P·H·Y·S·I·C·S
IMAGINATION AND REALITY

P·H·Y·S·I·C·S
IMAGINATION AND REALITY

P. R. Wallace

Department of Physics
McGill University
Montreal, Canada

World Scientific
Singapore • New Jersey • London • Hong Kong

Published by

World Scientific Publishing Co. Pte. Ltd.

5 Toh Tuck Link, Singapore 596224

USA office: 27 Warren Street, Suite 401-402, Hackensack, NJ 07601

UK office: 57 Shelton Street, Covent Garden, London WC2H 9HE

Library of Congress Cataloging-in-Publication Data
Wallace, Philip R. (Philip Russell), 1915–
   Physics: imagination and reality/P.R. Wallace.
     p. cm.
   ISBN-13 978-9971-5-0929-3 -- ISBN-10 9971-5-0929-6
   ISBN-13 978-9971-5-0930-9 (pbk) -- ISBN 9971-5-0930-X (pbk)
   1. Physics. 2. Astronomy. I. Title
   QC21.2. W35 1991
   530--dc20   89-22730
                                                 CIP

British Library Cataloguing-in-Publication Data
A catalogue record for this book is available from the British Library.

To those eager students whose searching questions influenced this book than they may imagine

CONTENTS

Processes 483

17.8 Nature of the Force between
Quarks 486

17.9 Other Quarks 487

17.10 Weak Interactions and
Unification 488

CHAPTER 18 THE INFLATIONARY
EARLY UNIVERSE 495

18.1 The "Homogeneity
Problem" 495

18.2 The "Smoothness Problem" 496

18.3 The "Flatness Problem" 498

CHAPTER 19 RADIATING BLACK HOLES:
WHERE QUANTUM
MECHANICS,
THERMODYNAMICS AND
RELATIVITY COME
TOGETHER 502

APPENDIX. An Informal Dialogue Concerning One
Not-so-new Science 511

INDEX 541

PREFACE

This book arises from a course which I gave for a number of years at McGill University, the goal of which was to convey to students in the humanities and the social sciences an account of the current state of knowledge and understanding of the physical world. The program of the course was ambitious — to guide the students through the "revolutions" which have taken place in our conceptions of that world since the early nineteenth century, culminating in the theories of relativity and quantum mechanics, and then to trace, up to as close to the present day as possible, their role in forming our current conceptions of the universe from the subnuclear scale to the cosmological one.

I undertook the course with the knowledge that public fascination with science was often tempered with a quite comprehensible unease concerning its social impact and also that there were in the public mind misconceptions about the nature of the scientific enterprise which were often rooted in outdated stereotypes. And indeed, these phenomena did manifest themselves in varying degrees. Two elements in particular were noted: first, there was a perception of science as mechanistic, and thus responsible for what M. Berman has characterized as "the disenchantment of the world"; secondly, science was identified with the technology which it has generated and for which scientists were held responsible.

In short, scientists were, in the view of some, technicians, and science itself utilitarian, morally neutral and without redeeming aesthetic qualities. To paraphrase March, it was not for poets.

This insensitivity to, or lack of awareness of, the fact that a scientific insight can generate the same sense of delight and exultation as a great work of music, literature, art or architecture constitutes an obstacle to cultural enrichment. There are many practical reasons why the intelligent layman should have some understanding of modern science; beyond that, why should the humanist be denied the cultural enrichment which can be provided by great scientific ideas while the scientist enjoys that provided by the arts?

The hope which lies behind the writing of this book is then this: that some who are not scientists will learn to experience the wonder of physics and come to understand what drives the scientist in his search for understanding, and that this will, in turn, encourage some kind of "unification" of diverse ways of looking at the world to the profit of us all.

The fact that this book has grown out of a university course (albeit an unusual one) does not imply that it is, or is intended to be, directed only toward students in such a course. Being a non-technical account of the development of modern physics, it should serve equally well any intelligent reader interested in understanding the modern physicists' view of the world, and the path that has led them to it over the past 150 years. Fortunately, the major ideas which have evolved have, over this time, become more accessible to the non-specialist than is generally realized, as the present account intends to show.

Where should one begin the story? If one starts back in classical times, or before, there is not likely to be time enough to reach our goal. A second possibility would be to start with the birth of modern science in the 17th century. However it is possible to identify the elements which distinguish contemporary physics from its precedents. Foremost among these elements is the recasting of physics in terms of "fields" rather than material elements. All the same, to appreciate fully what is novel in our new vision, we must first sketch the main features of the old one. This is also necessary so that we may keep track of the threads of continuity — the basic concepts of mass, force, energy and the rest — which link the new physics to the old.

The book therefore proceeds thus: first, I felt it necessary to distinguish in the sharpest possible terms between the intellectual bases of fundamental science and of technology; this provides the stage for the rest of the play. I have then given a very brief outline of the basic concepts of classical ("Newtonian") science.

We are then ready to get down to the business at hand — to show how the heretofore distinct (and non-mechanical) sciences of electricity and magnetism were found to be linked; how that linkage ("unification") came to be expressed in terms of the concept of fields; how this unification produced, in a manner which is typical of what was to follow, something new and unexpected, in this case the emergence of the conception of light as an electromagnetic phenomenon. While admitting a certain arbitrariness in the concept of "birth" in the face of evident historical continuity, it is this development of the unification of electricity, magnetism and light which can most appropriately be considered to mark the birth of modern physics. It was here that the long, slow process of erosion of the mechanistic Newtonian view of the world was set in motion.

The critical question posed by Maxwell's electromagnetic theory, and enunciated by Maxwell himself, was that of the effect of motion through the "aether" on observations of the speed of light — the "aether" being the medium through which light waves were supposedly carried. There is little of ultimate interest in the wasted decades of "aether physics" which followed, apart from the fact that they existed. The whole enterprise was wiped out with Einstein's extraordinary insight that the propagation of light required no medium, and his even more extraordinary postulate that nothing, including light, could distinguish between inertial frames of reference.

Chapter 5 is then devoted to exploring the physical consequences of Einstein's hypotheses. The discussion of various phenomena is necessarily somewhat ad hoc, due to forgoing the unifying power of mathematics; there is an advantage in this approach, however, in that it provides a sense of the physical processes involved. With some trepidation, I decided, after a year or two of experience, to try out, in a tutorial session, the technique of space-time diagrams. To my surprise, it was accepted enthusiastically. I have then integrated it into my discussion — and it does, in fact, represent a unifying factor in the treatment of the subject.

Surprisingly, the general theory proved to be easier to explain than the special. This is due to the fact that the principle of equivalence is, in this space age, fairly easy for students to grasp intuitively — and it forms a basis for all subsequent discussion. The idea of curved space flows rather naturally from the treatment of a non-inertial (rotating) "universe" from the viewpoint of an inertial frame of reference.

While the "classical" experiments verifying the general theory are usually considered to provide adequate evidence to support the theory, it has always seemed to me that the fact that it provided a framework for a viable cosmological theory was its most impressive testimonial — the more so when theory and observation converged with the observation of the recession of the galaxies by Hubble. Chapter 7 therefore deals with cosmology. In this and the three following chapters, the emphasis on relativity per se gives way to attention to the derivative problems in astronomy and astrophysics. The relativistic theme remains an issue throughout, however, since the discussion of the evolution of stars leads to the discovery of collapsed (strongly gravitating) objects, where the "Einsteinian" effects themselves no longer appear merely as small corrections to Newtonian gravitational theory, but produce large effects which provide, in effect, critical tests of relativity theory. This becomes especially evident in the discussion of black holes, first as theoretical predictions, and then as observable astrophysical objects (collapsed stars and quasars).

With Chapter 11, we roll back the calendar. In principle, this chapter could be inserted between the present Chapters 4 and 5, as it was when I first gave the course. This is obviously more appropriate historically, yet it breaks the intellectual continuity of the story at a crucial point. That consideration aside, I make no apology for including statistical physics as warranting its place in a survey of fundamental ideas. I do not regard thermodynamics and statistical mechanics as in any sense "poor relations" of relativity and quantum mechanics, but rather as deserving equal status. It seems to me that, when we reflect on it seriously, an understanding of the statistical nature of all the physics of macroscopic matter has profoundly affected our thinking; to put it in old familiar terms, it has taught us that the whole is much greater than the sum of its parts. We are too apt to forget, too, that the subject requires the introduction of an element additional to the ordinary "laws of physics", to wit, a hypothesis about à priori probabilities which we can make plausible, but not prove, from those laws. Nor is the idea of entropy, and its relation to the concept of information, in principle less sophisticated, subtle, or powerful than the ideas of relativity or quantum theory.

The rest of the book is devoted to quantum theory — its origins, its development and its flowering as perhaps the most powerful theory ever conceived by man. After a historical introduction in Chapter 12, I have tried to explore its essential ideas in Chapter 13. Here I will confess to a prejudice. I believe that there has been, in both the popular and the professional literature, a surfeit of "philosophical" word-spinning, and an obsession with supposed "paradox" and "queerness". Most of the issues raised in this context take on a different aspect if one poses problems in a quite different, but perfectly legitimate, way. For it is true that the mystery is largely dissipated if we recognize that it is generated by the lack of adequate vocabulary for dealing with a new reality. Our whole vocabulary has its conceptual foundation in "classical", i.e. pre-quantum, notions of physical reality. The problem is simply — how do we describe a revolutionary new view when we adopt as a framework of discussion the semantics of an earlier (discredited) one? The expression of our thought is forced into an inappropriate straightjacket; what were meaningful questions in the old scheme become meaningless in the new one. If we keep asking questions like "what is the taste of a Beethoven concerto?", the difficulty in answering may involve a "philosophical" problem, but it does not present us either with a paradox or a mystery. We do know how to express quantum ideas, with great precision, in the language of mathematics. Most physicists also have a keen intuition about quantum phenomena. What we need, if we are to talk about these things in ordinary language, are new words which are not "loaded" in favour of what are, in the context in question, obsolete concepts.

Another way of presenting the point is as follows: some features of the historical development of the subject have caused us to focus attention on what are considered to be peculiar new aspects of the problem of measurement. But the problem may be reformulated, not in terms of the "problem of measurement" but in terms of a proper understanding of what we are measuring. Questions which are meaningful when applied to intrinsically localized objects become misleading when applied to a non-localized one (e.g. "what is its precise location?"). If we start by accepting what is revealed to us by experiment, that quantum mechanics has forced us to adopt a new view of the nature of matter, many quantum "puzzles" disappear.

Often we also appear to have a strong egocentric bias. Why do we believe that our interaction with a physical entity is different in kind from the interaction with something inanimate? In the former case, we speak of "measurement". But if we were consistent, would we not have to admit that a quantum particle is constantly being "measured" by the totality of its physical environment? According to conventional arguments, the electron's state would have to be constantly "collapsing" to a "definite state" (whatever that means) from a "superposition of states". Where then has our "interpretation" of quantum mechanics led us? Is this not simply another way of saying that the "state" is constantly changing?

A last comment: why are we prepared to renounce what we know in favour of creating a seeming paradox? If we do not know that "Schrödinger's cat" either survived its ordeal alive, or died at some macroscopically determinable instant, what do we know?

All of this is a smokescreen which stops us from looking at what is really essential about quantum mechanics — that it provides us with a valid, meaningful and consistent theory of physical reality. That is what we should be conveying to students — not dissipating in a maze of word games. This is covered by the remaining chapters of the book. In its pattern, the treatment is conventional; first, to deal with atoms, then with agglomerates of atoms, whether small (molecules) or large (condensed matter systems). Regrettably, to do justice to the latter would involve an unacceptable inflation of the scope of this book; it is only possible to hint at the possibilities of the subject (my own!).

Invariably, students have been intrigued by Chapter 15 (particles and antiparticles). It is of course primordial in all that follows. They have sensed how revolutionary it is.

The problem with the presentation of particle physics is to know how far one can go. In so far as a descriptive account is possible, there is no problem. I have, however, not found what I consider to be a comprehensive account of the idea of gauge theories (much less of superstrings) at the level of this book. This may well be because my own understanding lacks the necessary depth. On the other hand, I have solicited help from colleagues with much greater expertise and have, so far, come up empty-handed. Beyond that, I am not sure that the purpose of the book is well served by the inclusion of material which can still be considered to be in the realm of speculation, and which may not survive the test of time.

Notwithstanding the above remark, I could not resist including the last chapter on radiating black holes and black hole thermodynamics. This has been done at the expense of taking a somewhat more mathematical approach than was deemed appropriate for the rest of the book. Despite that fact, I believe it is possible to extract some idea of physical mechanisms underlying the discussion. The seduction of the subject lies in the fact that it shows the interplay of all three themes of the book, and thus drives home the moral that, ultimately, the physical world is not conveniently sawed up into component bits, but reflects the play of all of its elements. To the timid reader I can only say, ignore this chapter if you must, though I would urge you to read it through rapidly, not trying to digest all of its details, just to get this one point: that more is possible by bringing different ideas together than can be accomplished with all of them separately.

Good luck!

Chapter 1

SCIENCE IN THE MODERN WORLD
— SOME OBSERVATIONS

For over half a century there has been a flow of books devoted to explaining aspects of the "revolutions" of modern physics to the lay public. The Darwinian thesis in biology, which aroused in many quarters a level of fear and hostility reminiscent of that of Galileo's time, created an atmosphere of polarization and confrontation between the advocates of the new science and proponents of traditional conservative dogmas. Although the conflict subsided, a residue of public suspicion of science remained. Sometimes arrogance and presumptions of near-omniscience among scientists reinforced that suspicion. Yet Einstein, whose impact on physics was unparalleled since Newton, evoked what was almost public reverence; his science was generally considered by the public to be nearly incomprehensible, but his evident humanity and personal humility sustained an image almost of saintliness.

Perhaps it was inevitable that the aura of glamour surrounding the new physics would be exploited to generate a more comfortable image of science — one based on a vision of the world more apt to inspire awe than fear. Thus the popular writings of Sir Arthur Eddington and Sir James Jeans exerted a strong influence, not only on the general public but on a new generation of young physicists in the making. (It is regrettable that only a few were subjected to the sobering influence of Susan Stebbing's critical "Philosophy and the Physicists".)

We will only note two important consequences of these developments. On the one hand, other sciences (and even areas outside the traditional boundaries of science of the time) were drawn to accept physics as a model for their own activities and attitudes, and this trend still persists to some degree today. On the other hand, much popular writing on science has followed the model of Eddington and Jeans in emphasizing the glamour and "mystery" of physical science, in playing on its strangeness and its "paradoxes", to the point of losing sight of the realities of actual scientific activity.

Both tendencies have had unfortunate effects on public understanding of what modern physics really is. The physics being mimicked in other areas of science is all too often the reductionist science of the nineteenth century, in which the behaviour of the real world is considered to be explicable on the basis of detailed study of its primitive components. The general public, on the other hand, is often distracted from the real significance of modern physics by an emphasis on supposedly incomprehensible "philosophical" paradoxes.

What has happened in physics in the last century and a half has often been misunderstood, not only by the general public but by the intellectual community outside the sciences. In approaching the study of physics, then, readers may have unjustified prejudices based on such misunderstanding. It is sad that this is true, but scientists too entertain many distorted and false views of humanistic studies, and these sometimes become evident and lead to a reciprocal suspicion of science by non-scientists.

C. P. Snow, in his latest book on "The Two Cultures" notes that:

> "Persons educated with the greatest intensity we know can
> no longer communicate with each other on the plane of their
> major intellectual concerns. This is serious for our creative,
> intellectual and above all, our normal life. It is leading us
> to interpret the past wrongly, to misjudge the present, and
> to deny our hopes for the future."

At the centre of the problem is the increasing specialization of intellectual endeavour. We carve out a bit of the world and assume it can be understood in isolation. It is, regretfully, a product of the scientific age, where *technique* has become our predominant concern, and technical expertise our most esteemed possession. This leads to compartmentalization, and we speak, as in the 19th century, of such things as the "conflict of Science and Religion", as though they were two powers disputing possession of a piece of territory.

With such an antithesis, we put arbitrary bounds on our perception of reality and we are unable to deepen our understanding of the world.

It is worth asking whether the greatest affliction of our society is not to divide life into arbitrary sectors, and to create in each a priesthood of expertise which reigns over it. It certainly creates feelings of helplessness and alienation in ordinary people.

Why do I emphasize this point? Because it is at the root of public misunderstanding of science. What we see, what affects us deeply in our daily lives is *technology*, which is an excrescence on the body of science.

I shall try to give you some notion of the issues that have dominated modern physics; they relate to the most fundamental questions of our conceptions of space, time, matter, and energy. One's first inclination is to say that such problems are purely theoretical and abstract, and have no impact upon the practicalities of our daily lives. Nothing could be further from the truth. Modern technology is derived from the revolution of concepts which occurred in the last century and a half of physics. The work of Faraday and Maxwell, which led to the theory of electromagnetic fields and the knowledge that light was an electromagnetic phenomenon, led to electric power generators and transformers as well as to the theory of relativity. This, in turn, led physicist Leigh Page to say, "THE ROTATING ARMATURES OF EVERY GENERATOR AND MOTOR IN THIS AGE OF ELECTRICITY ARE STEADILY PROCLAIMING THE TRUTH OF THE RELATIVITY THEORY TO ALL WHO HAVE EARS TO HEAR."

The awareness that visible, ultraviolet and infrared light, radio waves, X-rays, and γ-rays are simply different forms of electromagnetic radiation is behind all our modern communications technology. More profoundly, it has made us aware that our senses — in this case the sense of sight — have given us access only to a minute part of reality. Light, as we shall see, is essential for communication between the external world and us, as well as between each other. Yet our eyes are sensitive only to *visible* light. All around us are electromagnetic radiations to which we are largely insensitive, but which contain innumerable messages about our environment. When we understand this, we create extensions of our senses which enable us to pick up and translate these messages. The modern astronomer, conscious of the enormous forces of the cosmos, searches for signals in the whole range of electromagnetic waves, from microwaves to the γ-rays produced in nuclear reactions. The telescopes which are presently revealing to us the violent processes in the universe of which I have already spoken are *not optical* telescopes, but ones which use X-rays and γ-rays.

What we understand less well is how almost all of the technical marvels of our age are based on quantum principles, and so could not have been conceived in the era of Newtonian physics. The solar cell, the microchip, and the nuclear power plant — to say nothing of the nuclear bomb — are the purest manifestations of quantum principles.

Organic chemistry (the chemistry of life) becomes comprehensible only in the light of quantum concepts. The very stability of the atoms and molecules of which we are composed depends on them. So does life itself. It is the so-called uncertainty principle which gives life its stability, its colour, its almost infinite variety. The tools of scientific medicine, from radiation treatment of cancer to genetic manipulation depend on it.

In short, the technological structure of our society is a direct consequence of what we have learned about the most fundamental principles of the physical world.

It is this which has led Jacob Bronowski to say

"THE WORLD TODAY IS MADE, IT IS POWERED BY SCIENCE, AND FOR ANY MAN TO ABDICATE AN IN- TEREST IN SCIENCE IS TO WALK WITH OPEN EYES TOWARD SLAVERY."

That is a *practical* argument for scientific literacy, and one that I want to reinforce and develop, but it omits *another* aspect of science that may, in the long run, be even more important. For the view of the world which we have evolved is one which can and should also touch the human spirit. It has grandeur; it has beauty; it is awe-inspiring. In this sense it is profoundly *humanizing* rather than dehumanizing. It has brought us a view of the world which is no longer that of a vast machine, a mechanical world, a world of cer- tainties. It gives us a new understanding of the marvellously subtle structure of nature, and of our inseparability from it. It has left almost no domain of human thought unaffected. In this sense, it constitutes a fundamental part of our culture.

Yet almost everyone, I believe, has some sort of profound un- ease about science, and some tendency to mistrust it. Nor is this unease and this mistrust unjustified. The problem is not with our knowledge, our image of the world in which we live, but with what we can, and do, use our knowledge for. In short, we are concerned about the technology which springs from our knowledge of science.

It becomes, then, important to analyze clearly our dilemma, and to understand the difference between science and technology. Let me try to clearly show the difference.

Note first that the basic process of science is generalization. As Koestler said, science finds links between concepts and phenomena previously thought to be distinct. Feynman says the same in one of his Messenger lectures. For the true scientist, nothing stands in isolation; everything is linked together. The problems of science are the problems of synthesis.

This is in contrast to the specialist, the technical expert in one or another domain of knowledge. The technological problem is specific; the expert is an expert on a specific technique or phenomenon. *Isolation*, not generalization, is the key.

We can put it in another way. As we shall see throughout this course, the challenge to the scientist is to find the right way of looking at things — the perception of a pattern which will best reveal the interrelationships of phenomena. The questions a scientist asks are not pre-defined; he has to pose them himself. The decades of intensive scientific work in the last quarter of the 19th century when scientists occupied themselves with the properties of aether were wasted because they asked the wrong question. Instead of asking "what are the properties of aether?" they should have asked, "*is* there an aether?" or "why do we believe in an aether?" Science always progresses in this way; we are blocked because our questions are based on a preconceived, often false, impression of reality. Discovery only comes with the posing of perceptive questions.

For the technologist, the question is already given. How to make a flat television tube, how to increase the yield of wheat, how to get rid of mosquitoes, how to slow up the deterioration of food products, and so forth. He only has to solve the problem; success lies in its resolution.

The problem with technology is that as one solves a pre-determined problem, too often worse ones are created in its place. One does not look beyond pre-defined limits of space, time, or even the possible impact on life. Such an exclusive concern with solving the immediate problems has, for example, brought about pollution and, in turn, the ecological movement. It is also at the root of public concern with the implications of nuclear power development, and what we now so glibly call "genetic engineering".

From the viewpoint of the scientist, the mental framework which leads us to isolate problems tends to make us think of science as something apart from other aspects of human existence, or even implicitly to deny their existence. The technology-dominated mind, for example, has no use for history. It is no accident that it was Henry Ford, who gave modern society its favourite object of worship, the automobile, who said

HISTORY IS THE BUNK.

It is not amiss, however, to juxtapose this remark with the following observation of Dennis Gabor, who won the Nobel prize for the development of the hologram:

THE MOST IMPORTANT AND URGENT PROBLEMS OF THE TECHNOLOGY OF TODAY ARE NO LONGER THE SATISFACTION OF THE PRIMARY NEEDS (OR OF ARCHETYPAL WISHES) BUT THE REPARATION OF THE EVILS AND DAMAGES WROUGHT BY THE TECHNOLOGY OF YESTERDAY.

We ignore history at our own peril.

In a technological-industrial society, the world is continuously being destroyed and recreated; as we "advance" and become more "scientific", our cultural heritage is no longer relevant. We develop feelings of alienation.

Our society has a myth that technology is synonymous with *efficiency.*

Nowhere is the dubiousness of this proposition more evident than in agriculture. The historian G. P. Stavrianos, in "The Promise of the Coming Dark Age", makes the following observation about modern agrobusiness. Traditional agricultural methods, such as the use of natural fertilizers and a high proportion of human labour, uses only *one* calorie to produce 5 to 50 Calories of food. (1 Calorie = 1000 calories.) On the other hand, "advanced" methods, using synthetic fertilizers (petroleum products!) and sophisticated agricultural machinery, use between 5 and 10 calories to produce 1 Calorie of food. Which one is more "efficient"?

Stavrianos answers: this energy-production ratio makes the high-energy agricultural system of the industrialized world one of the least efficient in history.

Let's put it a bit differently.

"Under the new agricultural technology, the equivalent of 80

gallons of gasoline is needed to produce one acre of corn." No wonder
food prices are rising at record rates! In fact, if we tried to feed the
whole world with these agricultural methods, the world's oil reserves
would be exhausted in less than 5 years!

The problem is not confined to agriculture. Barry Commoner,
a distinguished biologist who has written extensively on ecological
problems starts his book "The Poverty of Power" as follows:

> "In the last ten years, the United States — the most pow-
> erful and technically advanced society in human history —
> has been confronted by a series of ominous, seemingly in-
> tractable crises. First there was the threat to environmental
> survival; then there was the apparent shortage of energy;
> and now there is the unexpected decline of the economy.
> These are usually regarded as separate afflictions, each to
> be solved in its own terms: environmental degradation by
> pollution controls; the energy crisis by finding new sources
> of energy and new ways of conserving it; the economic crisis
> by manipulating prices, taxes, and interest rates. But each
> effort to solve one crisis seems to clash with the solution of
> the others — pollution control reduces energy supplies; en-
> ergy conservation costs jobs. Inevitably, proponents of one
> solution become opponents of the others. Policy stagnates
> and remedial action is paralyzed, adding to the confusion
> and gloom that beset the country."

Commoner has recently pointed out in another article that:

> "What is new and profoundly unsettling is that thousands
> of separate entrepreneurial decisions that have been made in
> the United States in the last 30 years have, with alarming
> uniformity, favoured those that are less efficient in their use
> of energy and capital and more damaging to the environment
> than their alternatives."

We have seen how specialization leads to departmental barriers.
In the case of agriculture, it illustrates how limiting our perspective

to too narrow a geographical area distorts our perceptions. Concentrating problems within too narrow a time frame can also lead us to ignore important aspects of reality. An interesting illustration of this is given by exponential growth laws.

An exponential growth law is like the law of compound interest: the growth in a given period of time is a fixed fraction of what has been accumulated as a result of all past growth.

There are many areas in which we think in terms of *percentage* growth over given intervals of time. Financial experts talk of economic growth as a percentage from year to year. We chart the growth of prices (inflation). The rate of population growth and the rate of increase of energy consumption are also problems we notice. The consequences of constant growth can be explored in a way which is applicable in all these areas, and more.

A useful index of exponential growth is the "doubling time". Let us take, for example, the gross national product, in fixed dollars, of a nation. We may take any starting point as the base. If the growth rate is x per cent per year, there is a simple rule which says that, at this rate, the GNP will double in $70/x$ years. Thus, if $x = 5\%$, the doubling time is 14 years; if 7%, 10 years; if 10%, 7 years. No doubt most economists (and politicians) would be happy with a 5% growth rate (and even happier with a 7% one!). Let us, for purposes of illustration, take 5%.

Starting with the year 1950 as a base, which we designate 100, here is how the GNP increases with time:

1950	100
1955	128
1960	163
1965	208 ←←← It has doubled in 14 years!
1970	265
1975	338
1980	432

This looks very gratifying, and suggests prosperity. Let us, however, continue our table in bigger jumps:

2000	1147
2020	3035
2040	8042
2050	13,156

In only one century, at this rate, our GNP will have increased 131 times! We may now begin to wonder, first, whether we will be able to consume 130 times as much as we did in 1950, and, even more serious, where the raw materials for all this production will come from.

But an even greater shock awaits us. Let's look at things a bit differently. If we look at production in 14-year periods, we come up with the following pattern:

1st 14 yrs.	2nd	3rd	4th	5th	6th	7th
100	200	400	800	1600	3200	6400
Total to this point	300	700	1500	3100	6300	

If we look at the totals to a given point (bottom line) we observe that the production in any period is more than the *total* production in all previous periods!

Note also the effect of increasing the rate of growth from 5% to 7%. Going back to our original period 1950-2050:

	at 5%	at 7%
1950	100	100
1955	128	140
1960	163	197
1965	208	276
1970	265	387
1975	339	543
1980	432	761
1985	552	1068
1990	704	1497
1995	899	2100
2000	1147	2946
2050	13,150	86,772

As production increases, so must the amount of natural resources.[1] This may be seen in another way. There was a period, which happily ended in the early 70's, during which the consumption of petroleum in the West increased on the average by about 7% per year. Thus, the doubling time for consumption was 10 years. Few people at the time considered this to be a source of anything but gratification; after all, our "standard of living" was constantly improving (but so was the gap between the rich and poor of the earth). Consider, however, the long-range implications of this growth. Let's use a model calculation for illustration. Start again with some initial decade in which the process is in operation. Call the consumption in that decade one unit. Suppose, for the sake of argument, that at this rate of consumption, the world's petroleum resources were 7 units, i.e. sufficient to last for 7 decades. In fact, in the second decade the consumption would be 2 units, and in the third, 4. The resources would be exhausted in *three*, not seven decades! Still more striking is that to sustain the rate of growth of consumption for one more decade (the fourth) would require more than doubling the available resources; in each decade, remember, consumption is more than in all preceding ones.

Remember the point we set out to illustrate: if problems are approached within a limited time frame, and solutions found which are very gratifying in that time frame, one may lose sight of the fact that they have, in the longer term, disastrous consequences. We often appear willing enough, in our days, to satisfy our own immediate desires in this way, and to leave the inevitable resulting crisis to our children or our grandchildren. This, again, is the consequence of a preoccupation with a quick "technological fix". Here a final point must be made. Technical experts may be competent to tell us how to achieve certain ends, but they have no authority with regard to the questions of "whether" or "why". They have, in short, no authority with regard to the *values* which guide either individuals or societies.

---

[1] The figures in our table, therefore, can be taken as at least a rough indication of the despoiling of the natural environment.

There seems to exist, in our rapidly developing age of technology, a sort of technological imperative, whose rule is that whatever can be done, *must* be done. We shall illustrate this with two present-day examples.

One concerns what we have come to call "genetic engineering", and has to do with our technical ability, consequent to the unravelling of the mysteries of the genetic code, to create new life forms, or to reproduce indefinitely existing ones.

As with any discovery, this can be exploited for the pursuit of various goals. Since the driving force in our society is avowedly the pursuit of profit, this is surely one important context of exploitation. Other goals concern the human ego or the desire to combat disease. What is certain, however, is that no one will be unaffected by whatever "practical" use is made of new genetic knowledge and technology.

The environment of man will be permanently, and irrevocably changed. We have been through this before, with the unlocking of nuclear energy. There is little in our experience with nuclear energy that reassures us that technological advance serves the common interest; we are told that our "security" depends on the use of weapons capable of annihilating all life on earth. The mathematician, G. H. Hardy, who made no apologies for the fact that his mathematical discoveries had no "practical" applications, made the sardonic remark that "a science is said to be useful if its development tends to accentuate the existing inequalities in the distribution of wealth, or more directly promotes the destruction of human life."

Few people have written more forcefully or more extensively on the technological domination of modern life than Lewis Mumford. He has, in particular, called attention to the risk of human genetic manipulation and its implications. He cites an article by a social scientist, highly regarded as a population expert, in a book entitled "Genetics and the Future of Man", in which he states that deliberate genetic control is *bound* to occur, and, once begun, "it would soon benefit science and technology, which in turn would facilitate further hereditary improvement, which in turn would extend science, and so on in a self-reinforcing spiral without limit."

O Brave New World!

Mumford, however, notes that this statement begs the question of what constitutes "improvement" of the human species. Should the development of this bright new future be entrusted to technical experts? His answer is quite different:

> "One would hardly have guessed — that thousands of the wisest minds have meditated for thousands of years over what are the most desirable characteristics in human beings — "

He goes on to suggest that "the only effective approach to this problem is that taken long ago by nature; to provide the possibility of an endless variety of biological and cultural types, since no single one, however rich and rewarding, is capable of encompassing all the latent potentialities of man". It is a statement, in the human context, of the biologists' "law of hybrid vigour".

Finally, he returns to the point touched upon by Snow: that specialization, and a lack of awareness of the links of one's specialization with other areas of human concern, makes us blind to many of the dimensions of reality.

> "Those who speak with the highest authority upon some minute section of exact knowledge too often unblushingly claim the right to speak for mankind upon matters of general human experience upon which they can testify only on the same lowly basis as other human beings."

What it comes down to, in the end, is a question of values.

One last illustration of the problems with which advanced technology confronts modern society is the development of an almost unlimited capacity to store information in a readily accessible way. At first sight, this would seem an unmitigated boon to humanity; a second thought, however, generates not only doubts, but misgivings. Those who remember the rise of fascism in the 1930's, and its culmination in the bloody holocaust of 1939–45, may shudder to think what a powerful weapon a global information bank would be in the hands of a political dictator. Even in a democratic society, the potential for misuse is vast. Again, the question of whom such technological triumphs will serve, and for what purpose, is primordial. Again, too, a question of values is posed: what are the rights of the individual to privacy; what is his protection against the invasion of his most personal life by an impersonal bureaucracy?

The question is important: the more sophisticated the technology, the more likely we are to be enslaved by it because, by virtue of its very sophistication, it is outside our control. Quis custodiet ipsos custodies? Can systems become so complex that even experts, because of the limits of their specialization, no longer know how to control them?

The foregoing discussion may strike the reader as an unqualified condemnation of technology. One runs the risk of being characterized as a latter-day Luddite. Of course the *potential* value of modern technology to human welfare is immense, and beyond dispute. Yet the agonizing question is, how do we realize that potential and, at the same time, avoid having a powerful weapon turned back against us.

The thrust of this course is not the misuse of scientific knowledge in society. Our concern is trying to understand the world — the physical universe in which we live. This task is a profound affirmation of our humanity, and a challenge to both the intellect and the spirit. The vision which physical scientists have gained in the 20th century of the structure and evolution of the physical world is so unexpected and at the same time so wondrous that we marvel both at the ingenuity and subtlety of Einstein's god and at man's intellect for being able to read such simple yet intricate structures into it. Human imagination has been stretched to new limits, and produced a vision which is a work of art, pure poetry; the more exciting because it is the poetry of reality. Quantum mechanics, or Einstein's general theory of relativity, are among the greatest intellectual achievements of the human mind to date.

Yet they were not realized by a process of pure thought. By their very nature, our ability to penetrate the domains of the physical world which are not accessible to our direct sensory experience has depended on sophisticated instruments to act as extensions of our senses. These are by-products of the technological advances.

The following example illustrates this point. From the beginning, the advance of astronomy has depended on telescopes which have brought the planets, the stars and the galaxies closer to us. More recently, we have learned to look at the cosmos through many-coloured glasses which span the spectrum of the electromagnetic messages from distant objects. Radio astronomy revealed the pulsar, from which we learned that collapsed objects with enormous gravitational fields existed in nature as well as in theorists' imaginations. We become aware of an astronomical universe which is characterized by much instability and violent change; this belies the apparent constancy of its appearance to the naked eye. As a laboratory of physics, it manifested scales of energy far more extreme than we had imagined. But the logical place to look for evidence of these extreme energies was in the highly energetic/high frequency end of the electromagnetic spectrum; that of X-rays or even γ-rays, hidden from us by the earth's atmosphere, in which they are almost totally absorbed.

This world has now been opened up to us as a fallout from the space program. That program was regarded by some as an example of the misuse of technology arising from a distortion of governmental priorities; indeed, it did seem that the major powers, and the world in general, had more urgent problems than putting a man on the moon. Yet this same program made possible, and led to, the development of an X-ray astronomical observatory outside the earth's atmosphere — the High Energy Astronomy Observatory now appropriately named the Einstein Laboratory — which, since starting normal operations in January 1979, has given us vast quantities of new information about the remote reaches of the universe; information which could force a re-examination of many prevailing ideas in astrophysics, and radically alter our conception of the cosmos.

The fact is that, at the most fundamental level of science, existing concepts are being tested and new ones evolved as a direct result of space technology. Whatever the motivation of the space program, this by-product has made, and will continue to make, an important contribution to our intellectual and cultural heritage.

The question is not, then, whether technology is in itself good or bad, but how we use it — for whose benefit and for what purpose. What we must learn to do is to take technology out of its limited context and to examine it in its widest implications. Only then does it become an integral part of science.

One further warning is in order. Various modern philosophers — some physicists — have called our attention to the fact that the way we pose our questions about physical phenomena may affect the answers which we get and the way in which we interpret them. Foremost among those warning us of this phenomenon is J. A. Wheeler.[2] The issue is put rather vividly by W. Kaufmann in his book "Relativity and Cosmology":[*]

> "We are quick to realize that any of our modern ideas must incorporate all available data and observations, but how has our psychological orientation affected the concepts put forth in this book? The general approach of all of modern physical science is entirely mechanistic. It is mechanistic not in the crude sense of gears, levers, and pulleys, but rather in terms of trying to reduce all reality to concrete physical laws wherein the only truly important quantities are those that we can measure with machines such as spectrographs, galvanometers, and photographic film. Ancient man thought of the planets as gods and attributed spiritual qualities to what he saw in the sky. It is argued by the modern scientist that

[2] For a fashionable, but controversial view, see the article "What You See is What You Get" by David Harrison in *American Journal of Physics*, July 1979.

[*] Quoted with the permission of William J. Kaufmann.

such ideas contribute nothing of value to our understanding and indeed lead us astray. But when this same scientist walks up to his telescope or cyclotron, he carries with him a set of axioms that are just as much a part of his psyche as the unquestioned concepts in the mind of an ancient Babylonian astronomer climbing to the top of a ziggurat. It is obvious, therefore, that if we knew more about ourselves, we would be able to discover and understand more about the nature of the universe."

Chapter 2

FUNDAMENTALS OF NEWTONIAN MECHANICS

Let me, in form of a summary, sketch the main features of the classical or Newtonian world as physicists understood it in the early part of the 19th century.

At the core are Newton's mechanics.

It is beyond dispute that Newton was one of the greatest geniuses of all time. His physics was the physics of a world accessible to the senses of man. It summed up, in very simple and general laws, almost the whole of that experience. His laws of motion, which dominate our way of viewing the world of everyday experience, have become so ingrained in our experience that we are inclined to think them almost trivial. It is difficult to imagine, then, that they undoubtedly seemed as strange and contrary to common sense in those times as the most esoteric features of relativity and quantum mechanics are to the non-expert today.

The range of his intellect, and his capacity for seeing the connection between what had previously been considered isolated phenomena, was unmatched until recent times. Perhaps the most remarkable of his discoveries was the law of universal gravitation, of which Richard Feynman has said that, despite its 17th Century origin, it is remarkably modern in spirit. Nothing which Newton contributed to our knowledge, however, has had a more profound impact on modern thinking than his "methods of fluxions" — the infinitesimal calculus — which is nothing less than the mathematics of continuous change.

It is true that he shares responsibility for this discovery with a contemporary, Gottfied von Leibnitz. It is not fruitful to waste time trying to decide which, to use the current expression, deserves recognition for priority in the discovery. The fact is that the idea came to Newton in 1666 — the great year when the young Newton, then 23 years of age, fled to his mother's house from the university in Woolthorpe to avoid the Great Plague. Just as Einstein was to cram the essence of almost all his greatest contributions to science in the one year of 1905, so did Newton generate his greatest ideas in physics and mathematics in 18 months in 1666 and 1667. For both the greatest discoveries came in solitude, away from the academic milieu; for Einstein, in the Berne patent office; for Newton, in the countryside. Clearly, neither had any direct debt to their scientific peers.

Unfortunately, however, Newton was a difficult, suspicious and somewhat vain man, despite all his published avowals of humility. His personal characteristics led him to keep his discovery of calculus secret for some years, as a sort of "secret weapon" for the solution of problems not available to others. Thus, when von Leibnitz, working quite independently, published similar ideas in 1696, Newton was unable to believe that his rival had not had access to his own discoveries.

What is clear from all this, however, is that his "method of fluxions" was at the core of Newton's thinking from the beginning, and served as an indispensable tool for the detailed development of his physical ideas.

It is important to emphasize the above, since the modern physics is characterized by the rejection of Newtonian physics. But this "rejection" is based on the development of technical means of expanding our knowledge into areas beyond the scope of direct sensory experience, where, as we now know, Newtonian physics is no longer applicable. At the same time, the new physics does not deny, but rather incorporates, Newtonian physics as the physics of the macroscopic world.

From the modern viewpoint, however, it is important to recognize the very fundamental pre-suppositions of Newtonian physics, as well as their implications, because this is where modern physics has its roots.

The first, and most profound assumption, is in Newton's concepts of space and time. Basically, he considered them as a sort of divine stage — a background for the drama of nature.

"Absolute space," wrote Newton, "in its own nature, without relation to anything external, remains always similar and immovable." And again, "Absolute true and mathematical time, of itself, and from its own nature, flows equally without relation to anything external — just as relative time always more nearly approaches absolute time as we refine our measurements."

It is difficult to believe, for the layman, that the advances of modern physics have depended on our rejection of these two "common-sense" and seemingly innocuous hypotheses. Newton thought of space and time not as a *part* of physics, but as a *framework* for physics. Relativity incorporated them into the *structure* of physics. And the practical importance of relativity in our daily lives is illustrated in the words of Leigh Page, that "The rotating armatures of every generator and motor in this age of electricity are steadily proclaiming the truth of the theory of the relativity theory to all who have ears to hear".

Another underlying assumption, less specifically articulated, concerned the nature of "matter". In Roman times, both Democritus and Lucretius spoke of matter as being constituted of "atoms", and at all times it has been thought that matter was not infinitely divisible, but that there were some ultimate elements from which every material thing was composed. What was *not* imagined, however, was that these ultimate "particles" were different in nature from macroscopic matter itself; that is, they could not be characterized by the same properties that characterize bulk matter. Newton, therefore, and his followers, thought of each "particle of matter" as the ultimate object of the physical laws of motion. From the dynamics of such "particles", it appeared that those of macroscopic matter could be deduced.

The world of Newtonian physics was then quite machine-like. Particles of matter, and the bodies of which they were composed, moved according to well-defined laws, which could be studied by the methods of Newton's mathematics of change. So Simon de Laplace was able to say:

"Given for one instant an intelligence which could comprehend all the forces by which nature is animated and the respective positions of the beings which compose it, if, moreover, this intelligence were vast enough to submit these data to analysis, it would embrace in the same formula both the movements of the largest bodies in the universe and those of the lightest atom; to it nothing would be uncertain, and the future as the past would be present to its eyes."

Since the laws of matter were universal, everything, including the living organism, was subject to them. Since the structure of physics was, in the crudest sense, mechanical, the living organism became a machine, in principle totally predictable, subject to determinate law.[3] It may be claimed that the same sort of reductionism is still valid in the world of modern physics, and yet one is also permitted to believe that things are not as simple as that. Certainly, Laplace's simplistic statement is no longer meaningful; nor does it appear any longer that physical law is insensitive to scale. Our experience has shown us too many instances where quantitative difference merges into qualitative, leading Feynman to observe that "one does not, by knowing all the fundamental laws as we know them today, immediately obtain an understanding of anything much ..."

It is interesting, all the same, to observe that although, at the most fundamental level, our concepts have radically changed, elements of classical physics have been preserved more or less intact to the present day. Einstein has deepened our concept of the *mass* of

---

[3] From such considerations sprang the 19th century "Conflict between Science and Religion", whose ghost lives on.

objects, but it still exists as a measure of inertia in modern physics. Momentum, angular momentum and energy are still with us, and even largely retain the roles reserved for them in classical physics. Symmetry remains, and conservation laws, and causality. Each concept has been broadened and refined, and new relations found between them, but the essential features of the concept have remained.

It is important, therefore, that we talk about these concepts in the context of Newtonian mechanics, for only then will we be able to understand their subtler significance in modern physics.

2.1. Mass

Let us start with the concept of *mass*. A less terse definition for the *mass* of a body (or a "particle") is "coefficient of inertia".

Consider, for example, two objects made of the same material, one having twice the volume of the other ("twice as big"). Suppose that each is acted upon by an *identical* force for an identical length of time. It is found that the smaller body is accelerated to twice the speed of the larger one. The larger one has more resistance to acceleration, more *inertia* than the smaller. We say it has twice the mass.

Of course, inertia is only proportional to size for objects made of the *same* material. A ball of lead of 20 cm diameter has more inertia, and hence more mass than one of aluminum. How do we test their relative masses? In the same way as before, by comparing the relative accelerations which they are given by identical forces.

The above statements, which *define* mass (in relative terms at least) seem almost precisely equivalent to *Newton's Law of Motion*. Galileo had already formulated the rather strange law that every body, left to itself (i.e. not acted upon by any *force*) would continue indefinitely in its state of rest or uniform motion in a straight line. What is so difficult about this law is that it is hard to imagine something not acted upon by any force! It certainly has never existed in the real universe. It is an abstraction, a mental construct. We can, however, imagine it better today than in the time of Newton or Galileo, due to our increasing familiarity with space travel. Far enough from the earth and other gravitational bodies, and in the near-vacuum of space, objects may be subject to so little force that they *do* almost satisfy the ideal conditions envisaged.

2.2. Force

In any case, from this starting point, Newton affirmed that this ideal state of rest or uniform motion could only be changed by *force*, which would cause the body subjected to it to *change* its state of motion, i.e. to *accelerate*. In fact, in quantitative terms, Newton's law of motion says

$$\text{force} = \text{mass} \times \text{acceleration}$$

or alternatively,

$$\text{acceleration} = \text{force} \div \text{mass} .$$

The second version, however, is basically the *definition* of mass as a coefficient of inertia.

But not quite. If a force — any force — can be generated by a reproducible physical system or process, that alone will serve to determine the masses of objects in terms of some standard unit mass. Newton's equation then tells us how bodies will accelerate for *any* given force.

We must now ask whether the equation does not in fact simply define "force". The equation of motion can be qualitatively summarized by saying "acceleration is a manifestation of force", i.e. if something accelerates, we can postulate a force which is responsible. Clearly, if we are not to treat forces in such an "ad hoc" manner, we need another element in our theoretical structure — a theory of force(s), a physical law (or laws) governing the origin and nature of forces.

This is best illustrated by considering Newton's law of universal gravitation. Newton postulated a universal gravitational force with the following properties:

any two bodies are attracted to each other by a force which is proportional to the mass of each and *inversely* proportional to the square of the distance between them.

He was able to see that a force of such a nature could explain a large number of different physical observations — from the fall of an apple from a tree, to the orbit of the moon round the earth and of the planets round the sun. With such a *theory* of force Newton's law of motion is seen to be *truly* a law, whose consequences in many contexts can be predicted and tested by experiment!

It would be nice if we had a single law of force which governed everything in the universe. Unfortunately, such a law does not exist up to now. We have, in addition to the gravitational force, electric and magnetic forces (unified into a single law by J. C. Maxwell), the strong forces responsible for the binding of nuclei, and the so-called "weak" force manifested in beta-decay and other subatomic processes. Einstein tried, without success, to unify gravitational and electromagnetic forces; that is, to find a single law which included them both. Recently, progress has been made on a theory which states that weak and electromagnetic forces are different manifestations of a single theoretical structure. The unification of all forces, if indeed the forces *are* all related, is still far from realization.

The formulation of Newton's law which has been given above is not the most generally useful. Suppose, for example, that we want to work out the dynamics of a space rocket. Since it is constantly burning fuel, its mass is decreasing. The formulation which we have given is quite satisfactory when the mass is *constant*, but it is not immediately clear what to do when it is not. The correct formulation of the law in this case involves the concept of the *momentum* of an object.

2.3. Momentum

If an object of mass m moves with a velocity v, its momentum is defined as mv. (Note that this quantity, like the velocity, has a direction as well as a magnitude; such a quantity is called a vector.) This definition is in accordance with our intuitive notions of momentum. We can now redefine Newton's law of motion as

force = rate of change of momentum .

If the mass is constant, rate of change of momentum = mass times rate of change of velocity (acceleration), as before. The new version is valid even when the mass varies.

Note that if the speed of an object varies, but it changes direction, its *velocity* varies — i.e. it accelerates. This is necessarily so by virtue of Newton's law, since a force is necessary to change the direction of an object, even if its speed remains constant. Thus, if a tennis ball with momentum p is thrown horizontally against a wall, and bounces back with the same momentum in the opposite direction, we shall call the rebound momentum $-p$; the change of momentum is then $2p$. The force of the wall on it has then changed its momentum by this much.

This example illustrates another of Newton's principles, namely, that if I exert a force on something, it exerts an equal and opposite force back on me. The harder I push down on the table in front of me, the harder the reactive force which it exerts on my hand. If I swing a stone around on the end of a string, I must exert a force to keep it moving in a circle; the string exerts an equal and opposite force on my hand. Similarly, the earth exerts a gravitational force on the moon, but the moon exerts an equal and opposite gravitational force on the earth, and this is the force which is responsible for, among other things, the tides on the earth's oceans.

The law of conservation of momentum follows directly from this law of action and reaction. We note that the momentum of a closed system is conserved, since no external forces act upon it. There will, however, be internal forces between any two elements, A and B, of the system. The force F of A acting on B for a small interval of time t will change its momentum by Ft. The force of B acting on A is $-F$, so that the momentum of A will change by $-Ft$. The total momentum change due to the internal forces is therefore zero, i.e. the momentum is conserved.

2.4. Work and Energy

Let us turn now to two other concepts which play a central role in physics, work and energy. Both words we use in everyday life in a qualitative way; for the physicist, however, both are defined quantitatively, and their meaning is precise.

Consider first *work*. Work is done by forces on objects and is proportional to the force and to the distance which it displaces the object. Thus, if a force F acts on a body and displaces it by a distance x in the direction in which it acts, the *work* done on the body is Fx. But this work gives the body energy; in fact, the amount of work done is the amount of energy imparted; this, effectively, defines energy. That is,

$$\text{work done on body} = \text{energy given to it.}$$

This statement defines what we *mean* by energy. Consider, for example, a body falling freely under gravity. There is a gravitational force on it, which is proportional to its mass m, and which we shall therefore call mg. When the body falls a distance x, gravity has done work mgx on it, and has thus given it this much energy. (We neglect air resistance; if that is taken into account part of the work is done on air molecules in the path of the falling body, and some of the energy is therefore given to these molecules.)

The energy given to the body is energy of motion, which the physicist calls "kinetic" energy.

Suppose, on the other hand, that one uses a hoist to *raise* an object *against* the force of gravity. To raise it by a distance x requires that a force mg act over that distance, i.e. an amount of work mgx must be done, so that an amount of energy is *given* to the body. But this is not energy of motion, since at the end the body will be at rest. We say in this case that it has "gravitational potential energy". If it is released and again *falls* the distance x, it will lose the potential energy, but will *gain* an equivalent amount of kinetic energy.

Of course, in the previous example we have rather idealized things. To the extent that air resistance impedes the fall of the body under gravity, its kinetic energy is diminished; some energy is *lost* from the body, and appears as energy of motion of air molecules. Similarly, when it is raised against gravity, some of the work done may be against friction in the hoist. Thus, not all the energy will be given to the body; its potential energy will be less than the total work done. Some energy will be given to the hoist since the friction will produce *heat* in the hoist, and this is another form of energy. It seems that when work is done, there is some "waste"; some work has been done on, and some energy lost to, the surroundings. This is a very important phenomenon, which will become of vital concern in our later study of thermodynamics. For the present, however, we shall ignore this aspect of things. In some circumstances, this neglect will be justified to a high degree of accuracy.

We saw how potential energy can be quantified. If the body starts at rest at some position and we raise it against the gravitational force mg through a distance x, it gains potential energy mgx. What happens, however, to the *kinetic* energy which it acquires due to its being freely acted upon by a force? With a bit of simple algebra, we can determine it using Newton's law

$$\text{force} = \text{mass} \times \text{acceleration}.$$

Suppose a force F acts for a brief time Δt, increasing the velocity of the body from v_1 to v_2. Then, Newton's law says

$$F = m\frac{v_2 - v_1}{\Delta t} \, ,$$

where m is the mass of the body. Now the force will move it some distance Δx, so that the work done on it, and thus the energy which it is given, is

$$F\Delta x = m\frac{v_2 - v_1}{\Delta t}\Delta x \, .$$

But $\frac{\Delta x}{\Delta t}$, the distance travelled by the time taken, is the average velocity. If the time is short, it is $\frac{1}{2}(v_1 + v_2)$. Thus, the energy given to the body is the difference of its initial and final energy:

$$m(v_2 - v_1)\frac{1}{2}(v_2 + v_1) = \frac{1}{2}m(v_2^2 - v_1^2) \, .$$

We conclude that the kinetic energy of the body is $\frac{1}{2}mv^2$. This is an important result to remember. If the speed of a body is doubled, the kinetic energy is increased by a factor of 4 and so on.

2.5. Angular Momentum

Angular momentum is to rotational motion what momentum is to linear motion. Whereas *momentum* is only changed by applying a force, so that bodies move in a straight line with constant velocity (or momentum) unless acted upon by a force, they move with constant *angular* momentum about any point F unless a *torque* is exerted around that point. Consider a body with velocity v at point A (Fig. 2.1). Call its velocity *transverse* to FA, v_t. Its transverse momentum is mv_t if m is its mass. Its angular momentum is defined as rmv_t (transverse momentum × distance from F).

Let us now show that if a body is acted upon only by forces *toward* a fixed point, its angular momentum *around* that point does not change.

$$P'Q' = v_t\Delta t$$

$v_t \equiv$ transverse velocity

$\Delta t =$ time between points P and Q in the orbit

Distance $FR = r$

Area swept out in the orbit between P and Q is $\frac{1}{2}rv_t\Delta t$; area swept out per *unit* time $= \frac{1}{2}rv_t =$ angular momentum$/2m$.

Next, let $v =$ velocity at A; v_t and v_ℓ the transverse and longitudinal components respectively.

Imagine a body moving in a straight line through A, in the same direction as our orbiting body at A, and at the same speed. At A, then, they are for the moment moving together.

Now the triangle of velocities at A is similar to the triangle BAF (their sides are in the same proportions). Thus

$$\frac{z}{r} = \frac{FB}{FA} = \frac{v_t}{v} \qquad \text{so that} \qquad rv_t = zv \ .$$

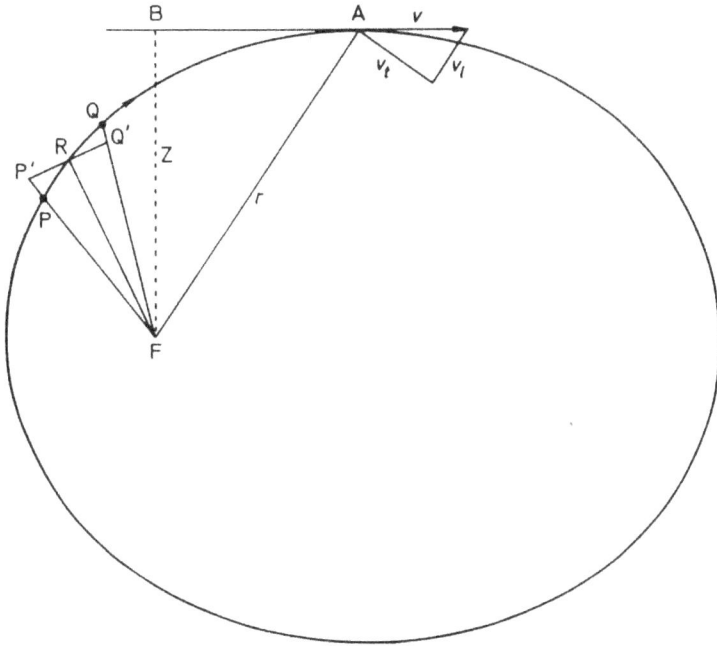

Fig. 2.1. The figure shows an elliptical orbit of a planet about a star. The velocity at A is seen to have a longitudinal component v_l and a transverse component v_t. Imagine a body moving freely with the same velocity as the planet at A. Let B be the point of its trajectory nearest to the sun at F.

Looking at another part of the orbit, let the planet go from P to Q in some interval of time. In that time, its orbit will trace out an area $FPQ = FP'Q' = \frac{1}{2}r \times P'Q' = \frac{1}{2}r \times PQ$.

But for the co-moving body, z and v are constant; so then is zv and *therefore* rv_t. That is,

rv_t **is not changing at the point A.**

The same will be true for the orbiting body if its transverse motion is also free, that is, if no transverse force is exerted about F so that v_t is *changing* at A. A *torque* on the orbiting body at A is defined as ($r \times$ transverse force). A force *along* FA, *toward* or *away from* F, exerts no turning effect (torque) about A, and does not change its angular momentum.

Although it is not too difficult to understand angular momentum for a single body, like a planet revolving around a sun, it is more difficult to envisage, or predict, the consequences of angular momentum for a macroscopic object. From the law of conservation of angular momentum of a single particle one can see that that for rigid bodies, say, follows. This results from two facts: one, that the angular momentum of a composite body is the sum of the angular momenta of its parts, and two, that the internal forces between elements of the body, being equal and opposite, do not exert any torque on the body as a whole.

The *consequences* of this conservation, however, or those of the behaviour of a rigid body acted upon by a torque, are much more complicated. We are familiar with gyroscopes and tops, and also with the phenomenon by which skaters, spinning with their arms outstretched, can speed up their spin by drawing in their arms. We know that gyroscopes are used to stabilize spacecrafts against undesirable rotations. It is important to realize that angular momentum has a *direction* associated with it, that of the axis of spin, and that conservation of angular momentum implies that the *axis* remains fixed unless a torque is applied to it. In space no such torque exists, so the axis of a gyroscope aboard remains fixed. If the spacecraft is "locked" to it, its direction will also remain fixed. On the other hand, interaction between the gyroscope and the ship can permit transfer of some angular momentum, and thus provides a means of orienting the ship. No such interaction, however, can *change* the total angular momentum of the system.

A close parallel exists between momentum and angular momentum. Momentum remains constant, and bodies travel in a straight line with constant velocity, *unless a force is exerted* from the *outside*. The force may, however, change the *direction* of momentum without changing its magnitude (as, for example, in a perfectly elastic collision with a fixed object). That is, however, a *change* in momentum; at any moment of the collision this change is proportional to, and in the same direction as, the force being exerted.

Similarly, it takes a torque to change the *axis* of rotation of a body, even if its rate of change is not affected. That is why a top is more stable the faster it spins.

2.6. Law of Universal Gravitation and Kepler's Laws of Planetary Orbits

As we mentioned earlier, Newton's mechanics is not complete without postulating laws of force. The most famous such law was Newton's law of universal gravitation, which stated that *all* objects in the universe attracted each other, that is, there was a force along the line joining them that pulled them together; and that this force was proportional to the mass of each and *inversely* proportional to the square of the distance between them.

The most important thing about this law is that it is *universal*; everything in the universe attracts everything else. Between small objects, the force is weak; between electrons it is of the order of 10^{-42} times the electric force between them at any distance. 10^{-42} means $1/1,000,000...$ (42 zeros in all) $...000$. If we consider protons rather than electrons the ratio is closer to 10^{-36}. For a fixed charge, the forces become comparable for a mass of a millionth of a gram. On the other hand, for massive bodies, and especially astronomical ones, gravity dominates over all other forces. Finally, the universe itself is held together by gravity, which determines the very structure of the universe.

The formulation of Newton's law of gravitation created the theoretical basis for astronomy. Johannes Kepler had already formulated the laws of planetary motion, on the basis of empirical observation, as follows:

1. The planets move in ellipses with the sun at a focus.
2. Their orbits trace out equal areas in equal times.
3. For the different planets the squares of their periods in their orbits are proportional to the cubes of the major axes of those orbits. Thus, knowing the distances of the planets, the lengths of their years can be determined, and vice versa.

But the distance of Mars from the sun is 142 million miles, compared with 93 million miles for the earth. The ratio of the cubes of these quantities is 3.56. It follows that

$$\text{Mars year}/\text{earth year} = (3.56)^{1/2} = 1.89 \ ,$$

exactly as observed.

The second of Kepler's laws is, as we have already seen, in effect equivalent to the law of conservation of angular momentum, and arises from the fact that the gravitational force is exerted *toward* the point around which the orbit rotates. It puts into quantitative form the intuitive expectation that planets move slowest when they are farthest away.

It is instructive to consider planetary orbits from the viewpoint of *energy*. If one wanted to move a planet or anything else in the sun's gravitational field outside the influence of that field, one would have to do work, i.e. to provide energy. The *amount* of energy needed to remove it is called the *binding* energy. The body has *less* energy when it is bound by gravitational force than when it is not. This idea also applies to the chemical forces which bind atoms and molecules together, and the nuclear forces which hold protons and neutrons together in nuclei. The general principle is that the attraction of bodies decreases the energy of the system which they constitute.

The angular momentum law also has consequences beyond that concerning motion in planetary orbits.

Consider, for example, galaxies which are agglomerations of some hundreds of billions of stars, as well as other interstellar matter. Evidence shows that, while our galaxy and many others have the form of flattened spirals, younger galaxies are more spherical in form. This might be a consequence of gravitational interaction in a system with angular momentum. The gravitational attraction causes the matter of the galaxy to pull together. However, the contraction is inhibited in directions perpendicular to the axis of rotation, and particularly at greater distances from that axis, because of a "centrifugal force". It will then tend to flatten in the direction of the axis, but not so much in the plane perpendicular to it.

Another striking illustration of conservation of angular momentum is of quite recent interest. Ordinary stars of sufficient mass, when they have exhausted their supply of fuel, may collapse under their own gravitational force to incredible densities, so great that a cubic centimetre of their matter has a mass of approximately 1.4 hundred million tonnes. The mass of a sun of a million kilometres radius will then be compressed to a radius of the order of ten kilometres.

Since the collapse is accompanied by the "blowing off" of considerable amounts of matter, some angular momentum may be lost in the process. However, a substantial fraction of the angular momentum of the original sun should be retained by the resulting collapsed object. Because it is smaller, it must rotate much faster than its parent sun, so much so that, despite its huge mass, it may rotate many times a second. The first such object found, in 1964, rotated 30 times a second! These objects are "pulsars" or "neutron stars" which have been intensively studied by astronomers in recent years.

While the concept of motion changed radically with the advent of quantum mechanics over 50 years ago, the conservation laws, of energy, momentum and angular momentum, remain in the new mechanics as in the old. We shall therefore use them throughout this course.

We have, up to now, talked of the laws governing the motions of *particles* of matter. Modern physics tells us that even very tiny macroscopic objects (e.g. a cube of sugar) contain incredible numbers of component particles. A cubic centimetre of solid matter will contain some million million million millions of electrons, protons and neutrons. The simplest object seems to present problems of incredible complexity, if our starting point is with the ultimate components of matter and the fundamental laws. Such an approach is quickly seen to be impractical, so we must take instead a point of departure not so remote from our observation and study *macroscopic* properties. This is the program of modern statistical physics. It provides surprises, however. The properties of complex systems made up of fundamental particles do not appear to be simply related to the properties of their components; in fact, they display new qualitative features. This is the failure of reductionism; the whole is more than the sum of its parts; novelty appears to have its origin in their interactions.

It is this sort of question which will occupy us when we deal with the theory of heat and thermodynamics.

Chapter 3

TWO SCIENCES MADE ONE:
ELECTRICITY AND MAGNETISM

The evolution of science is a continuous process. Thus, it is difficult to choose a starting point. Yet there has been, in this century, a revolution in scientific ideas; our framework of concepts has radically changed. What we can try to do, then, is to trace the new concepts back to their origin. The new ideas are embodied particularly in the theories of relativity and of quantum mechanics. The new concepts have to do with how we regard the fundamental elements of space, time and matter. The mechanical world which was embodied in the work of Newton and which dominated the thinking of scientists for several centuries has now been replaced by a world of "fields". A reasonable starting point for our discussion of modern physics is therefore to see how the image of a world of material particles in motion has been replaced by one of the flow and interplay of these more intangible entities called "fields". This enables us to choose a starting point: the work done, starting in the first half of the 19th century, on electricity and magnetism, and the discovery of the unexpected relationships between them. This was the beginning of discovering a unified view of the physical world. It is still our main preoccupation today.

3.1. Electricity

Electricity and magnetism have been known since classical times. We have a primitive experience of electricity when we rub a comb over a piece of cloth and find that bits of paper will then stick to it. At a slightly less primitive level a battery powers a flashlight or a transistor radio.

In the rubbing experiment, we explain what happens by evoking two kinds of "electric charge" — positive and negative (quite arbitrary designations) — such that like charges (positive-positive or negative-negative) repel each other but unlike ones attract (positive-negative). We believe that, in the universe as a whole, positive and negative charges are in complete balance (we shall discuss later the radical consequences if this were *not* the case). However, in the cloth-comb experiment, charges are transferred from one object to the other upon rubbing, so that the comb has an excess of one kind of charge, the cloth of the other. Similarly, you can amuse children by rubbing an inflated balloon on your hair. Afterwards, the balloon may be made to adhere to a wall, and your hair may stand on end. What causes this? To understand, it is not enough to say that like charges are attracted to each other while unlike ones repel. One must say how strong these attractions or repulsions are, and what they depend on. They depend evidently on *how much* charge is involved — the greater the *amount* of charge, the greater the force. But they depend also on the *distance* between the charges. In fact, experiments show that, for given amounts of charge, the force between them is 1/4 at twice the distance, 1/9 at three times the distance, 1/16 at four times the distance, etc. We call this the "inverse square law", or sometimes "Coulomb's law" after its discoverer. This means that the force becomes *very* strong at very short distances.

Imagine now the balloon rubbed on hair. The balloon acquires an excess of one kind of charge (let's call it negative; it doesn't really matter). We now bring the balloon close to a wall. The latter we consider like everything else to be made up of equal quantities of positive and negative charge. The balloon will attract the positive charges on the wall, and repel the negative ones. Its negative charges will then be attracted by the near positive charges on the wall, and repelled less strongly by the more remote negative charges. The attractive force dominates, and the balloon sticks to the wall. The action on the wall by which one sort of charge is pushed toward the surface and the other is pushed away is called the "electric polarization".

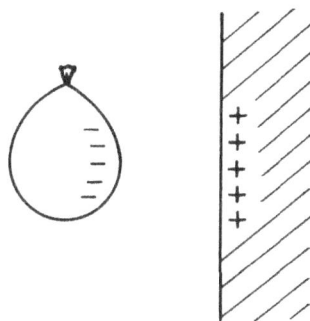

Fig. 3.1. The phenomenon of polarization. Negative charges on the balloon attract positive charges on the wall. Result: the balloon is attracted to the wall.

This concept of matter being made up of two kinds of charges is very important, because it gives us a basis for thinking about another aspect of electricity, namely, that it can "flow". We call a flow of electricity an *electric current*. The most primitive source of electric current is a battery. The first batteries were known as Voltaic cells (after the Italian physicist Volta). These consisted of plates of two different metals in a weak acid solution. Positive charges in the liquid tended to accumulate on one plate, negative charges on the other. If a wire (which we call a conductor because it is capable of transporting electric charge) is then used to connect the two plates, charge will then flow through the wire in such a way as to neutralize the accumulated charges.

Maxwell tended to think of the flow of electricity in a wire as the flow of liquid through a tube, i.e. electricity was regarded as a kind of fluid.

Why do we believe that positive and negative charges in our world are in complete balance? To answer this question we first point out that, just as there are forces of attraction or repulsion between *charged* particles, there are *attractive* forces of gravitation between *all* particles of matter in the universe. We shall see that gravitation plays a rather paradoxical role in nature, appearing at times to be negligibly weak and at others to be so strong that it can bring about the total collapse of systems in the universe or even of the universe itself.

If we compare the electric and gravitational force between two elementary particles of matter – electrons – the electric force is about 100,000,000,000,000,000,000,000,000,000,000,000,000,000,000 times the gravitational force. So when we discuss problems at the atomic level, we certainly do not need to concern ourselves with the gravitational force!

On the other hand, gravity dominates our daily life, holding us to the earth, guiding the planets in their orbits about the sun and holding the stars together in the galaxies. In fact, as Einstein showed, it even determines the evolution of our universe and may end up crushing it into a small, dense formless mass. We believe that there are astronomical objects so condensed that the mass of a mountain would be a billion times too small to be seen under a microscope — objects capable of crushing all surrounding matter into formlessness. These objects are known as "black holes", and we shall have more to say about them later.

How do we reconcile the weakness of the gravitational force in atomic processes with its overwhelming power in astronomical or cosmical phenomena? It is simply that in the gravitational case, *all* matter attracts all other matter. Each atom in the sun may exert a weak gravitational force, but the *number* of such atoms is incredibly large. Even our little earth contains about 10^{42} (this is 1 followed by 42 zeros) of them. Gravity is important because its effect is cumulative — the more matter there is, the stronger the gravitational attraction. If we take a large object like a star, every bit of it is pulled by all the rest of its matter. If it is near the outside, it is pulled *in*.

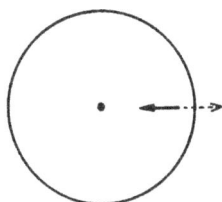

Fig. 3.2. There's a lot of matter pulling *in*. But there's very little pulling *out*.

Consider the gravitational force at point O outside the region bounded by the dashed circle (Fig. 3.3). All forces are pulling toward that region.

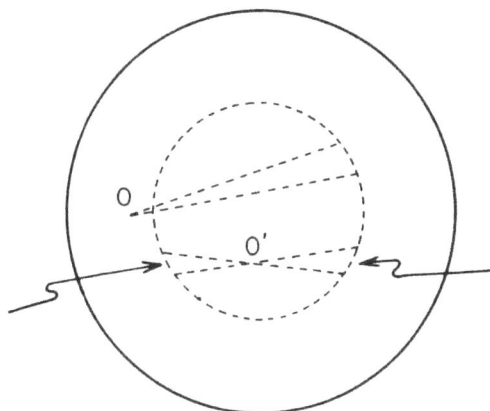

Fig. 3.3.

At O', on the other hand, the material to the left of O' is pulling one way, and that on the right the other. On the left, there is less material at a shorter distance; on the right, more material at a greater distance. The net effect is that the forces cancel each other out.

Why, then, doesn't everything collapse under gravitational force? Only because some other force is balancing it. When such other forces are exhausted or overwhelmed, unrestricted gravitational collapse *will* take place — hence "black holes".

Now, let us come back to our fundamental question — why is it that, if they are so much stronger than gravitational forces, electrical forces do not have more catastrophic effects on our universe? The answer is found in the existence of *two* types of charged matter, positive and negative — with the possibilities of both *attraction* and *repulsion*. An atom of hydrogen, for example, is held together by electrical attraction, being made up of a positive charge — a proton — and a negative charge — an electron. The result is that it exerts no net electrical force on another distant electron; its proton attracts it, and its electron repels it with exactly the same force. Attraction and repulsion balance each other, and no net electrical force is felt. But if there were even a slight imbalance in the amounts of positive and negative charges in our world, the unbalanced electrical force which it would exert would be much greater than the gravitational force, and would play a decisive role in astronomy. In fact, as has been known from the time of Newton, Galileo, *et al.*, the behaviour of the solar system seems to be completely understandable on the assumption that gravity alone governs the way things evolve.

For this reason we have confidence that the electrical charges of the matter of the world around us are in complete balance; the amount of positive charge is in balance with the amount of negative charge.

3.2. Fields and Lines of Force

We recognize that gravitational, as well as electrical, forces act "at a distance", i.e. through space. A charged object then has the *capacity* to exert a force on another charge (a "test charge"), according to where that charge is placed. We can conceive of that "capacity to exert a force" as a quality distributed in space even when the test charge is absent. Each point of space is then characterized by a force defined in magnitude and direction. A more modern way to put it is to say that the charged object creates a "field of force" in the space surrounding it. The direction of that field varies from point to point, and this fact may be displayed in an ingenious way, whose virtue is that it gives us a visual image of the field. This is the way it is done for the electric field: fine pollen grains are suspended in a shallow dish filled with an insulating fluid. If an electric field is created in the fluid, the pollen grains will point in the direction of the electrical force. Why? Firstly, grains are long and thin. Secondly, they contain positive and negative charges which can be displaced relative to each other by an electrical force (actually, we know that the positive ones are more or less fixed, while the negative ones can move).

In the accompanying diagram, negative charges are pulled to the end E_2 of a grain nearer to an external charge e^+. This leaves an excess of positive charge at the other end (E_1). The result is that the field produced by the external charge pulls the end E_2 of the grain toward it, and pushes the end E_1 away. Consequently the grain will be lined up in the direction of the electric force — i.e. toward e^+.

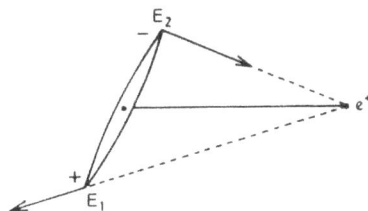

Fig. 3.4.

With this in mind, we can interpret the following pictures. In all cases, a field is created in the insulating fluid by introducing charges into it.

In Fig. 3.5(a), the tip of an electrically charged needle is inserted. (Assume the needle's charge to be positive.) Negative

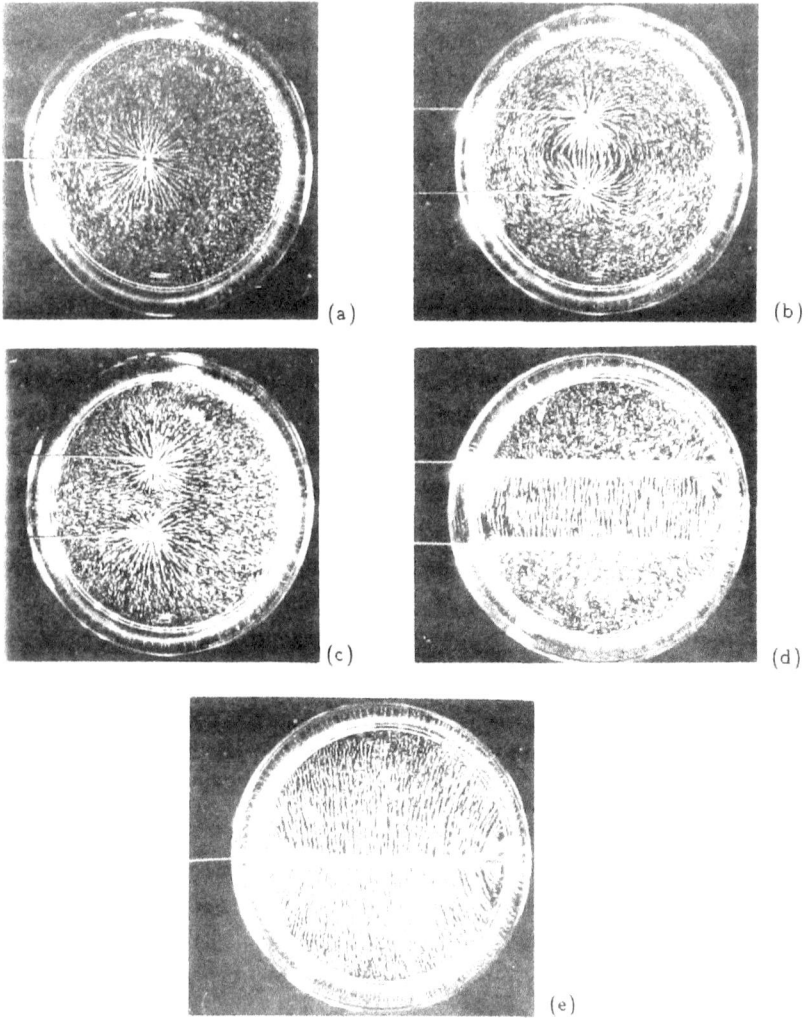

Fig. 3.5. Electric force field patterns formed by grass seeds in an insulating liquid. (a) a single charged rod; (b) two rods with equal and opposite charges; (c) two rods with the same charge; (d) two parallel plates with opposite charges; (e) a single charged metal plate. (Physical Science Study Committee, *Physics*, 2nd edition, 1965; D. C. Heath and Co. with Education Development Center, Inc., Newton, Massachusetts.)

charges are attracted to the ends of nearby grains which are nearest the needle; positive charges are pushed to the other end (polarization). The two forces on the needle, the attraction of the negative charges on the near end, and the repulsion of the positive charges on the more distant end, tend to turn the needle so that it aligns itself in the direction of the electric force of the needle. Thus, in (a), the grains all point toward (or away from) the tip of the needle.

In Fig. 3.5(b), two such charged needles are inserted; the force on the charges in the grains leads them to align themselves in the direction of the net force, a combination of the force of the two charges. The fact that the grains between the two charges point from one needle tip to the other shows that the needles are oppositely charged. Note now that the "lines of force" shown by the needle orientations go from one pole to the other; this system is called a "dipole".

In (c), the two poles are similarly charged. Near the mid-point between them the lines of force tend to point away from both, while retaining symmetry between the two.

In (d), the edges of two charged objects are inserted. (They might for example be charged razor blades.) They must again be oppositely charged, because the lines of force lead from one to the other.

Finally, in (e), a single charged blade is inserted, and lines of force run out perpendicularly from it. Their bending near the edges is also easily understood, since there is more charge influencing the grains on the side toward the centre than on the opposite one.

What these demonstrations have done is to show how we may, so to speak, make the force visible; the pollen grains point along the direction of the net electric force at each point, the *lines of force*.

3.3. Magnetism

How about magnetism? The classical magnet is a mineral called lodestone, which can be found in nature. It has the "magic" property of attracting certain metallic objects. But it also "magnetizes" (i.e.

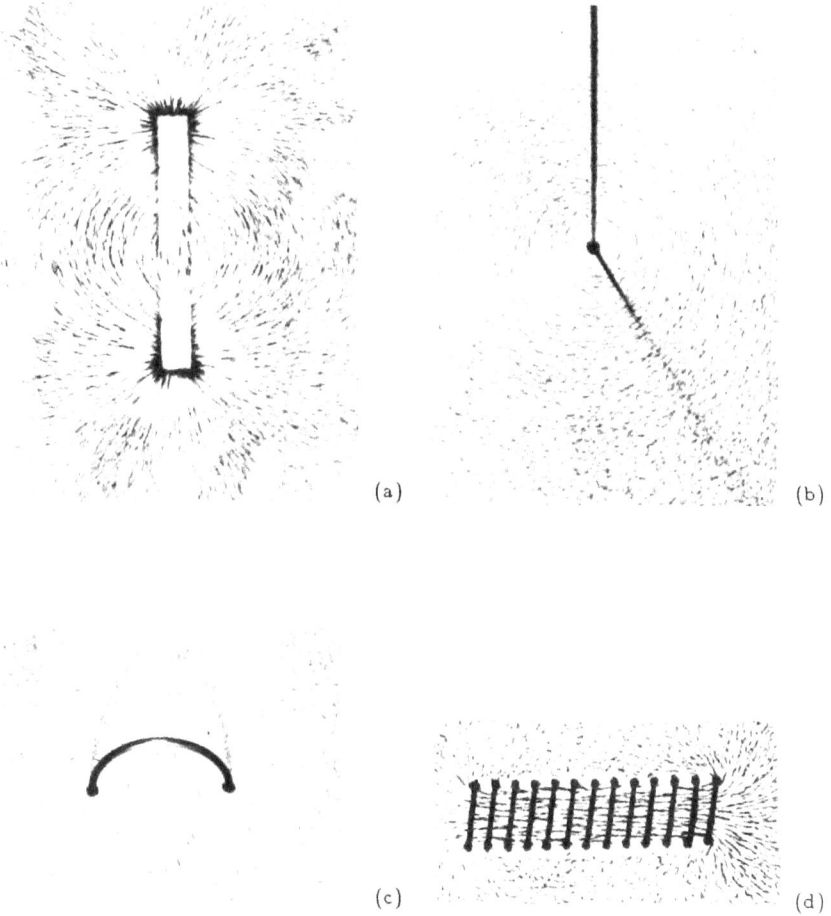

Fig. 3.6. Magnetic field patterns due to: (a) a bar magnet; (b) current flowing in a long straight wire; (c) current flowing in a loop wire; (d) current flowing in a solenoid. (a–c: Physical Science Study Committee, *Physics*, 2nd edition, 1965; D. C. Heath and Co. with Education Development Center, Inc., Newton, Massachusetts. d: Reprinted with permission from R. Kronig, ed., *Textbook of Physics*, copyright 1959, Pergamon Press PLC.)

converts into magnets) these objects themselves. The most common magnets are, of course, made of iron.

At this point I should like to mention a simple observation about magnets. This observation had great impact on the thinking of Michael Faraday. Faraday had little formal schooling and never learned formal mathematics. Therefore, instead of casting physical laws in mathematical form and "reasoning" with mathematical calculations, he depended on mental pictures of physical processes to guide his thinking.

Let's then look at this simple experiment. Take a bar magnet and place a thin piece of paper over it. Then sprinkle it with iron filings. These iron filings tend to have a needle-like shape, and what we observe is that these needles orient themselves in an easily distinguishable pattern (Fig. 3.6a). They seem to point along lines running from one end of the magnet (one "pole") to the other. Here is how we interpret this observation: we say that every magnet has two poles (arbitrarily called "north" and "south", not unreasonably since the familiar compass tells us that the earth itself is a giant magnet). Consider now two magnets: we may verify that, much as in the case of electricity, like poles repel each other and unlike ones attract. Thus if we place two magnets like this,

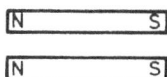

Fig. 3.7a.

they repel each other. But if we place them like this,

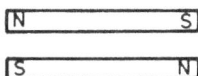

Fig. 3.7b.

they attract each other.

Suppose that matter is made up, not only of charges, but also of elementary magnets. Imagine that we take a needle, and run the north pole (say) of a magnet over it:

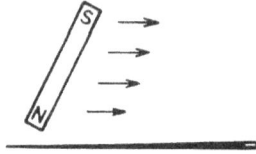

Fig. 3.8a.

The *south* poles of the elementary magnets of the needle will be pulled to the right, so that the needle will become magnetized thus:

N ━━━━━━━━━━━━━━━━━━ S

Fig. 3.8b.

As in the case of electricity, we again call this process "polarization" – in this case, *magnetic* polarization.

Now imagine an iron filing placed near a bar magnet. If it is equi-distant from the two poles of the magnet, it will orient itself as shown (south pole attracted to north pole and vice versa).

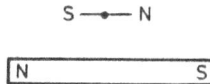

S ─•─ N

| N   S |

Fig. 3.9.

You may say, but will it not be pulled *toward* the magnet? In fact, there *will* be a force tending to act in such a way. But this force is relatively weak compared to that tending to turn the filing, and the friction with the paper is sufficient to prevent it from having any effect.

Faraday, in observing this phenomenon of orientation of the iron filings, saw "lines of force" emanating from the magnet. The iron filings served to make them visible to the eye. The filings, the tiny magnets, were constrained to align themselves in the direction of these "lines of force".

Note a difference between electric and magnetic "lines of force". If there is an *electric* charge at a point P the force on the other charges is along the lines pointing directly toward or away from it. They radiate into space:

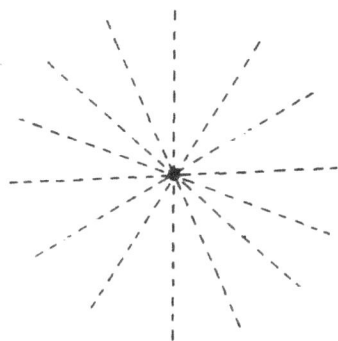

Fig. 3.10.

But with magnets the situation is different. If one breaks or

Fig. 3.11.

cuts a bar magnet, one does not make single poles, but two new magnets, each of which has a north pole and a south pole. The lines of force start and end at the magnet.

For many centuries, although both electricity and magnetism

were known, no one suspected a connection between them. Arthur Koestler, in his film on the psychology of creativity, contends that the key to the act of creation is to make an association between apparently independent things or concepts. It is certainly the key to *scientific* discovery, since its purpose is the pursuit of general laws and its aim is to reveal connections between all aspects of the natural world.

The first evidence of a *connection* between electricity and magnetism was discovered by Oersted (1777–1851), who found that a compass needle was deflected by a nearby wire carrying an electric current. The force on the compass needle was greater the closer it was to the wire, and the greater the magnitude of the current flow. Most curious, though, was the direction of the force on the compass. The lines of force circulated *around* the wire (Fig. 3.6b); the arrows indicate the direction in which the compass needle would point at the position indicated. Previously known forces had acted along the line joining the interacting objects (charges, gravitating bodies). This force acted *perpendicular* to that direction!

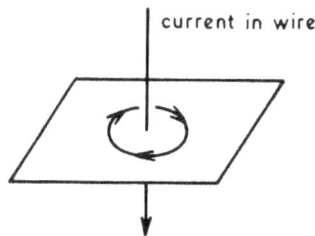

Fig. 3.12.

This leads us to an interesting observation. Suppose we take a horseshoe magnet of a well-known type and place it under a piece of paper, pole faces upward, and again test its field with iron filings.

What we find is something like this: the lines all lead from north pole to south pole (Fig. 3.13a). Now let's imagine a current flowing in a system like that shown in Fig. 3.13b. Using iron filings again, a rather similar pattern of lines of force can be observed. It is true that lines of force close to the separate wires may be circulated around one of them. However, we cannot be certain that similar closed lines of force do not exist *within* the two poles of the bar magnet, so it is still evident that coils of wire may simulate magnets. Is it not then possible that magnetism is a manifestation of circulating electric currents?

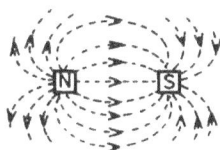

(a)

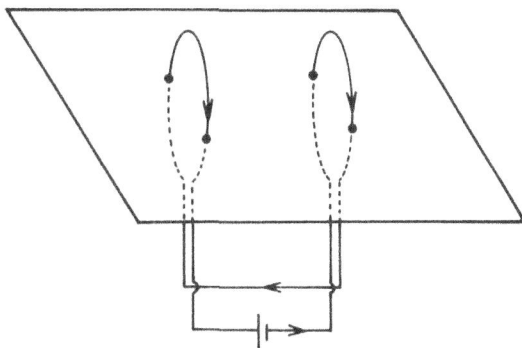

(b)

Fig. 3.13.

Chapter 4

FARADAY, MAXWELL AND
ELECTROMAGNETIC THEORY

4.1. The Contributions of Michael Faraday

First, let me reiterate the motivation behind the work of Faraday which led to his discovery of the law of induction. Moved by a belief in the symmetry of nature, and by Oersted's discovery that an electric current produced a magnetic force which, at any point, circulated around the wire, Faraday expressed the feeling that "it appeared very extraordinary that as every electric current was accompanied by a corresponding intensity of magnetic action at right angles to the current, good conductors of electricity, when placed within the sphere of this action, should not have any current induced through them, or some sensible effect produced equivalent in force to such a current".

The apparatus he used is shown in Fig. 4.1. You will recall that a consequence of Oersted's law is that a solenoidal coil acts like a magnet, the magnetic forces of the separate turns of the coil adding along the axis. If one puts a bar of iron inside the coil, it in turn becomes magnetized. If the iron is in the form of a ring or doughnut, closed lines of force run through it in circles.

What Faraday did was this: he wrapped a coil of wire, the ends of which would be attached to a battery, around such a circular ring of iron. Then, separated from it by insulation, he wrapped another coil the ends of which were connected to a galvanometer. For our purposes, this is simply an instrument for measuring an

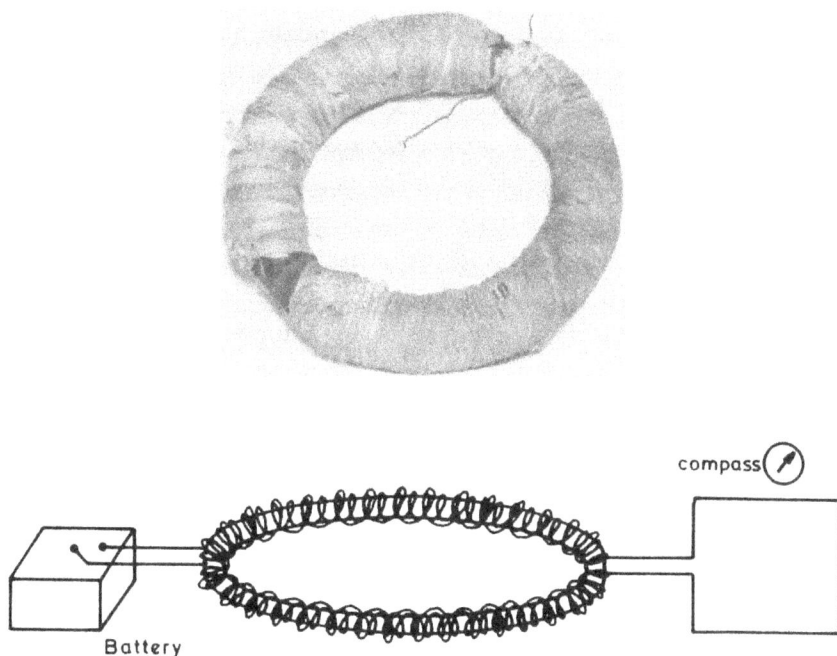

Fig. 4.1. Top: One of the coils used by Faraday in his experiments. Bottom: Schematic representation of an induction experiment.

electric current.

His idea in this experiment was the following: when the battery coils were connected, a current would flow through the first coil; this would not only produce its own magnetic field, but that field would magnetize the iron. He would then see whether this magnetization would create a current in the other wire, which would be detected by a swing of the galvanometer needle.

In his first observations, no current whatsoever was detected in the second coil. On closer observation, however, when the terminals of the first coils were connected, there was a small temporary deflection of the galvanometer; on cutting the current a similar deflection in the opposite direction was seen.

While a less meticulous scientist than Faraday might well have missed this effect or, having observed it, might have dismissed it as spurious, Faraday immediately recognized this as the sort of effect he was searching. Nothing happened when a steady current flowed. The effect was associated with a *change* in the current flowing in the primary; it was evidently a *changing* magnetic field inside the secondary coil which had induced the current.

Faraday proceeded to study this effect in detail in a series of experiments designed to obtain more precise information on the qualitative and quantitative effects of his law of induction. He summarizes his conclusions in terms of his own image of lines of force.

"If a terminated wire moves so as to cut a magnetic curve, a power is called into action which tends to urge an electric current through it; but this current cannot be brought into existence unless provision is made at the ends of the wire for its discharge and renewal.

If a second wire moves with a different velocity or in some other direction, then variations in the force exerted take place; and if connected at their extremities, an electric current passes through them."

Evidently, then, Faraday had found evidence for the sort of symmetry between electricity and magnetism which he was convinced ought to exist.

Let us for the moment use the word "field" to summarize where things stood as a consequence of the experiments of Oersted and Faraday.

(i) An electric current produces a magnetic field which circulates around it;

(ii) A *changing* magnetic field produces a current in a wire which passes across the lines of magnetic force.

It seems that the situation is not symmetric, since a *steady* electric current produces a magnetic "field", but only a changing magnetic field produces a current in a wire.

As pointed out by Einstein and Infeld in "The Evolution of

Physics", however, the American physicist Rowland showed that this lack of symmetry was only apparent. By rapidly rotating a disc around the circumference of which the charges were placed, he was able to produce the same effect on a magnet as an electric current. In fact, now that we know that an electric current consists of a *flow* of electrons, we see that it is the *motion* of charge, and not charge at rest, which produces a magnetic field. Similarly, the *motion* of a magnet produces electric effects, as may be seen by plunging a bar magnet through the core of a solenoidal coil and observing that a current is produced in the coil.

Another argument clarifies the situation somewhat. It is based on the conservation of energy. If a constant magnetic field could produce a current, and since current carries energy (as evidenced in the fact that it produces heat), we would have a mechanism for the perpetual production of energy. The same would of course be true if a fixed charge produced a magnetic field, since the latter could be used to pick up (and thus give energy to) iron nails, etc. It must be noted, however, that at the time when Faraday was conducting his experiments, the law of conservation of energy had not been given an explicit general formulation.

Although Faraday gave evidence for "the unity of the forces of nature", it was Maxwell who formulated in an explicit and general way the relation between electricity and magnetism.

Before going on to discuss his contribution, though, it is useful to summarize the contribution, not only of Faraday's experiments, but also of the conceptual framework in which he placed them. To what extent can we give Faraday credit for explaining electricity and magnetism and their interrelations in terms of "fields"?

As Koestler noted, the "lines of force" associated with electric and magnetic fields were for Faraday as real as solid matter. Faraday was influenced by the 18th century mathematician, physicist and philosopher Boscovitch (1711-87), who propounded the idea that material particles were essentially centres of force, each such centre being connected to every other by the forces they exerted on each other. The young Faraday wrote, in 1818, with regard to the dictum of the time that "bodies do not act where they are not", "Is not the reverse of this true? Do not all bodies act where they are not and do any of them act where they are?" and again,

"Final brooding impression that particles are only centres of force; *that the force or forces constitute matter*"; which almost seems to presage Einstein's equivalence of mass and energy.

Tyndall, a successor of Faraday at the Royal Institution, tried to convey Faraday's point of view as follows:

"Let it be remembered that Faraday entertained notions regarding matter and force altogether different from the view ordinarily held by scientific men. Force seemed to him an entity dwelling along the line in which it is exerted. The lines on which gravity acts between the sun and the earth seemed figured in his mind as so many elastic strings .... . Such views, fruitful in the case of magnetism, barren as yet in the case of gravity, explain his efforts to transform this latter force. When he goes into the open air and permits his helices to fall, to his mind's eye they are tearing through the lines of gravitating power, and hence his hope and conviction that an effect would and ought to be produced. It must ever be borne in mind that Faraday's difficulty in dealing with these conceptions was at bottom the same as Newton; that he is, in fact, trying to overleap this difficulty, and with it probably the limits prescribed by the intellect itself."

In 1851, Faraday submitted a paper "On the Lines of Magnetic Force" in which he says:

"From my earliest experiments on the relation of electricity and magnetism, I have had to think and speak of lines of magnetic force as representations of the magnetic power, not merely in the points of quality and direction, but also in quantity. The necessity I was under of a more frequent use of the term in some recent researches, has led me to believe that the time has arrived, when the idea conveyed by the phrase should be stated very clearly, and should also be carefully examined, that it may be ascertained how far it may be truly applied in representing magnetic conditions and phenomena; how far it may be useful in their elucidation; and how far it may assist in leading the mind correctly on to further conceptions of the physical nature of the force, and the recognition of the possible effects, either new or old, which may be produced by it.

A line of force may be defined as that line which is described by a very small magnetic needle, when it is so moved in other directions correspondent to its length, that the needle is constantly a tangent to the line of motion."

In fact, the last sentence comes very close to our current definition of the magnetic field, except that the latter defines not only the *direction* of the magnetic force at any point, but also the *magnitude* of that force. Faraday also represented this in a pictorial way. He imagined that the number of lines of force emanating from a magnetic pole was proportional to its magnetic strength. Thus, if there was a stronger magnet, which exerted a correspondingly stronger force, there would be more lines of force passing through a given small region of space, that is, the lines of force would be more dense.

A few paragraphs on, Faraday writes:

"I desire to restrict the meaning of the term line of force so that it shall imply no more than the condition of force at any given place as to strength and direction; and not to include (at present) any idea of the nature of the physical cause of the phenomena; or to be tied up with, or in any way dependent on, such an idea. Still, there is no impropriety in endeavouring to conceive the method in which the physical forces are either excited, or exist, or are transmitted."

It seems clear that what Faraday is looking for is a way to "understand" the new phenomena of electromagnetism in terms of a *mechanical* model, that is, in terms of old concepts. This tendency carried on, as we shall see, in the work of Maxwell. In fact, electromagnetism would ultimately be seen to be separate from, and on an equal footing with, mechanics, rather than a derivative of it; the two together would be united at a higher theoretical level, when matter and field were seen to be different aspects of the same thing.

It is a similar tendency to wish to "explain" the new in terms of the old which has made it difficult, from the beginning, for some physicists to fully accept quantum mechanics. Einstein, even, was always unhappy about the intrinsic impossibility, within the quantum scheme, of giving simultaneous precise values to all of the quantities used to describe the states of classical systems. He was convinced that "God does not play dice with the universe" in allowing us only to prescribe the probabilities of certain values occurring for physical quantities. But in arguing thus, is one not simply displaying a resistance to discarding an old picture of nature and replacing it by a new one? Is not the inherent impossibility of characterizing a physical system in the same way as in classical physics the very essence of quantum mechanics, which insists that we look at nature in a new way?

K. Mendelsohn of Oxford has expressed the problem this way: "... it is very difficult to avoid introducing — quite unconsciously — analogies with everyday experience which have no basis in observation. The dualism of wave and particle would never have arisen except that it seemed quite legitimate to the physicists of the 19th century to compare light with ripples seen on the surface of water and molecules with ivory balls on the surface of a billiard table. But these are images which go clearly far beyond observational justification."

Faraday is clearly not totally unaware of the hazards of his models and pictures. Like Maxwell, he realizes that his pictures are not strictly necessary consequences of experiments, but are merely aids to "understanding" — understanding the new, that is, in terms of the old.

Before leaving Faraday, it is interesting to call attention to a letter he wrote to Maxwell in 1857 (he was 66 years old, while Maxwell was 26).

"There is one thing I would be glad to ask you. When a mathematician engaged in investigating physical actions and results has arrived at his own conclusions, may they not be expressed in common language as fully, clearly and definitely as in mathematical formulae? If so, would it not be a great boon to such as we to express them so — translating them out of their hieroglyphics that we might also work upon them by experiment. I think it must be so, because I have always found that you could convey to me a perfectly clear idea of your conclusions which, though they may give me no full understanding of the steps of your process, gave me the results neither above nor below the truth, and so clear in character that I can think and work from them."

As R A.R. Tricker points out, Faraday's comment overlooks the contribution which mathematics makes to the process of thinking itself. It was Faraday's lack of mathematical training which prevented him from drawing the sort of precise conclusions from his discoveries which Kelvin and Maxwell were later able to make.

You may recall that Feynman said that the effort to express fully the essence of physical law in ordinary language could never be completely successful "because mathematics is not just another language. Mathematics is a language plus reasoning: it is like a language plus logic. Mathematics is a tool for reasoning". The latter statement is certainly true, but leaves some questions unanswered. We shall discuss this proposition in a later chapter.

Of course, much ingenuity can go into the development of the tools of mathematical analysis. Often, in this process, one is guided by one's intuitive feeling for the physical processes being studied, that is, the mathematical techniques used are a function of one's way of looking at the physical problem.

Sometimes, too, mathematics has another important function in providing a framework for what might be called "disciplined speculation". Mathematical models of nature can be constructed which conform to broad guiding principles of symmetry, of relativistic invariance, etc. Conclusions may be drawn from these speculations, and may lead to predictions of new phenomena. Examples of this sort of discovery by mathematical speculation are to be found in

Yukawa's prediction of the meson, or Dirac's of antiparticles such as the positron.

4.2. The Contribution of J. C. Maxwell

The foregoing discussion of the role of mathematics in physics is a useful introduction to James Clark Maxwell's contribution to the theory of the electromagnetic field. The books of Everitt and of R.A.R. Tricker provide biographical information on Maxwell, which we summarize briefly:

Maxwell was born in 1831, when Faraday was already forty years old and embarking on his most important research. Unlike Faraday, he came from a prosperous family, and had the best formal education available at the time. His grandfather was an East India Company captain whose brother was a baronet, his father was of the landed gentry who inherited an estate in Dumfriesshire in the Scottish border country. He received private tutoring at his father's country estate till the age of ten, when he went to Edinburgh to live with an aunt and became a student at Edinburgh Academy. He sometimes spent holidays at the home of William Thomson (Lord Kelvin) and was encouraged by his father, who had an amateur's interest in science, in his studies in mathematics and physics. At the age of 16 he entered the University of Edinburgh; while a student there he had two papers published, one on a geometrical problem and one on the elasticity of solids.

At the age of 19 he went to Peterhouse College, Cambridge, and later transferred to Trinity College, where he passed the famous mathematical tripos in 1854, ranking second in the class.

Like most scientists of that time, his education and his interests were broad. He read the classics, studied philosophy, and wrote a considerable quantity of English verse. He was, rather like Faraday, a conservative and religious man, and engaged in the polemic against Darwin's theory of evolution. His argument was that the atoms of which matter was made could not have evolved!

In 1855, at the age of 24, he obtained a fellowship at Trinity College, Cambridge. A year later, a chair became vacant at Marischal College, Aberdeen; he applied for and received the appointment. All the while he spent as much of his time as possible at his country estate of Glenair in Dumfriesshire. He had a tendency, it is said, to talk well over the heads of most of his students.

In 1860 Maxwell was appointed to King's College, London, where he did his most important work on electromagnetism. During his time there he gave evening lectures to working men. He also made the acquaintance of Faraday, by now well past his prime. Maxwell had a lasting admiration for Faraday. In addition to Maxwell's contribution to electricity and magnetism, he made contributions to the theory of colour, to dynamics and to the theory of gases, the last being work of major importance to which we shall return later.

His last post was at Cambridge, where he became the first Cavendish professor of physics and was instrumental in setting up the Cavendish Laboratory, which became world famous and remains so to this day.

In his later years Maxwell wrote two great treatises, one on electricity and magnetism. It was the culmination of his life's work in this field. The other work was his "Treatise on Heat". He died at the age of 48 in 1879 of intestinal cancer.

At the beginning Maxwell's work was firmly based on that of Faraday. His first major paper on electromagnetism was entitled "On Faraday's Lines of Force". In it he set about to translate Faraday's ideas into mathematical language. It is interesting that he did this with very little reference to electromagnetism itself; rather, he attempted to develop mechanical analogies which he felt made Faraday's lines of force easier to visualize. The analogy which Maxwell chose in this case was that of motion of an incompressible fluid. He explains his goal as follows:

"It is not even a hypothetical fluid introduced to explain the phenomena. It is merely a collection of imaginary properties which may be employed for establishing certain theorems in pure mathematics in a way more intelligible to many minds and more applicable to physical problems than that in which algebraic symbols alone are used."

The analogy to the lines of force is that of lines of fluid flow. What the paper provided was the basic mathematical apparatus with which electric problems could be approached. He is, however, modest in his claims for the work. He says "I do not think it contains even the shadow of a true physical theory; in fact its chief merit as a temporary instrument of research is that it does not, even in appearance, account for anything."

In a later paper on "A dynamical theory of the electromagnetic field" Maxwell addressed himself to another problem. The general notion of the time regarding the interactions of two objects (e.g. electric current in a wire and magnet) was that the interactions reached out, so to speak, across space, and did not depend on what, if anything, was between them. That certainly seemed to be the case with gravitation. However, in Oersted's experiment, the interaction was not only not along the line joining the interacting elements, but the force of interaction actually depended on the velocity of one element — that of the electric charges or "elastic fluid" which carried the electric current. Maxwell remarks:

"The mechanical difficulties, however, which are involved in the assumptions of particles acting at a distance with forces which depend on their velocities are such as to prevent me from considering this theory as the ultimate one, though it may have been, and may yet be, useful in leading to the coordination of phenomena.

I have therefore preferred to seek an explanation of the fact in another direction, by supposing them to be produced by actions which go on in the surrounding medium as well as in the excited bodies ...

The theory I propose may therefore be called a theory of the electromagnetic field, because it has to do with the space in the neighbourhood of the electric or magnetic bodies, and it may be called a dynamical theory, because it assumes that in that space there is matter in motion, by which the observed electromagnetic phenomena are produced."

We see that Maxwell has here made two steps — first, he has attributed electric and magnetic properties to space itself, the very essence of the theory of fields; but secondly he has filled space with an invisible "matter in motion", which we shall henceforth call the aether. Ironically, it was the consistent development of the first idea which was ultimately to undermine the second, but not till some forty years later.

It was at this time that Maxwell introduced an extremely important modification of Oersted's law, without which the prediction of electromagnetic waves could not have been made. We previously summarized Faraday's and Oersted's laws thus:

(i) Oersted: an electric current produces a circulating magnetic field and

(ii) Faraday: a changing magnetic field produces an electric current in a wire.

It is evident that, in this form, the situation is not really symmetric between electric and magnetic forces. We speak, on the one hand, of magnetic *fields* (which permeate space) but on the other hand, of electric *currents* (which flow only in matter). Should not electric *fields* occur in a true field theory, rather than material electric currents?

Now it was already understood that a current was produced in a wire by an "electromotive force" — an electric force or field in which charges moved. In the case of Faraday's law, a changing magnetic field produces an electric field, and this field can be detected by the current produced in a wire. But the essential difference is that the electric field is produced *whether or not* a circuit is present in which to produce a current. The wire only serves the secondary purpose of detecting the electric field.

We therefore write Faraday's law of induction in the following form: a changing magnetic field produces an *electric* field which circulates around the lines of magnetic force.

With respect to Oersted's law, consider the simple system indicated in Fig. 4.2.

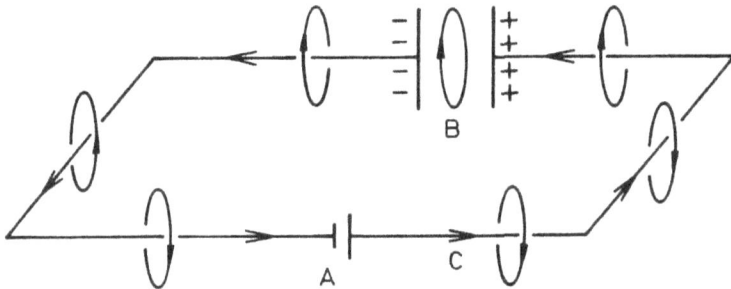

Fig. 4.2.

A is a battery which causes a current to flow in the circuit in the direction shown by the arrow C. The circuit is broken by a condenser, a pair of parallel metal plates, at B. As the current flows, a magnetic field will be set up around the wire, as shown by the loops indicated. According to our present understanding, electrons, carrying negative charge, will accumulate at the left hand plate. The resulting accumulation will repel the electrons from the right hand plate, leaving a net positive charge. The flow of current is in the same direction around the circuit. Lines of magnetic force at all points circulate around the wire.

What happens at the region between the plates? Surely the magnetic circulation around the circuit will not suddenly disappear. Maxwell said that as the charge on the condenser plates accumulates, a constantly increasing electric field is set up between the plates. This changing electric field can give rise to a transverse magnetic field around the electric lines of force. The magnetic lines of force ringing the circuit would be continuous around the circuit, just as if the circuit were closed and the current would circulate through without interruption.

This interpretation has two attractive features. First, it avoids the troublesome question of the arbitrary "disappearance" of the magnetic field in the region between the plates, and restores the continuity of the magnetic field. Secondly, even more important, it puts the electric and magnetic fields on a more equal footing, in that the electric aspects of the system no longer depend on the presence of a material carrier of electricity. Now a *changing electric field in space* produces the circulating magnetic field. Just as a changing magnetic field produces an electric field, now a changing electric field produces a magnetic field.

Once this step is taken, the possibility of the propagation of electromagnetic waves becomes possible. For a changing electric field produces a (changing) magnetic field, which in turn produces a changing electric field, etc.; one has a self-sustaining mechanism which maintains these fields in space!

It is a bit difficult, without mathematics, to show how this leads to the propagation of a wave in space, but with the aid of the accompanying diagram (Fig. 4.3) and a bit of imagination, one can get some idea of how it works. The simplest procedure is to describe how the wave is propagated, and then to see how the principles we have evolved permit such a wave to be sustained. The mechanism is described in the text accompanying the diagram. The relationships between the fields are exactly as described in Oersted's and Faraday's laws. All mechanical objects have disappeared from the picture.

A detailed calculation yields still another piece of information, the speed of the wave, the rate of advance of its crests or valleys. This is expressed in terms of the constants of proportionality in the Oersted's and Faraday's laws; one constant determines the strength of a magnetic field due to a given current at a given distance, and the other constant determines the strength of the electric field due to a given rate of change of the magnetic field. Both quantities are determined experimentally. Using these values, Maxwell calculated the velocity of propagation of the wave to be approximately 186,000

Physics: Imagination & Reality

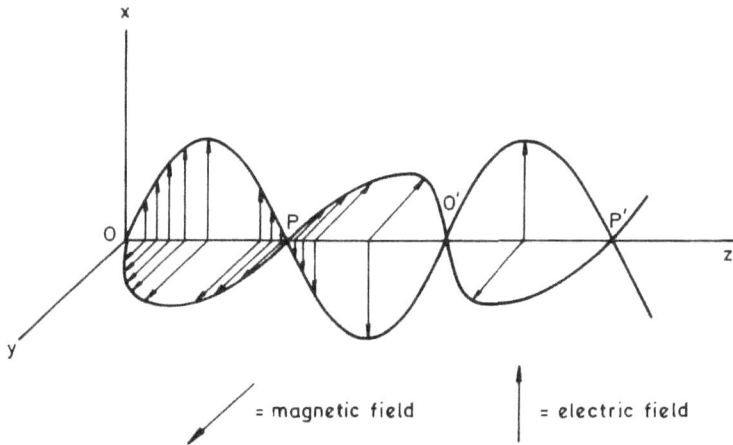

Fig. 4.3. Propagation of an electromagnetic field. The electric field always points *up*, and the magnetic field always points out. The whole pattern moves forward to the right.

The electric field is changing most rapidly at O, O', P and P'. At O it is changing from up (positive) to down (negative). The *magnetic* field is zero at O, and increases towards the right. This is equivalent to rotating, clockwise, about the vertical z axis, i.e. rotating about the direction of the electric field.

Conversely the magnetic field is also changing most rapidly at the same points. At O the magnetic field is changing from positive to negative, and the electric field is rotating about its direction (the y-direction outward) in an *anticlockwise* direction.

At P and P' the fields are changing in the opposite direction to that at O, O' (from negative to positive) and the direction of rotation of the fields is also reversed.

miles per second, which was the already known value of the velocity of light. Thus, the velocity of light can be determined, if not truly measured, by two rather simple experiments which can be carried out by a physics student, and which require only a very modest expenditure on equipment.

The argument that leads to the prediction of electromagnetic waves does not make any reference to the properties of the medium through which they are propagated. In fact it implies that they can be propagated in a vacuum. Nevertheless, Maxwell and other physicists of his time were so conditioned to mechanical models that it seemed inconceivable that a medium was not needed for the propagation of the waves. Certainly, all other sorts of waves had an identifiable medium: waves on water, sound waves in air, etc.

In one of his papers, Maxwell says:

"The electromagnetic field is that part of space which contains and surrounds bodies in electric or magnetic conditions.

It may be filled with any kind of matter, or we may attempt to render it empty of all gross matter ...

There is always, however, enough matter left to receive and transmit the undulations of light and heat, and because the transmission of these radiations is not greatly altered when transparent bodies of measurable density are substituted for the so-called vacuum, we are obliged to admit that the undulations are those of an aethereal substance, and not of gross matter, so as to heat it and affect it in various ways ... Professor Thomson has argued that the medium must have a density capable of comparison with that of gross matter and has even assigned an inferior limit to that density.

We may, therefore, receive as a datum derived from a branch of science independent of that with which we have to deal, the existence of a pervading medium of small but real density, capable of being set in motion and of transmitting motion from one part to another with great but not infinite velocity."

Still later, in his 1873 Treatise, he reaffirms his belief in the aethereal medium:

"A theory of molecular vortices, which I worked out at considerable length, was published in the *Phil. Mag.*"

"I think we have good evidence for the opinion that some phenomenon of rotation is going on in the magnetic field, that this rotation is performed by a great number of very small portions of matter, each rotating on its own axis, this axis being parallel to the direction of the magnetic force, and that the rotations of these different vortices are made to depend on one another by means of some kind of mechanism connecting them."

Of course, the "datum received from another branch of science" had no factual basis whatsoever, nor was there *any* evidence of rotations of molecular vortices, such as Maxwell wished to associate with magnetic fields. The basis of Maxwell's convictions was not scientific evidence, but merely a reflection of deep-rooted prejudices founded on a view of nature which had existed since Newton and had been accepted as self-evident by almost all scientists at that time

Figure 4.4 illustrates the mechanical model Maxwell had in mind, one of rotating hexagons separated by smooth ball-bearings. There is a problem – one cannot fill a plane with circles, while rotating hexagons run into obstruction. Lord Kelvin, in 1889, put forth a more elaborate model for an aether which could carry electromagnetic disturbances:

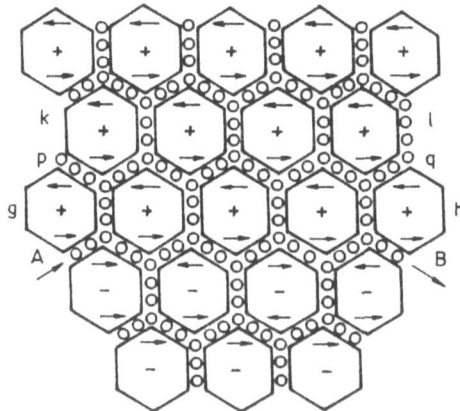

Fig. 4.4. Maxwell vortices. (With permission from R. A. R. Tricker, *The Contributions of Faraday and Maxwell to Electrical Science*, copyright 1966, Pergamon Press PLC.)

"[a structure] formed of spheres, each sphere being in the centre of a tetrahedron formed by its four nearest neighbours. Let each sphere be joined to these four neighbours by rigid bars, which have spherical caps at their ends so as to slide freely on the spheres. Such a structure would, for small deformations, behave like an incompressible fluid. Now attach to each bar a pair of gyroscopically mounted flywheels, rotating with equal and opposite angular velocities and having their axes in the line of the bar; a bar thus equipped will require a couple to hold it at rest in any position inclined to its original position and the structure as a whole will possess that kind of quasi-elasticity which was first imagined by McCullagh."

Here a stubborn adherence to the mechanical model of the world has led to almost comic extremes. Yet is it more far-fetched than the efforts of modern geneticists to find a basis for ethics in the chemical structure of the gene?

It seems to me that there is a lesson in this, which should give pause to workers in newer and less exact sciences, a lesson which was spelled out by Kaufmann in a reference I quoted earlier. It is very easy to fall into the trap of accepting as obvious viewpoints or theories which have no basis in fact whatsoever but which nevertheless carry the weight of general acceptance and sanction by the highest authority. Such ideas may be unquestioningly held as a result of a generally accepted way of looking at the world and of posing questions which are false and misleading. The motto of the scientist is "question all things". The idea is not very new. It was enunciated by Pierre Abelard in the 12th Century. Of course one cannot constantly question everything but one *can*, at least, cultivate a degree of humility ... about what we take to be certainties, and to keep our minds open to the possibility that we may be most mistaken precisely in those instances where we are most certain that we have grasped unshakable truth. Jacob Bronowski in "Ascent of Man" makes this point cogently in the chapter on certainty and uncertainty.

It is perhaps astonishing that physicists of Maxwell's time were so ready to accept the existence of aether, given the strange assortment of properties which it would have to possess. It has to be very rigid so that waves can travel through with high velocity. At the same time it has to be thin enough to permit bodies to pass through without resistance. And it has to be capable of exerting forces at right angles to the line joining interacting objects. To devise a mechanical model with such seemingly contradictory properties takes great scientific ingenuity. If the mark of genius is to solve extremely complicated problems, the theory of aether spawned numerous geniuses. Maybe we should be cautious about giving high marks for engendering and mastering complexity. The ideas of Einstein, on the other hand, were much simpler, much more radical and much more profound.

Decades of work devoted to developing an acceptable aether theory only led to new problems each time an existing one was solved by ad hoc hypotheses. In such a case we should have known by experience that we are on the wrong track entirely. Great theoretical advances enable us to dispense with baffling complexity by means of a simple principle. Were it not so, we would have been buried long ago under a mass of unrelated detail too voluminous to be absorbed.

In fact, faced with the volume of current scientific literature, present-day scientists often feel overwhelmed. This may well be a measure of our failure.

To do justice to Maxwell, it must be admitted that he was aware that his theory did not require aether. He states in one of his later papers:

"I have on a former occasion attempted to describe a particular kind of motion and a particular kind of strain, so arranged as to account for the phenomena. In the paper, I avoid any hypothesis of this kind; and in using such words as electric momentum and electric elasticity in reference to the known phenomena of the induction of currents and the polarization of dielectrics, I wish merely to direct the mind of the reader to mechanical phenomena which will assist him in understanding the electrical ones. *All such phrases in the present paper are to be considered as illustrative, not as explanatory.*"

Despite such qualifications, Maxwell's last words on the subject, which appeared in an article on aether which he wrote for "Encyclopedia Britannica" and which still appeared in an 1898 edition in my possession, are as follows:

"Whatever difficulties we may have in forming a consistent idea of the constitution of the aether, *there can be no doubt* that the interplanetary and interstellar spaces are not empty, but are occupied by a material substance or body, which is certainly the largest, and probably the most uniform body of which we have any knowledge."

In the same *Encyclopedia Britannica* article Maxwell cites the results of "calculations" of the detailed properties of the aether:

Energy/cc = 1.886 ergs (one cc of sunlight has 4×10^{-5} ergs of energy);

Coefficient of rigidity = 843 (comparable to gelatin; that of steel is 8×10^{11});

Density = 10^{-18} g/cc or 10^{-12} g/m^3 (in space, there is about one hydrogen molecule per cc, about 3×10^{-24} g/m^3).

Aether cannot be like a gas, since the amplitude of transverse oscillations of a gas decreases by a factor of 500 in one wavelength, i.e. transverse waves cannot be propagated in a gas, since the energy would rapidly be dissipated into heat.

The above table makes evident the problems of aether theory. How can something with the density of a good vacuum have the rigidity of gelatin? Mechanical waves travelling at the speed of light would require inconceivably high rigidity. And how could planets pass through a material with such properties with no effect on their motion?

There is clearly no conceivable way in which such a combination of requirements could be met.

4.3. Prospect and Retrospect

The discoveries we shall now discuss put Maxwell's theory in a different perspective.

Since light (electromagnetic waves) is a form of energy and Einstein demonstrated the equivalence of mass and energy, light, far from being propagated through matter, *is* a form of matter. (Caution! we must not think of this matter as a *medium* for the propagation of electromagnetic waves; it *is* electromagnetic waves.) The work of Planck and Einstein at the same time showed that these waves might have a quantized ("particle?") character. Later, with the general theory of relativity, Einstein showed that matter (including electromagnetic waves) determined the fabric of space and time rather than merely being propagated through them. To use Wheeler's metaphor, then, energy is more than an actor in the drama of the physical world, it is the fabric of the stage.

But quantum theory shows us that not only do electromagnetic waves display some properties of matter, but "matter" (e.g. electrons, protons, etc.) may be represented by waves of a field. In the end, only fields exist, distributed over space and at the same time shaping that space (or more accurately, space-time).

What a marvellous tapestry, and how unimaginable in the time of Maxwell! Maxwell's discoveries, among the greatest in the history of science, nevertheless were made in a framework which misrepresented the basic relationships of space, time and matter.

Even at the peak of our greatest scientific triumphs, we may not yet understand some of the most fundamental features of the fabric of nature.

Nevertheless, without the work of Faraday, Maxwell and their contemporaries, we would not have progressed along the path of discovery. This story still remains to be told.

Chapter 5

RELATIVITY — THE SPECIAL THEORY

Maxwell derived the speed of light from electricity and magnetism. It occurred to him that a measurement of the speed of light on earth might provide us with a means of determining the earth's speed through space. The principle is simple. Imagine two cars travelling at a steady speed in the same direction along a highway, one going 40 km/h and the other 100 km/h. As the faster car passes the slower one, it would appear to the occupants of the slower one to be going at 60 km/h. Similarly, to the extent that we are moving in space, we should measure a speed of light different from that of Maxwell, the difference being our own velocity.

The problem with this, as Maxwell realized, was that the speed of light is so great that the measurement would be an extremely delicate one; too delicate for the resources of his time.

"If it were possible to determine the velocity of light by observing the time it takes to travel between one station and another on the earth's surface, we might, by comparing the observed velocities in opposite directions, determine the velocity of the aether with respect to these terrestrial stations. All methods, however, by which it is practicable to determine the velocity of light from terrestrial experiments depend on the measurement of the time required for the double journey from one station to the other and back again, and the increase of this time on account of a relative velocity of the aether equal to that of the earth in its orbit would be only about one hundred millionth part of the whole time of transmission, and would therefore be quite insensible."

It was Michelson and Morley at Case School, Cleveland, Ohio who carried out an experiment to measure our velocity through aether. In the experiment, light was to be sent along two different paths at right angles to each other and reflected back to a fixed point, where the two beams could interfere with each other. The way in which the two light waves interfered would show the different speeds of light in the two directions.

How can you decide this from the interference pattern? By looking at one pattern, you can't. But if you rotate the apparatus, so that the directions of propagation of the two waves relative to the motions of the earth are changed, the interference pattern should also change, and this is what the experiment is designed to detect.

Since we are measuring a change, it is not necessary to make the length of the two paths exactly the same. Anyway, this would be inordinately difficult.

In detail, the apparatus consisted of a source of light A, a half-silvered mirror MM', two mirrors C and D to reflect the light-beams in the two perpendicular directions, and an interferometer at E (Fig. 5.1). The path of one beam, which first reflected at MM', is ABCBE; the other passed through MM' and follows the path AB-DBE. The whole apparatus was mounted on a massive slab of stone floating on mercury, which insulated it from external disturbance.

The experiment was, in terms of its goals, a failure; *no* displacement of the interference pattern was found, although the effect anticipated should have been easy to measure. To rule out the possibility that the null result might be because the earth was *not* moving appreciably at the time of the experiment, the experiment was repeated after six months, when the earth's motion around the sun was reversed in direction. The null result persisted.

One explanation was that the earth was dragging the luminiferous aether with it. As theories of the aether became increasingly untenable, this explanation soon lost credibility.

Shortly before the turn of the century Lorentz suggested that

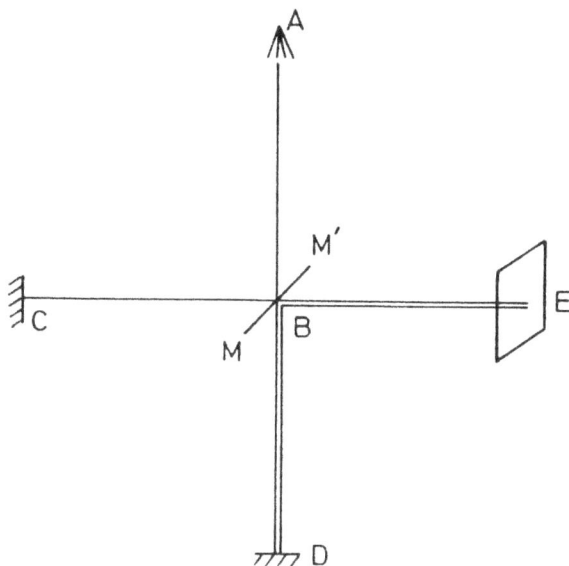

Fig. 5.1.

the length of any moving object might change due to its motion through aether; in particular that it might shrink in its direction of motion. The explanation was to be found in Maxwell's electromagnetic equations, by now well established and accepted. One feature of these equations was that charges at rest produced only an electric, and not a magnetic, field. In motion, they also generated a *magnetic* field. Thus, the electromagnetic forces existing inside the body were modified by its motion. The idea was then to see whether the effect of these additional fields might not be responsible for the contraction.

To establish that this was in fact so was a formidable task, requiring numerous uncertain assumptions. For example, he assumed electrons to be spheres of a definite radius, which themselves contracted by the means postulated. He also had to assume that *all* forces in the body, and not only explicitly electromagnetic ones, behaved under the motion of the body just like electromagnetic ones. He thus evolved a physical picture of a moving body which was able to explain *precisely* a contraction which nullifies the effects of motion (at the time, considered a motion through aether).

What an extraordinary conspiracy of nature! This was purported to be a real physical phenomenon, yet the laws of nature made it unobservable. Other attempts were made but none was successful. Henri Poincaré remarked that such a conspiracy itself must be a law of nature. But it was Einstein who suggested a simple and fundamental way of resolving the problem. The idea was very radical despite its simplicity. Before introducing it, let us first discuss the notion of "frames of reference", and in particular, "inertial" or "Galileian" frames of reference (this terminology of course derives from Galileo).

A reader, sitting in a chair in his office or a classroom, is observing the world from a particular "frame of reference". But the "frame of reference" is not peculiar to *him*; everyone else who is at rest relative to him is in the same frame of reference, which means that each interprets physical phenomena in precisely the same way.

But someone going by in a car on the street outside is in a *different* frame of reference. Things that are moving relative to him are fixed relative to us. If the car is accelerating, that acceleration has physical consequences for him which we do not share. His frame of reference is one which is accelerating relative to the earth's surface. Someone accelerating at the same rate on a parallel street some distance away is in the *same* frame of reference.

Frames of reference are important for interpreting *Newtonian* mechanics. In fact we can say that Newton's laws hold only in certain frames of reference. We know that the rotation of our earth, its motion around the sun, and the sun's rotation in our galaxy all represent acceleration, and this acceleration affects the dynamics of events on earth; for instance, the difference in the directions of the prevailing winds in the northern and southern hemispheres is an effect of the earth's rotation. But acceleration is always *with respect to something*, some basic frame of reference in which objects which were not acted upon by any force would move in straight lines with constant velocity. This frame of reference is usually taken to be one in which fixed stars are at rest. Newton's laws are formulated for such a frame.

But Newton's law of motion, mass × acceleration = force, is in no way altered if one uses a frame of reference which moves with constant speed relative to that of the fixed stars. This is because acceleration is *change of velocity*; thus, acceleration is the same in all frames of reference moving with constant velocity relative to each other. All such frames of reference, which are completely equivalent for Newton's equations of motion, are called "inertial" or "Galileian" frames of reference. We see that Newtonian mechanics is already a sort of relativistic theory; dynamics is identical in all such frames. We might call such relativity a *Galileian* relativity.

But Maxwell's theory threatened the foundations of Newtonian dynamics. Maxwell had predicted a constant speed of light, but constant in what frame of reference? It did not seem to be true that Maxwell's equations were unchanged if one went from one Galileian frame of reference to another moving with constant speed relative to the first. A charge at rest does not produce a magnetic field, one in motion *at constant velocity* does. So electromagnetic theory challenged Galileian relativity. This is precisely the problem to which Einstein addressed his new theory. The approach was to *accept as fundamental* the basic requirement of electromagnetism, embodied in the negative result of the Michelson-Morley experiment, that the speed of light was the same in *all* inertial frames of reference; the problem then was to make mechanics conform to this new requirement. Einstein's resolution was embodied in the following two basic postulates:

1. The speed of light is the same in all inertial frames of reference.
2. *All* laws of physics have the same form in all inertial frames.

The special theory of relativity is built on these two hypotheses. What is important about these propositions is not only their simplicity but their generality. They constitute the basis of an understanding of all phenomena taking place in nature at high speeds (i.e. speeds comparable to the speed of light, c) and thus at high energies.

In fact the theory does even more. With some mathematical ingenuity, Maxwell's theory of electromagnetism can be expressed in a form which is relativistically invariant, that is, a form which is the same in all inertial coordinate systems, thus accommodating Einstein's second postulate. This is equivalent to saying that the theory conforms, without any modification, to the postulates of relativity. This revelation is perhaps not so surprising as might at first appear, since it was the consequences of the theory of electromagnetism which forced us along the path to relativity. Having assured that Maxwell's theory would be the cornerstone of the new theory, it remained to be seen what consequences it would have for classical dynamics. Those consequences turned out to be very profound indeed, requiring a complete revision of our concepts of space and time. These could no longer be absolute and universal, as they had been for Newton; henceforth, we have to accept that the very measures of space and time differ from one inertial frame of reference to another.

Let us now explore the most obvious consequences of Einstein's hypotheses. First, what does it state about the concept of time?

5.1. Relative Time and the Relativity of Simultaneity

Once we abandon the concept of an absolute time, we are compelled to view clocks as integral parts of the physical world, obeying themselves the laws of physics. This is reasonable, since man has always used physical processes to measure the passage of time. First, there were the motions of the heavenly bodies which gave us the day, the month and the year. These can be further subdivided and measured with the pendulum clock. Finally came the "atomic" (or, more properly, "molecular") clock, for which the unit of time was an elementary molecular vibration. The virtue of this last is its almost total independence of macroscopic influences of the environment.

In view of the negative result of the Michelson-Morley experiment, which shows that the speed of light is the same in inertial
frames in relative motion, it must be assumed that length and time
intervals are different in different inertial frames of reference.

A quite simple method exists which examines how time intervals
in different frames of reference may be related. It is based on the
concept of a "light-beam clock", a fictional entity which could be
used to measure time intervals in any reference frame. It consists of
two parallel mirrors separated by some fixed distance L. Light may
be reflected back and forth between the mirrors. The basic time
interval measured by the clock is the time taken for a light beam to
make a return journey from one mirror to the other and back.

Of course, this is not a prescription for a practical clock, since we
are left with the question of how we know when the light leaves one of
the mirrors and when it returns. We do not have to bother ourselves
with such technical details, however, for we are only interested in
performing one of Einstein's "thought experiments".

The situation is illustrated in Fig. 5.2. The rectangle on the left
contains the light-beam clock assembly. If the clock is stationary in
our frame of reference, the periodic interval is $2L/c$. This is the unit
in which time intervals between events are measured.

Suppose now that this light-beam clock moves to the right in
our frame of reference with speed v. While the light-beam signal
is travelling between the mirrors, the whole clock is moving to the
right. To an observer moving with the clock, the clock is still at
rest, and its elementary time interval remains $2L/c$. But from the
viewpoint of the outside observer, (i.e. the original frame of reference
relative to which the clock is now moving) the light will follow the
diagonal path from P to P″ and then back again to the new position
of P.

But of course as measured in this original rest-frame, the speed
of light is c, just as in the moving frame S′. We conclude that since
for S the light must follow a longer path to make the return journey,

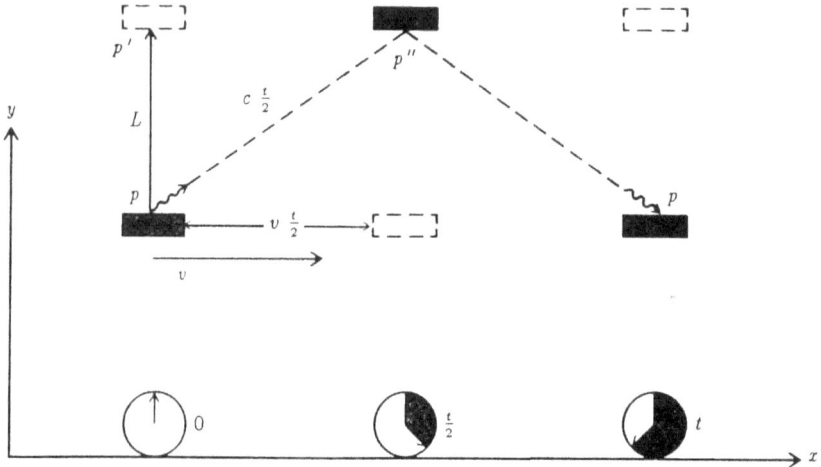

Fig. 5.2. The light-beam clock aboard the spaceship, as observed by ground-based observer. a ground-based observer. (Reprinted with permission of Macmillan Publishing Company from *The Fabric of Reality* by S. K. Kim. Copyright © 1975 by Sung Kyu Kim.)

the clocks of S' must have a longer time interval. Whatever happens to any *resting* object in S' must appear to S to be happening more slowly. This is not merely a subjective impression; it can be verified by clocks (of another sort, e.g. pendulum clocks) fixed in each of the two frames.

Let us make this quantitative. Using the geometrical theorem of Pythagoras, the sum of the squares of the lengths of the sides of a right-angled triangle which form the right angle is the square of the length of the hypotenuse. This means that

$$L^2 + (\frac{1}{2}vt)^2 = (\frac{1}{2}ct)^2 \,,$$

t being the time taken relative to the frame S, for light to get from P to P''.

It follows that

$$4L^2 = (c^2 - v^2)t^2 = c^2(1 - v^2/c^2)t^2 \,.$$

If we write

$$\gamma = \sqrt{1 - v^2/c^2} \, ,$$

this equation states that

$$4L^2 = c^2\gamma^2 t^2$$

or

$$t = \frac{2L}{c\gamma} \, .$$

Thus the time interval of the clock, which in its own rest-frame is $2L/c$, is increased, in the frame relative to which it is moving, by a factor

$$\frac{1}{\sqrt{1 - v^2/c^2}} \, .$$

Of course, as the velocity v of the clock approaches c, the velocity of light, the clock appears to the stationary observer to slow down more and more, till at last nothing in the moving frame seems to change at all; everything becomes, to all intents and purposes, frozen in time.

You may ask: how do we know that the *length* of the clock may not depend on the frame of reference? If it does, our argument breaks down. Imagine, however, that two parallel measuring rods, equal in length when they are both at rest, are put in relative motion (Fig. 5.3). Suppose also that, if rod L_2 is moving perpendicular to its length relative to L_1, it would appear to the inertial frame of reference of L_1 to be *shortened*. Suppose that the ends of the rods were equipped with felt pens, so that they could mark each other as they passed. Due to the shortening of L_2, its pens would leave marks on L_1, and not the reverse.

Consider now the situation from the point of view of the rest-frame of L_2 (Fig. 5.4). In this frame, L_1 would be moving, and it should appear to be shorter than L_2. The marks would then be left on L_2.

But of course it cannot be true that L_1 is marked and not L_2, and at the same time L_2 is marked and not L_1. Since the assumption

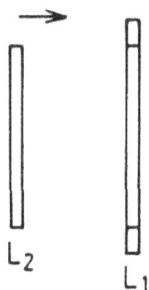

Fig. 5.3.

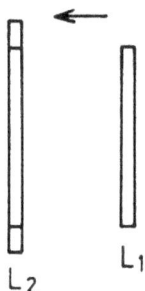

Fig. 5.4.

of a relative shortening of the "moving" rod has led to an absurdity, we must conclude that it does not exist; thus, the lengths of two parallel identical rods moving in a direction perpendicular to their length must be the same.

One question remains: can we be sure that time measured by light-beam clocks is identical with that measured by *other* clocks? It is obvious that it must be, since we must use another sort of clock to determine the speed of light; the fact that light has a *constant* speed relative to these clocks ensures that they must measure the same time.

5.2. The Mount Washington Experiment and Time Dilation

Physicists will not often turn from publishable research to do an experiment, particularly a tedious one, with a purely pedagogical purpose. The experiment on the verification of time dilation through decay of μ-mesons by J. H. Smith and David Frisch is a notable exception. The film based on this experiment is extremely well done. It gives a clear and convincing demonstration.

The idea is that particles called μ-mesons enter the earth's lower atmosphere at a very high energy, and with speeds very close to the speed of light. Many of them decay as they pass through the atmosphere toward the earth. They may be slowed down and captured and the mean lifetime for decay measured in the laboratory. This lifetime is known to be about 1.5×10^{-6} second ($1\frac{1}{2}$ microseconds).

In this time, a particle moving at the speed of light travels a distance of about 450 metres.

An important feature of radioactive decay is that the instant at which it will take place is completely unpredictable, but the *probability* that it will decay in a given interval of time is constant. The life of a meson is not like that of a living organism. A human being is much more likely to die in the next year if he is 80 years old than if he is 30. This is not true for a meson. If, by chance, one survives, not for 1.5×10^{-6} second but for 10^{-5} second, it is no more likely to decay in the next microsecond than in the same period of time after its birth.

Imagine, then, that one could detect some mesons high in the atmosphere – in the experiment in question at the top of Mt. Washington in New Hampshire, which is 6500 ft. (about 2000 metres) high. Suppose, that we followed them in their descent toward the earth, and charted how far each of them would go before decaying. What proportion can be expected to reach ground level?

One might think that half would decay for each 450 metres of descent; after 1800 metres only $(\frac{1}{2})^4 = \frac{1}{16}$ would remain. After 2000 m, only 4.8 percent would survive.

The flaw in this argument lies in the assumption that mesons in flight, at speeds near that of light, will decay at the same rate as mesons almost at rest. The meson should be thought of as a sort of subatomic clock which will destruct at a random time, but on the average in 1.5 microseconds. (This seems very short, but we shall see, when we discuss particle physics, that the natural period of this clock is about 10^{-23} second; on this scale, its lifetime is in fact very long.)

Now, if the phenomenon of time dilation predicted by relativity is true, a clock travelling close to the speed of light should be slowed down by a large factor (if the velocity is 99.5% that of light, by a factor of 10). It should live 10 times longer or about 15 microseconds, and on the average should travel 4500 m, not 450 m, before decaying. Most of the Mt. Washington mesons should live to reach ground level.

The experiment tests the validity of this theoretical prediction of relativity.

Of course it cannot be done as described above. We cannot detect a meson at the top of Mt. Washington, and then follow it down to sea level to see when it decays. But some facts make it possible to accomplish the goals of the experiment in a simpler way. The most important of these is the fact that the flux of μ-mesons through the atmosphere is quite constant in time, and also over a large area. Furthermore, all mesons are identical in their properties, so we need not track particular mesons, but may instead use a statistical approach.

The experiment may therefore be redefined as follows:

First, we set up an experiment in a laboratory at the top of Mt. Washington, which allows us to slow down incoming mesons in a block of *iron* and let them decay. The decay lifetime is measured by determining the interval between their detection on entering the apparatus and their decay. At each observation, one determines how far down from the top of Mt. Washington the meson would have gone if we had not captured it *on the assumption* that its lifetime in flight was the same as that at rest.

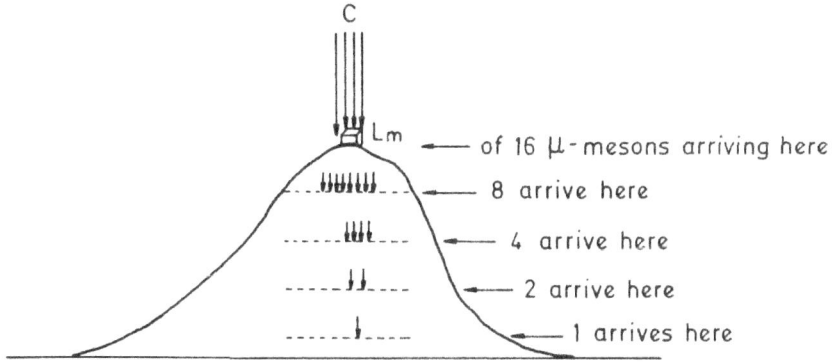

Fig. 5.5. The μ-mesons have a half-life of 1.5×10^{-6} second; in other words, half of them decay in that time. The distance between two successive levels shown is that which can be traversed by light in that time. In fact, however, nearly 3/4 of the mesons reach sea level.

Observations over several hourly periods showed that between 500 and 600 mesons have decayed. Let us consider a typical hour's observation, in which 563 mesons were found to have decayed. Translating their lifetimes into distances travelled at the speed of light, we deduce that the number which should reach the earth's surface was 4.8% of 563 = 27.

One now takes the apparatus down to sea level (at M.I.T.), sets up the experiment again, and observes how many mesons do decay in the apparatus in an hour (an average number over several runs of an hour, actually). What is found is not 27, but 408!

Using this, we can reargue the actual half-life of the fast-moving mesons. Mathematical tables tell us that

$$\frac{408}{563} = \left(\frac{1}{2}\right)^{0.46} .$$

So the trip down was only 0.46 of the actual half-life of the very fast meson. The time needed for something to travel at the speed of light from the top of Mt. Washington is easily calculated as 6.7

microseconds, or 4.4 times the half-life of a meson at rest. On the other hand, these 6.7 microseconds represent only 46% of the half-life which the meson displays when it is moving near the speed of light; this true half-life is thus $\frac{6.7}{0.46} = 14.4$ microseconds, or nearly 10 times that measured for mesons at rest.

There is, therefore, a *time dilation* factor of nearly 10, which corresponds, according to Einstein's theory, to a speed 99.5% that of light.

To state it differently: the lifetime of the meson as measured in our earthbound reference frames is 10 times greater than that which it would have in the frame of reference in which it is at rest.

Note the significance of this statement. If one could construct a laboratory moving *with* the meson, one would measure a meson lifetime in that frame of 1.5 microseconds.

But now we are confronted with another difficulty. Using the frame of reference co-moving with the meson, ground level is rushing up toward us at 99.5% of the speed of light, so that sea level should reach us in a time equal to

$$0.46 \times 1.5 \times 10^{-6} = 0.69 \text{ microseconds}.$$

We would then be forced to conclude that sea level would only appear to have gone

$$0.69 \times 10^{-6} \times 3 \times 10^{10} \text{ cm} = 207 \text{ m}.$$

Thus, from the viewpoint of an observer moving with the meson, the distance between the top of Mt. Washington and sea level is only 207 m, rather than the 2000 m measured in a frame of reference fixed to the earth. This is the phenomenon of the Lorentz contraction of length. From the viewpoint of the meson's frame of reference, Mt. Washington is *shrunk* by a factor of almost 10.

It is very interesting that a phenomenon can manifest itself as a time dilation in one frame of reference, and as a length contraction in another. This is an indication of the entanglement of space and time so characteristic of relativity. The fact that the same phenomenon can be interpreted as a temporal one in one frame of reference and a spatial one in another suggests that space and time are as intimately linked in the physical world as were electricity and magnetism in Maxwell's theory.

5.3. Length Contraction

Argument 1

This argument has its origin in the famous "Mount Washington Experiment" on the decay of μ-mesons. In this experiment, μ-mesons penetrate the earth's atmosphere at speeds of more than 99% the speed of light. Thus, their "natural clocks", which determine their decay rate, run slower by a factor of about 10 in the stationary frame of reference of the earth laboratory than if they were at rest in that frame. Consequently many more arrive at the surface of the earth than would do so if this time dilation did not exist.

If, on the other hand, the situation is viewed from a frame of reference moving with the mesons, the decay time (the "natural" one) is ten times shorter. Nevertheless, the number of mesons reaching the earth's surface remains the same. How can this be explained using the aforementioned co-moving frame of reference? The time in flight must be ten times shorter, so the distance travelled must be ten times shorter as well. In the meson's frame of reference, the earth is flying toward it at more than 99% of the speed of light, and all distances must be reduced by a factor of ten (Lorentz contraction).

Thus, the phenomena of time dilation and length contraction are manifestations, in different reference frames, of the *same* phenomenon.

While this argument is perfectly valid, one must ask whether, given the equivalence of the two phenomena, one must use *one* of them — time dilation — to explain the other, the contraction of length. Is there an argument for the latter based on considerations of length alone? The answer is positive, as we now see.

Argument 2

Imagine a platform mounted on wheels on a straight track (such as the one we envisaged above to demonstrate the time dilation) on which we place a source of light. Imagine also a mirror mounted on the same platform vertically above the source (Case A) (Fig. 5.6).

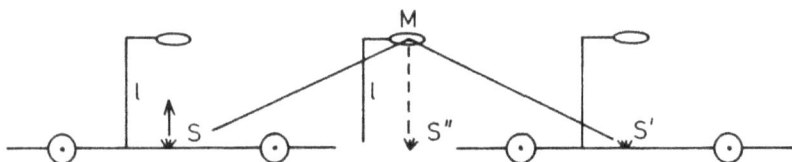

Fig. 5.6.

$$SM = ct_0$$
$$SS'' = vt_0$$
$$S''M = l$$
$$(ct_0)^2 - (vt_0)^2 = l^2$$
$$t_0 = (l/c)\frac{1}{\sqrt{1-v^2/c^2}}$$

If the platform is in motion, a beam of light, emitted from source S and reflected at the mirror M back to S is seen by the external viewer to have a longer trajectory than that seen by an observer moving *with* the platform. This is the time dilation in a moving frame of reference. Given the constancy of the speed of light in all frames, it is evident that the total time of the trajectory must appear greater to the external observer.

Suppose now that the mirror is mounted at the same height as the source, and at the same distance l from it as before so that the trajectory is in the direction of motion of the platform (Case B) (Fig. 5.7).

In the frame of reference of the platform, if light signals are sent out simultaneously from S in the horizontal and vertical directions, they will return to S simultaneously.

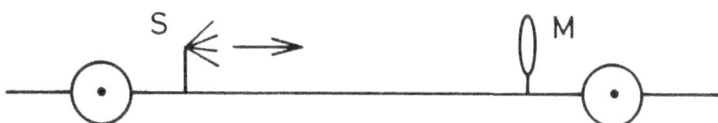

Fig. 5.7.

The same must then be true in the frame of reference relative to which the platform is moving with speed v.

We know that in the direction *perpendicular* to the motion, lengths measured in the two frames of reference must be identical. We *cannot* assume the same for those measured in the longitudinal direction. Thus, we shall assume that in the rest frame the distance is measured as l'.

Let us measure distances in the direction of motion from the positions of the source at the instant of the emission of the signals ($t = 0$). Then, as seen by the stationary observer, the signal will reach the horizontal mirror (M') at a time t_1 given by

$$l' + vt_1 = ct_1$$

or

$$t_1 = \frac{l'}{c - v} \ .$$

The time for the return journey will, on the other hand, be given by t_2 such that

$$l' - vt_2 = ct_2 \ ,$$

so that $t_2 = \frac{l'}{c+v}$. The time for the signal to go out and back is then

$$T = l' \left(\frac{l}{c-v} + \frac{l}{c+v} \right) = \frac{2l'c}{c^2 - v^2} = \frac{2l'}{c(l - v^2/c^2)} \ .$$

In the same frame of reference, what will be the time required for the signal to go to the *vertical* mirror and back (a distance $2l$)?

This we have already shown to be $\frac{2l}{c\sqrt{1-v^2/c^2}}$. But since the signals arrive back simultaneously, we conclude that

$$\frac{2l'}{c(1-v^2/c^2)} = \frac{2l}{c\sqrt{1-v^2/c}}.$$

In words, the argument is: since a journey of a given length made *horizontally* takes longer than a journey of the same length *vertically*, if the signals are to arrive back simultaneously, the horizontal journey must have been *shorter* than the vertical one. The journeys are of the same length in the frame of reference of the platform, but relative to the rest frame lengths must be contracted along the direction of motion.

It may be useful to illustrate the above argument with numbers. Let us suppose that $v = 3/5c$; then $\sqrt{1 - v^2/c^2} = 4/5$, so lengths in the direction of motion are reduced by 20%.

If we take a length of 10 m, the time for the signal to go out and back vertically, at the speed of light $(3 \times 10^8$ m/s) is

$$\frac{20}{3 \times 10^8 \times 4/5} = 8.33 \times 10^{-8} \text{ s}$$

or 4.17×10^{-8} s each way.

In the longitudinal direction, the time for the outward leg is

$$\frac{10k}{3 \times 10^8 \times 2/5} = 8.33 \times 10^{-8} k \text{ s}$$

(which reflects the fact that the light signal is chasing the receding platform), and

$$\frac{10k}{3 \times 10^8 \times 8/5} = 2.08 \times 10^{-8} k \text{ s}$$

for the inward leg, where k is the Lorentz contraction factor. The total time taken for the signal to travel out and back horizontally is therefore $10.41 \times 10^{-8} k$ s. Since the return distance travelled vertically is 8.33×10^{-8} s, the contraction factor is

$$k = \frac{8.33}{10.41} = 0.80$$

so that the 10 m is contracted to 8 m. To be precise $k = \sqrt{1 - 9/25} = 4/5 = 0.80$ exactly.

5.4. Other Relativistic Phenomena: Relativistic Mass Increase

One of the striking features of the phenomena of time dilation and length contraction is that both become catastrophic when the relative speed of the frames of reference is c. In this case all time intervals are dilated to infinity, which is to say that nothing happens! On the other hand, all lengths are contracted to zero, which implies that there is no finite world for things to happen in anyway.

These comments are pertinent to the famous question which Einstein purportedly asked himself when he was a 15-year old schoolboy, a question which, he said, set him on the train of thought which was to culminate in the special theory of relativity: what would the world look like if one could travel on a beam of light? The answer appears to be that time and space would cease to exist.

There is a further question which can be (and in recent times has been) proposed, namely, can anything travel *faster* than the speed of light? While there has been much speculation on the possible existence of faster-than-light particles ("tachyons"), one thing is clear: one cannot start with particles *slower* than light and accelerate them to speeds greater than light. This hardly seems surprising, since the abolition of space and time stands between the two regimes. However, this in itself does not resolve the question of the *existence* of tachyons.

What we shall show, though, is that the mass of an ordinary particle becomes greater as its speed approaches the speed of light. Because its inertia keeps increasing, no amount of energy conveyed to it will accelerate it to a speed equal to — much less greater than — the speed of light. We shall in fact see that an acceleration from 99% of the speed of light to 99.99% involves increasing the particle energy by a factor of 10, while to accelerate it to 99.9999% of c requires another increase of a factor of 10 in its energy.

The speed of light is the ultimate speed.

How is all this a consequence of relativistic principles?

The demonstration depends, once again, upon the analysis of a given phenomenon from the viewpoint of two different frames of reference. The situation to consider is the following: suppose there is a glancing collision of two different entities (which could be either "particles" of physics or billiard balls). One object is at rest (B), and another (A) collides with it in such a way that it is deflected almost sideways (B') while A goes off at an angle. Let A', B' be the positions of the objects an instant after the collision. By conservation of transverse momentum (momentum perpendicular to AQ) the distance BB' = QA' if A and B have the same mass. This is because momentum = mass × velocity = mass × distance ÷ time: for a fixed time and equal masses the subsequent distances of A, B from the original path of A must be almost equal.

If we look at the same collision from the viewpoint of a frame of reference fixed to B, and in which A is initially at rest, the corresponding diagram for the collision is as shown in Fig. 5.8a; it is the reflection, in the point of collision, of the original, with A and B interchanged. Now AA' = Q'B.

Because of conservation of momentum and the symmetry of the situation between objects A and B, we may argue as follows:

If we first consider the frame of reference in which B is not in motion in the direction of the collision, its transverse momentum is $m_o v_t$, where v_t is its recoil velocity and m_o its mass at low velocity ("rest-mass").

Consider next the frame of reference in which A is initially at rest, and B is initially moving horizontally and to the left with speed v_o (Fig. 5.8b). This frame is then moving with speed v_o relative to

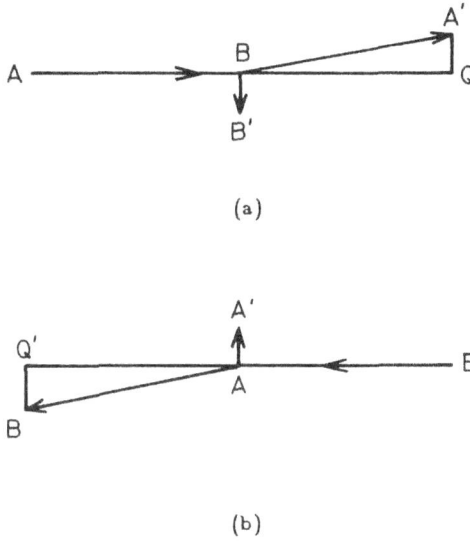

(a)

(b)

Fig. 5.8.

the preceding one. B is then in a frame in which times are dilated, i.e. everything that happens in it is slowed down accordingly. Thus, as seen by A, B's recoil velocity is reduced to $v_t \sqrt{1 - v_o^2/c^2}$. But the recoil *momenta* are the same in the two cases, so that the *mass* of B in its moving frame must be correspondingly *increased* by a factor $\frac{1}{\sqrt{1 - v_o^2/c_2}}$. The *mass* of B is therefore

$$m = \frac{m_o}{\sqrt{1 - v_o^2/c^2}} .$$

Note that the argument depends on the fact that *distances* perpendicular to the direction of initial relative motion are not affected by that motion, so that the transverse velocity of B in the frame which is moving relative to the target particle A of the collision is decreased only because of time dilation.

Time dilation, on the other hand, is related only to the relative motion of frames of reference, but is quite unaffected by the *direction* of that motion.

5.5. On the Famous Relation $E = mc^2$

Of all the contributions that Albert Einstein made to physics, none is better known to the world at large than the equation which expresses the equivalence of mass and energy. But what precisely does it mean?

We have already noted that mass appears independently in two different physical contexts. On the one hand it is a measure of inertia, of resistance to motion. On the other hand it is what determines the strength of gravitational force. Extremely delicate experiments have however established that these two sorts of mass are in fact the same.

If, then, energy is equivalent to mass, energy must display inertia and must be a source of a gravitational field. These statements must be subject to experimental verification. We shall show that they are.

Energy then must be equivalent to mass. But is mass equivalent to energy? What are the particular characteristics of "energy" that mass must share?

The most fundamental thing we know about energy is that it is a conserved quantity; it may be transferred from one form to another, but it may never be created or destroyed; its total quantity is always the same. The equivalence of *mass* to *energy* must therefore mean that this conservation law holds only when the energy equivalent of mass, as given by $E = mc^2$, is taken into account.

This statement is also subject to experimental verification.

The famous Einstein relation is therefore verifiable. It is also a logical deduction from the general principles of relativity.

There is a simple "thought experiment", which Einstein himself used to justify mass-energy equivalence. To understand it, however, one must first become familiar with the concept of "centre of mass".

Suppose we consider a collection of point-like masses. Their centre of mass is defined in the following way: it is a point such that the sum of the products of their masses and their distances from this point in any direction is zero, distances to one side being measured as positive and to the other side as negative.

Take, for purposes of illustration, three masses of 1, 2, and 3 units lying in a plane as shown in Fig. 5.9. If we take the horizontal direction, in which distances are designated as x, the sum of mass $\times$ distance from O is $-1 \times m + (-1) \times 2m + 1 \times 3m = 0$. Similarly, for vertical distances we have $-1 \times m + 2 \times 2m + (-1) \times 3m = 0$. Thus, O is the centre of mass of the three masses.

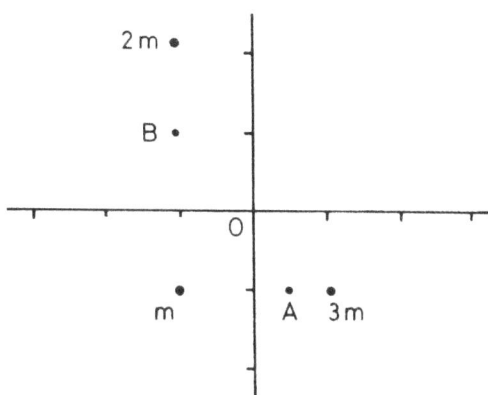

Fig. 5.9.

Note that the centre of mass of m and $3m$ alone is at A. Suppose then that we replace them by a single mass equal to their combined mass $(4m)$ at A; the centre of mass will remain the same, at O. Or, if we consider the centre of mass of $m, 2m$ alone it is seen to be at B. Again if we replace the two by a single mass $3m$ located at B, nothing is changed.

A further interesting result is the following: imagine an internal force between m and $2m$, which acts on them in equal and opposite directions (the force could be gravitational, or electric, or they could be attached by a spring, etc.). Now since $2m$ has twice as much inertia as m, it will only move half as far in a given time as m. Clearly, then, their centre of mass will remain stationary. Thus, the centre of mass of two particles (or more) acted upon only by internal forces between them will not change as they move. That is why we can say that the centre of mass of a rigid bar or ruler, for example, remains at the same place when no external forces are acting on it, despite the fact that it consists of countless atoms or molecules which are jiggling about under the influence of their internal forces.

The point of this digression is the following: we shall show that the centre of mass of a system of massive particles on the one hand and electromagnetic radiation on the other is determined by attributing to a quantity of radiation of energy E a mass equivalent of E/c^2.

With Einstein, imagine a railway car with completely frictionless wheels sitting at rest on an absolutely horizontal straight track (remember, this is a *thought* experiment so that we are free to postulate this rather fictitious entity!). Suppose that at one end of the car a powerful source of laser radiation is directed toward a perfectly absorbing screen at the other end.

Suppose now that this source is turned on and a pulse of energy E is emitted. According to Maxwell this radiation has a momentum of E/c. If the car has mass M, it will, by momentum conservation, recoil with velocity E/Mc.

The important thing is that the car will have moved!

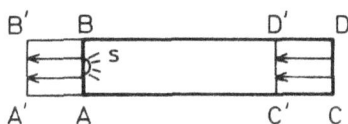

Fig. 5.10. The heavy lines show the original position of the car (ABCD), the light ones the final position (A′B′C′D′). S is the source of the radiation.

Now let us ask for how long a time (t) it will recoil. During this time the signal covers a distance ct. But the end of the car will come toward it by vt or $\frac{E}{Mc}t$. The sum of these two distances must be L, so

$$\left(c + \frac{E}{Mc}\right)t = L$$

and

$$t = \frac{L}{c + E/Mc} = \frac{McL}{E + Mc^2}$$

The *distance* it will recoil is then

$$vt = \frac{E}{Mc} \cdot \frac{McL}{E + Mc^2} = L\frac{E}{E + Mc^2} \ .$$

Thus the mass of the car × the distance recoiled is $\frac{MLE}{E+Mc^2}$. This must be balanced by the same quantity for the radiation (since the centre of mass cannot have moved). But the *distance* the radiation has moved is

$$ct = \frac{Mc^2 L}{E + Mc^2} \ .$$

Therefore the effective mass of the radiation must be M_r such that

$$M_r \frac{Mc^2 L}{E + Mc^2} = \frac{MLE}{E + Mc^2}$$

from which we deduce that $M_r = E/c^2$!

So what have we shown? We have shown that, for purposes of inertia, the radiation acts as though it had a mass E/c^2. The radiative energy is equivalent to mass.

What is remarkable about this result is that we have already shown that nothing with a mass can possibly move with the speed of light. Yet light, which *does* have that speed, nevertheless has an *inertia* associated with its energy.

Because of the equivalence of gravitational and inertial masses, it is a logical consequence of this statement that *energy* also is a source of gravitational fields.

This fact creates the complications of Einstein's theory of gravitation, with which we shall deal later. All fields, including the gravitational field, contain energy. The energy of the gravitational field can be the source of more gravitational fields.

This situation sharply distinguishes the gravitational field from the electromagnetic one. The sources of the electromagnetic field are charges, but charges do not produce more charge. Electromagnetic fields are not themselves charged, and thus cannot produce further electromagnetic fields.

5.6. Does the Existence of Mass Imply Energy?

We have demonstrated that energy shows the attributes of mass, but what about the reverse? Is energy associated with mass essential to the law of energy conservation?

The following two experimental indications show that it is.

First, antiparticles have this property: a particle (an electron) can annihilate in the presence of its antiparticle (positron) to create two quanta of radiation, i.e. a pure electromagnetic field. Conversely, under the proper conditions electromagnetic radiation (more precisely the quanta thereof) can produce pairs of massive particle and antiparticle. Thus, particles are transmutable to electromagnetic radiation and vice versa.

This phenomenon is a manifestation of quantum principles as well as relativistic ones. We shall say much more about this in a later chapter.

Secondly, we know that atomic nuclei are made of neutrons and protons. The masses of these two particles can be measured. Naively we might expect the masses of the nuclei to be the sum of the masses of its constituent neutrons and protons. However, the masses are less.

If two particles are bound to each other by forces, the *energy* of the resulting combination is less than the sum of the *energies* of the particles when separated. This is because work has to be done, i.e. energy expended, to separate them.

A nucleus has a negative binding energy due to the attractive interactions of its components. It also has positive *kinetic* energy. The fact that the nucleus remains *bound* is proof that the negative binding energy is greater than the positive kinetic energy. This *negative* energy is manifested as a decrease in *mass* of the nucleus with respect to the sum of the masses of its components. It appears that *potential energy* and *kinetic energy* are also equivalent to mass. This seems much more astonishing than the fact that radiative energy, in the form of quanta and thus showing particle-like behaviour, should behave as though it had mass.

Examples

(a) The helium nucleus consists of two neutrons and two protons. Using an "atomic unit" of mass ($\frac{1}{16}$ of the oxygen mass), the masses are as follows:

2 neutrons	2×1.0087	2.0174
2 protons	2×1.0078	2.0156
	Total	4.0330
Helium nucleus		4.0026
	Difference	0.0304

The difference represents the sum of (i) the binding energy of the four nucleons *minus* their total kinetic energy, both of which have mass equivalents.

(b) The nucleus of carbon-12 consists of 6 neutrons and 6 protons. The balance sheet in this case is

6 neutrons	6.0520
6 protons	6.0470
Total	12.0990
Carbon-12	12.0026
Difference	0.0964

Up to this point we appear to have discussed unrelated and independent phenomena, such as time dilation, Lorentz contraction, and mass-energy equivalence. Theory casts them as different manifestations of a single principle; they are seen as a *single* phenomenon. This is accomplished by showing that they are all consequences of Einstein's basic relativistic postulates. This can only be done by giving a mathematical expression to the principles of relativity and then using the reasoning power of mathematics to draw diverse conclusions. Only thus can we demonstrate the unity of many phenomena.

5.7. A Digression on the Relation of Mathematics to Physics

We have here a vivid illustration of the role of mathematics in physics. The basic principles of physics can in general be expressed and understood without mathematics. So may various consequences (particularly simple ones) of those principles. The overall unity of the phenomena of nature, however, requires mathematics for its expression. One can be even more explicit. Mathematics is a unified language for the expression of physical ideas. The ideas may usually, however, be expressed in ordinary language. But mathematics is also a structure of relationships, an efficient tool for reasoning. Feynman has defined mathematics as language *plus* reasoning. It is more apt to describe mathematics as a language *for* reasoning. One then admits that *all* languages are tools for reasoning. What makes mathematics distinct is that it serves as a means of codifying images, i.e. creations of the human imagination, while at the same time providing a tool for the expression of, and logical deduction from, relationships between these creations. It thus links our perceptions of the physical world with our powers of abstract reasoning. Its special significance lies in its efficacy in uniting these two functions.

This perspective permits us to better understand the gap which has progressively widened in recent times between the mathematician and the physicist. The pure mathematician tends to regard the evolution of a logic of relationships, of structure, as transcending the linguistic function of mathematics, that is to say, the function of expressing concepts derived from our experience of the world external to our minds. Thus, mathematics becomes more and more the development of a theory of the structure of relationships, and tries to attain greater abstraction. Its goal is to cast off all links to physical reality (an exercise which may well be founded in delusion).

For the scientist, on the other hand, the unification of the expressive and deductive powers of mathematics rouse his sense of wonder.

The author, however, rejects the notion that physics can only be understood in the language of mathematics. The basic ideas of the special and general theories of relativity, for example, are expressible in non-mathematical form, and can be understood without mathematics (debating this point could involve endless arguments about the definition of "understanding". However that question may be settled, we suggest that a more fruitful follow-up question is, what does mathematics *add* to "understanding"?) Without mathematical training one can understand a wide range of phenomena, in the sense of being able to interpret them as manifestations of general principles. It is only through mathematics, however, that one can hope to appreciate the full ramifications of these principles. It is our purpose in this book to show nonetheless that many important deductions from these principles are still within the grasp of the non-mathematician.

5.8. The Relativistic Doppler Effect

The Doppler effect is well known in classical physics. The simplest example is the whistling of a rapidly moving train passing by a station. As the train approaches, the pitch of the sound is raised in proportion to the speed of the train. As it recedes, the pitch is lowered. Sound is a wave phenomenon in the atmosphere; small compressions and rarefactions of the air succeed each other periodically with a given frequency (for simplicity we shall imagine that the whistle has a single frequency). Suppose that successive compressions are normally separated by a time interval T. They will then be separated *spatially* by a distance VT, where V is the speed at which successive compressions move through the air.

Suppose, however, that the train is moving toward us with speed v; in the same interval T, it will move a distance vT. Thus the distance between successive compressions will be reduced to $(V - v)T$. This wavelength is therefore shortened by the factor

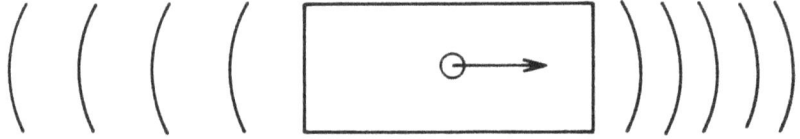

Fig. 5.11.

$1 - v/V$. Its frequency will then be raised by the same fraction. If the train is moving *away* from us, the wavelength increases in the ratio $(1 + v/V)$, and the frequency is *reduced* by the same amount.

The same argument holds for a source emitting *light* which is moving toward or away from us with speed v. When relativity is taken into account, however, there is an extra complication. Due to time dilation, though in its own frame of reference the periodic time of the light signal is T, in our frame of reference it will be lengthened to $T/\sqrt{1 - v^2/c^2}$ (on the other hand, the *frequency* f_0 of the light, i.e. the number of periodic times per second will be the inverse of that i.e. $\sqrt{1 - v^2/c^2}\,f_0$, where f_0 is the frequency in the rest-frame). This effect must be *superimposed* on the one due to motion. Thus, for an approaching source, the Doppler-shifted frequency will be

$$f_D = \sqrt{1 - v^2/c^2}\,\frac{f_0}{1 - v/c} = \sqrt{\frac{1 + v/c}{1 - v/c}}\,f_0 \ .$$

For a source *receding* at the same speed, the sign of the speed v will be reversed, and the Doppler frequency will be

$$f_D = \sqrt{\frac{1 - v/c}{1 + v/c}}\,f_0 \ ,$$

the factors $\sqrt{\frac{1+v/c}{1-v/c}}$ or $\sqrt{\frac{1-v/c}{1+v/c}}$ are known as the Doppler shift (more properly, the *longitudinal* Doppler shift, since they apply when the motion of the source is directly toward or directly away from us).

When the source is moving *perpendicular* to the line from us to it, only the time-dilation effect applies and

$$f_D = \sqrt{1 - v^2/c^2}\, f_0 \ .$$

Let us put in some numbers. We take for purposes of illustration $v = 7/25\,c$ (i.e. the speed is 28% that of light). Then, for an approaching source

$$f_D = \sqrt{\frac{1 + 7/25}{1 - 7/25}}\, f_0 = \sqrt{\frac{32/25}{18/25}}\, f_0 = \frac{4}{3} f_0 \ ,$$

i.e. the frequency is increased by 1/3. For a *receding* source

$$f_D = \frac{3}{4} f_0 \ ,$$

the frequency is *reduced* by 1/4. In the transverse case

$$f_D = \sqrt{1 - \frac{49}{625}}\, f_0 = \frac{24}{25} f_0$$

and the Doppler shifting is 4%.

Since in the optical spectrum the frequency of light *increases* from red to blue, a *decrease* in frequency is called a red-shift, and an increase a blue-shift. The above figures make it clear that (because of the time dilation effect) red-shifts are more common than blue-shifts.

5.9. On the Addition of Velocities

The fact that relativity implies an ultimate speed past which nothing can be accelerated raises the following interesting problem: imagine three frames of reference, which we shall call S_0, S_1, and S_2. Suppose we are at rest in S_0; that S_1 is moving with speed u relative to S_0, and that S_2 is moving with speed v relative to S_1, always in the same direction. In classical physics speeds are additive; if u were

60% of the speed of light, and v the same, S_2 and everything at rest in it would have a speed 1.2 times the speed of light relative to S_0. This obviously cannot be so in relativity, so we must ask, how *do* we determine the speed of S_2 relative to S_0?

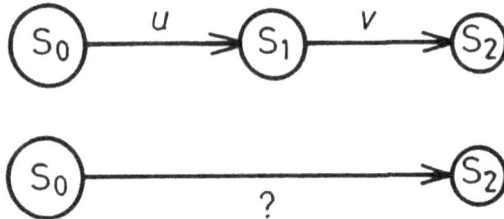

Qualitatively, we can see how the apparent paradox might be avoided. Because of time dilation, the observer in S_0 will observe the speed of S_2 relative to S_1 to be lessened (everything appears to move more slowly). In this way the speed of S_2 as observed by S_0 may be reduced to a value below that of light. We may however, with a little algebra, solve the problem quantitatively by considering the Doppler effect. If an oscillatory source of light of frequency f_0 is placed at rest in S_2, the frequency as observed in S_1, due to the Doppler effect, will be

$$f_1 = \sqrt{\frac{1 - v/c}{1 + v/c}} f_0 \ .$$

This frequency may also be *recorded* in S_1, and the recording sent back to S_0. This will result in further Doppler shifting by a factor of

$$\sqrt{\frac{1 - u/c}{1 + u/c}}$$

so that the frequency observed by S_0 is

$$f_2 = \sqrt{\frac{1 - u/c}{1 + u/c}} \sqrt{\frac{1 - v/c}{1 + v/c}} f_0$$

$$= \sqrt{\frac{1 - \frac{u+v}{c} + \frac{uv}{c^2}}{1 + \frac{u+v}{c} + \frac{uv}{c^2}}} f_0$$

$$= \sqrt{\frac{1 - \frac{(u+v)/c}{1 + uv/c^2}}{1 + \frac{(u+v)/c}{1 + uv/c^2}}} f_0 \ .$$

But this is precisely the Doppler shift observed in S_0 if S_2 is moving with a speed relative to it of

$$w = \frac{u + v}{1 + uv/c^2} \ .$$

We see then that the speeds u and v do not simply add, as our qualitative argument suggested. We would in fact expect w to be less than c. And in fact this is so, because if we consider $w/c = \frac{(u+v)/c}{1 + uv/c^2}$ we can deduce that it must be less than 1 because

$$\left(1 - \frac{u}{c}\right)\left(1 - \frac{v}{c}\right) = 1 + \frac{uv}{c^2} - \frac{u + v}{c} \ .$$

Since the quantity on the left side of this equation is clearly positive, $\frac{u+v}{c}$ must be less than $(1 + uv/c^2)$ and $w < c$.

Let us put in some numbers to see what emerges:

(i) if $u = v = \frac{1}{2}c$, $w = \frac{c}{1 + 1/4} = \frac{4}{5}c$,

not c;

(ii) if $u = v = \frac{4}{5}c$, $w = \frac{8/5\,c}{1 + 16/25} = \frac{40}{41}c$,

which is getting close to c. But if we push it much further, and put $u = v = 0.99\,c$,

$$w = \frac{1.98}{1 + 0.9801}c = \frac{1.98}{1.9801}c \ .$$

Note, though, that a small increase in speed may give rise to an enormous increase in energy. Going from 99% the speed of light to 99.5%, we find that

$$\frac{1}{\sqrt{1 - v^2/c^2}} = 7.09 \qquad \text{at} \quad v = 0.99c$$

whereas it is 10.01 at $v = 0.995c$. The increase in speed is half a percent; in energy, over 41%. There is a moral for accelerator builders in this; they must add large amounts of energy to the particles they accelerate (almost three times their rest energy) in order to increase their speed by half a percent! "Accelerators" might better be called "energizers"!

At this point, those readers who have an allergy to mathematics (algebra) will feel they are drowning in symbols.

How can one "understand" a fact demonstrated by means of mathematical formulae? First, you must understand what the mathematics is saying, that is, you must comprehend the language. But "understanding" is essentially qualitative, not quantitative. The derivation just proposed for the addition of velocities in relativity is quantitative and precise; it provides a numerical answer to the question: if B moves relative to A with speed u and C moves relative to B with speed v, what is the speed of C relative to A? But perhaps this numerical answer is not very interesting to anyone but a professional physicist. What *is* interesting, and what follows from the formula, is that however great the speeds u and v are, the speed of C cannot be greater than the speed of light, since nothing can exceed that speed. But do you understand why this is so?

One can understand some reasons why it *might* be so, once you have accepted the reality of the phenomena of length contraction and time dilation. For instance, the speed of C relative to B will not appear the same to an observer at rest in A as to one at rest in B or C. That is because the distance traversed between B and C, as well as the time interval taken to traverse it will, when viewed in A, appear to be modified by these effects. So the speed of C relative to

B cannot simply be added to that of B relative to A, because these are not speeds measured in the same frame of reference; to add them is to add unlike entities.

But of course this doesn't show why the speed of C relative to A cannot be greater than the speed of light; only that it cannot be determined as simply as one might at first have thought. The demonstration that it is so can all the same be seen from a general argument based on the Doppler effect. We use a process which might be called "chain reasoning". The idea is to start from something which we have already understood, and then show how it can be used to make further deductions, but this time qualitative rather than quantitative ones.

So let us refresh our memories about the Doppler effect. It states that any periodic phenomenon in a physical system is seen to run more slowly when that system is receding than when it is at rest, or more rapidly when it is approaching. To be specific, take a light signal coming from a periodic source with a definite frequency; say, blue light of wavelength 4500 Å and frequency 6.7×10^{14} Hz. If the source is receding, the observed frequency of the light will be shifted toward the red end of the spectrum; more specifically, it will always be shifted to a finite value, which will approach zero as the recession speed approaches the speed of light. If there is another frame of reference F relative to which *ours* is moving, it will see light further shifted to the red than we have observed, but it will still of finite frequency. If the frequency of the light is shifted relative to the original value, that shift can be regarded as a Doppler shift. It will correspond to a finite recession velocity, which must then be the velocity of frame F relative to the rest frame. It follows that F must be moving relative to the source frame with a speed less than the speed of light, whatever the relative speeds of the different frames of reference may be.

In other words, if the speed of F relative to the rest frame were greater than the speed of light, light from the rest frame would never reach an observer fixed in F, but our argument from the Doppler effect says it does, albeit with a reduced frequency.

We can illustrate this with specific numbers. Suppose that the source recedes from us with a speed of $\frac{4}{5}c$. In this case the Doppler shift factor is 3, which means that the frequency of the light reaching us is 2.2×10^{14} Hz. Its wavelength will be 13,500 Å, which places it well into the infrared region of the spectrum.

Suppose now that we are in turn moving with speed $\frac{4}{5}c$ relative to F. The "light" reaching F will have a frequency of about 7.3×10^{13} Hz, and a wavelength of 45,000 Å, and thus is still deeper in the infrared. Still, this is the Doppler shift for a finite velocity less than c, viz. $\frac{40}{41}c = 0.976c$ approximately.

We arrive then at the surprising result that in relativity, *compounding* velocities is not the same as adding them, as we would do in classical Newtonian physics. In fact, if we take a sequence of frames of reference, each moving with a high speed relative to that which precedes it, the speed of the last relative to the first will still be less than that of light.

5.10. Relativity of Simultaneity

This section might also be called the *relativity of time*. The title chosen, however, is adequate in distinguishing it from time dilation. It has to do with the basic question posed by Einstein — how does one decide whether things happen "at the same time" when they are at different places? There is no absolute answer.

It is best to start by considering how we measure time. We use *clocks*, that is, instruments to measure periods of time. The clocks of ancient man were the daily rotation of the earth detected by the position of the sun, the moon and the stars, the orbit of the moon around the earth and the orbit of the earth around the sun. These gave us the day, the month and the year, which could then be appropriately subdivided. That our instruments of measurement have vastly improved with technological development is dramatically shown by the fact that downhill ski races are now won, or lost, by one-hundredth of a second.

What constitutes a "good" clock? Primarily, independence of environment; we want something which will work as well on the ice of the North Pole as in the equatorial heat; on a rotating earth as on a spaceship. We want a *universal* standard.

In recent times we have learned to exploit the periodicity of atomic or molecular processes which, by virtue of the laws of quantum mechanics, are independent of macroscopic conditions; thus, "atomic clocks". Orbits and rotations of the earth, the moon and the sun may change over astronomical time; pendulum clocks are subject to variations of temperature, air pressure and the like which may affect the regularity of their motion. Spring-based clocks are subject to expansion and contraction with temperature; over long periods, to metal fatigue. The atomic clock, based on the periodicity of atomic or molecular processes, may be taken anywhere and assumed to behave in a constant way.

We always think of time as something absolute; that is, that the *times* of events exist independent of their measurement. Newton made this one of his fundamental postulates. A major effect of the Einsteinian revolution is to lead us to the realization that time is determined by the processes of nature, a conviction strengthened by our knowledge of thermodynamics. We shall see that the Einsteinian theory of gravitation tells us that the very structure of space itself is determined by the distribution of energy in the universe. We also learn from Einstein that space and time are inseparable. When we look at an astronomical object a billion light years away we are also looking back a billion years in time. So it is reasonable to assume that time, like space, is a *property* of the physical world. Space and time are not the stage on which the drama of nature unfolds; they are part of the drama itself.

Even if we have a "perfect" atomic clock (and the μ-mesons of the Mount Washington experiment qualify as such) we have discovered that the intervals of time they define may appear different in different frames of reference. One is then tempted to say that time is relative. Feynman berates "cocktail party philosophers" who tell us that the essence of the theory of relativity is to say that "everything is relative". This is not the essence at all. We know, of course, that the speed of light is absolute, not relative. But if we are to give any meaning to the measurement of time, we must have another absolute: the atomic clock. To make this statement more precise, in the light of Einstein: a perfect atomic clock, *read in a frame of reference in which it is at rest*, gives an absolute measure of time. This is not at all in conflict with time dilation. If, wherever we go, whatever we do, we measure time by an atomic clock that goes with us, it will always "keep the same time". It will always measure the same behaviour in time of atomic and molecular events: radioactive decays, for example, or the processes of the organic molecules of our bodies. We ourselves are governed by this absolute time. Physicists call it "proper time". Einstein tells us that it is invariant with respect to the frame of reference. In a given frame of reference, atomic clocks at different places keep the same time, i.e. they may be used to measure unambiguously the times of events at different places.

(A word of caution: all the above is true for the special theory of relativity. In the general theory, where acceleration and gravity become indistinguishable, gravitational effects on clocks are the same as the effects in accelerating frames of reference. But the special theory only deals with frames of reference *not* accelerating with respect to each other. This is a warning of things to come.)

A frame of reference, then, must not be thought of as existing at a particular place, or having a particular character. It is rather a set of identical measuring instruments all at rest with respect to each other.

The essence of the phenomenon of time dilation is this: when a standard clock is in motion in a certain frame of reference, the time it measures is dilated relative to that measured by an identical clock at rest.

We now return to an earlier question: can it be established whether two events taking place at different locations are simultaneous, i.e. "happen at the same time"? The answer is, in a given frame of reference, yes. But if two events are judged simultaneous in *one* frame of reference, they will not appear so in another frame of reference moving with respect to the first one. Let us see how this comes about.

Suppose that I wish to establish the simultaneity of an event in my rest frame of reference and one at a distance $a(x = a)$. What I can do is station myself halfway between the places where the two events take place and watch signals coming from the two sites (O and A). Clearly, signals arriving from these sites at the same time are simultaneous.

Now consider another frame, moving with respect to me with speed v. I shall take the origin (base-point) of this frame to coincide with mine at a time which I shall call zero. In both frames, then, the initial event will be observed to take place at $x = 0$ and $t = 0$.

In the subsequent argument I shall make v sufficiently small that length contraction between the two frames can be neglected.

Let us consider how I will view the arrival of the signals from these two events at the *midpoint* M in the other frame. The signal from point O will reach this point at time t_1. Suppose this second frame is moving *away* from my origin; thus that signal will travel a distance $\frac{1}{2}a + vt_1$ to get to M. This must be ct_1, so

$$\frac{1}{2}a = (c - v)t_1$$

and

$$t_1 = \frac{a}{2(c - v)} \ .$$

The signal from the *other* end will only have to travel a distance $\frac{1}{2}a - vt_2$ if t_2 is the time for the signal to reach its destination. This must be equal to ct_2 so

$$\frac{1}{2}a = (c + v)t_2$$

and

$$t_2 = \frac{a}{2(c + v)} \ .$$

Thus, the signal from A will arrive ahead of that from O by an amount

$$\frac{1}{2}a\left(\frac{1}{c-v} - \frac{1}{c+v}\right) = \frac{av}{c^2 - v^2} \simeq \frac{av}{c^2}$$

where $\simeq$ means "is approximately equal to".

The second frame of reference will *not* recognize the events as simultaneous. The one at A will be presumed to have occurred first.

There is a simple graph to show this (Fig. 5.12). Draw, for my frame of reference (S), two perpendicular axes along which we measure respectively distance and time. (Actually we shall use x and ct, which makes a more convenient scale, since for light x and ct change at the same rate.) My two events are marked by circles on the diagram, i.e. they both take place at $t = 0$.

Now let us trace the origin of the other frame of reference (S'). Since it moves much slower than light, its x increases slowly with time. Since the path of light again corresponds to $x' = ct'$, the x-axis of this frame (x') must be such that the path of light bisects the angle between the x' and ct' axes.

In its own frame of reference, only time changes; therefore, the path of its origin is simply the time-axis of S'. Ox' then designates the space axis.

Let us take an event represented by P. All lines parallel to Ox' are lines on which time is constant, i.e. they designate simultaneous events in S'. PQ is such a line. The *time* at P is then given by the distance OQ (or PR) and the place (coordinate x') is given by the length QP.

We now see that the events O and A, which are simultaneous in S, are *not* so in S'. In the frame S', the time of event O is zero, but that of A is the negative value ON, where AN is drawn parallel to Ox'. Thus, in frame S', the event A takes place *before* event O, as stated in an earlier argument.

Suppose next that the origin of the frame of reference S' were moving in the *opposite* direction (increasing *negative* x) with respect to S (Fig. 5.13). Its path is then $O - ct'$, which is the time-axis in

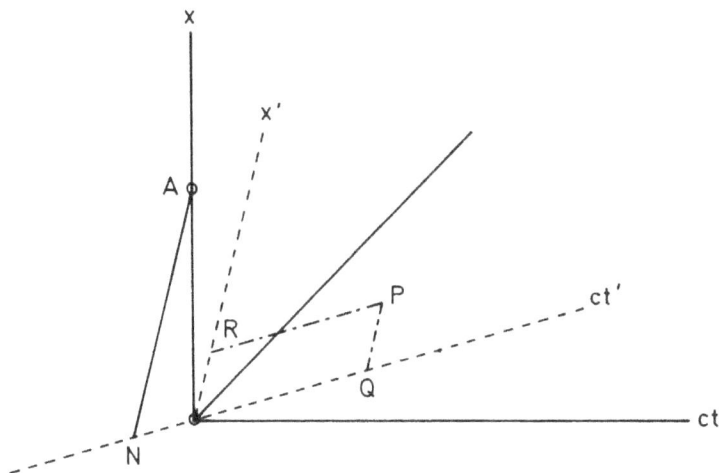

Fig. 5.12. Consider first the perpendicular axes; an event at a particular place (*z*) taking place at a definite time (*t*) may be represented by a point. Measure *z* along the vertical axis, *ct* along the horizontal one. The horizontal distance from the *z* (vertical) axis is *z*; the vertical distance from the *ct* (horizontal) axis is *ct*.

Since light has a velocity *c*, the path of a ray of light is a line making equal angles with the two axes.

Suppose that the origin of another frame S′ is moving relative to the first. Its path will be represented by the lower dashed line. Since the origin is at rest in that frame, the line will be its *ct′* axis. Since the speed of light is the same in all frames, the *z′* axis will make the same angle with the light line in the new frame. We call it *z′*.

All lines in the direction of the *ct′* axis (e.g. RP) have the same coordinate *z′* in frame S′. All lines in the direction of the *z′* axis (e.g. QP) have the same *ct′*, so the events lying on them are simultaneous in the frame S′.

The event A is at time *t* = 0, and so is simultaneous with the origin O in S. But it is simultaneous with N in S′, and so takes place *before* the event O in that frame.

frame S′. The space-axis is then O*x′*, where the angle *x* − O − *x′* is the same as the angle *ct* − O − *ct′*.

Again, we indicate by O and A the original two events simultaneous in S. Again, however, they are *not* simultaneous in S′. A takes place at the time corresponding to point N. This time event A corresponds to a *positive t′*, so it is seen to take place *after* the event O (Fig. 5.13).

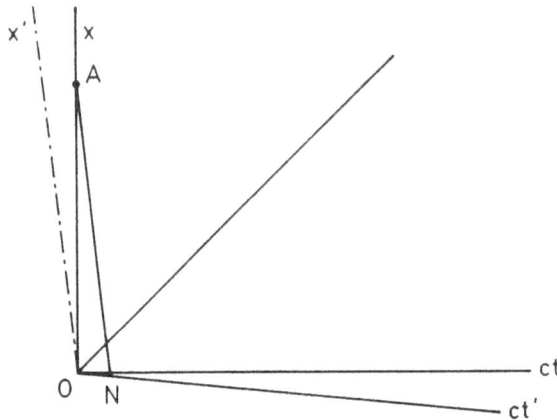

Fig. 5.13.

Thus not only are events O and A not simultaneous in any frame but S, but A may appear to *precede* O in some frames and to follow it in others! Even time ordering is not absolute!

At first glance, this seems to violate the law of cause and effect. The reason for this "violation" is simple. The events A and O, which take place simultaneously in frame S but are spatially separated, cannot possibly have any cause and effect relationship, since for one event to influence the other that influence would have to be exercised instantly (i.e. in no time at all) at a distance. But this violates the law that nothing travels faster than light. No information can go from O to A (or vice versa) faster than the speed of light.

Conversely, events in a causal relationship to each other cannot be simultaneous in *any* frame of reference.

If we again draw mutually perpendicular time and space axes for a frame S, we can put in the light paths KL and K'L' (Fig. 5.14). Then any events which can be influenced at O must lie in the region L'OL, while any events which could have influenced O would have

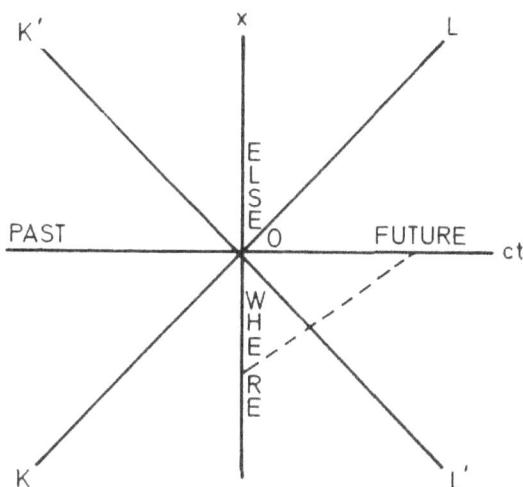

Fig. 5.14.

had to lie in the region KOK'. The first region is the region of the *future* for O, the second that of the past. Events in the other regions (KOL' and K'OL) are outside the world of the observer in S just as events taking place *now* on a distant star are outside my world. They are "elsewhere".

Of course, a signal sent out *now* from an event F in my "elsewhere", at the speed of light, could arrive in my world some later time. Astronomers are now seeing objects in the sky billions of light years away. The messages from the time of my predecessors of billions of years ago are now entering my world, the world of their successors.

A final remark — the non-simultaneity calculated earlier, $\frac{av}{c^2}$, is small compared with c; it depends on the ratio $\frac{v}{c}$. But it is not as small as the time dilation or length contraction, which depend on $\left(\frac{v}{c}\right)^2$. If, for example, $\frac{v}{c} = 0.01$, $\left(\frac{v}{c}\right)^2 = 0.0001$. Non-simultaneity is therefore the biggest relativistic effect. What is not evident without mathematical analysis is that time dilation and length contraction, and in fact all other relativistic effects, are *consequences* of non-simultaneity. This is the source from which everything else springs.

5.11. More on the Use of Space-time Diagrams

We can use space-time diagrams to illustrate time dilation and length contraction.

5.11.1. *Time dilation*

Imagine two frames of reference S and S', S' having a velocity of $\frac{4}{5}c$ relative to S, along their common x-axis. As seen by S, the world-line of the origin of S' is Ox'. This must be the time-axis of S', since the spatial coordinate of the origin of S' remains fixed. But since the velocity of light is c in all frames of reference, the time and space axes of S' must make equal angles with the light line OL (see Fig. 5.15).

We do not know yet how to measure distances and times along these axes, in relation to corresponding distances and times in S. A possible method would be: let a signal be emitted at the origin of S at time 1. By virtue of the Doppler effect, one time interval in S is "seen" as three in S', that is, the signals emitted at the beginning and end of one of S's intervals arrive at S' three intervals apart. Time intervals appear "stretched out" by a factor of three.

The moment the signal is received at the origin of S', a signal is sent back saying "your signal arrived here at time 3". When does this return massage arrive back at S? The Doppler effect now works in reverse; every interval in S' appears to S to be stretched out into 3. Thus, the return signal is received at time 9. The observer at S now reasons "I sent out my signal at time 1 and got a reply at time 9". Since the signal took the same time to go in each direction, it must have been received at S' at time 5 — the signal taking 4 units of time each way. But the observer at S' says he received it at time 3 by his clocks. What is happening in the frame of reference of the observer at S' appears to the observer at S to be slowed down; time is *dilated* by a factor 5/3.

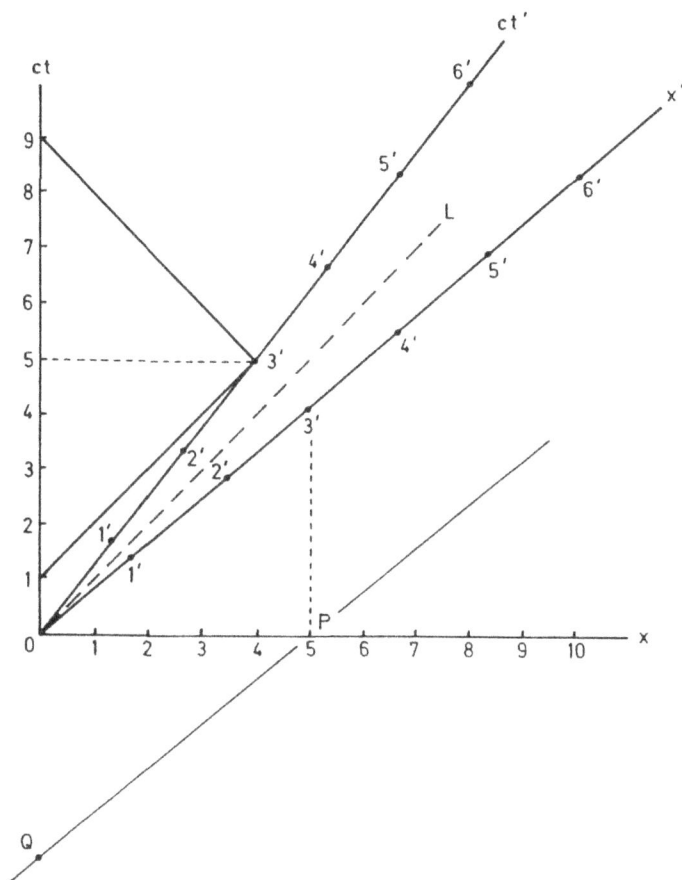

Fig. 5.15. x', ct' are the axes of a frame of reference S', which is moving at speed 4/5 c relative to the "fixed" frame S.

The scale markings on these axes are deduced as explained in the text.

Lorentz contraction is revealed in the fact that the end-points of the 5-unit rod at rest in the frame S (OP) have coordinates (O, 3') at a fixed time in S' (3').

The effect is of course reciprocal. To S', everything in S appears to have been slowed down. Is a contradiction apparent? Not at all, because S and S' cannot confront each other to determine who is "right". In fact, both are right, so no contradiction is involved!

5.11.2. *Space (Lorentz) contraction*

The first thing we have to note is the symmetry of the diagram with respect to space and time. This is simply a consequence of the constancy of the velocity of light. The discussion of time dilation enabled us to mark out the time scale of S′ on the diagram. By virtue of the symmetry, the length scale may be marked out correspondingly.

We can use the same diagram, but in a different way. Suppose there is a measuring rod in S, 5 units long, between $x = 0$ and $x = 5$. The ends are two *events*, O and P, which occur at the same time in S. What we mean by "the length as measured in S′" is: what is the spatial separation of the two ends of the rod *at the same time in S′*? When the question is posed in this way, we see that an important element is the relativity of time. What we mean by "at the same time" is different in the two frames. In S, we are concerned with the separation of two events *simultaneous in that frame; in S′, we are concerned with events simultaneous in S′*. But *simultaneity is relative*. That, in the end, is the source of time dilation and length contraction.

How then do we determine the length contraction from the diagram? Events simultaneous with P in S′ lie along the line PQ. The event at time zero in S′, at the position O, is Q. Thus, the length of the rod in S′ is the distance QP measured in the direction of the space-axis of S′. This is obviously the same as the distance O3′, i.e. the length in S′ is 3 units. This is the space contraction observed in an object moving with respect to the frame of reference in which the observer is at rest.

Note that what we have done here is deduce the space contraction from the time dilation. They are two sides of the same coin.

5.12. Are Relativistic Phenomena Real?

Do objects moving past us at high speeds really contract in their direction of motion? Do clocks moving very fast with respect to us really slow down? One can reply with a question. What do you mean by *really*? What I *mean* is, are relativistic effects some kind of conjurer's trick, some sort of illusion, which lead us to see things, or interpret things, which are not objectively happening? If what we observe in nature depends on our frame of reference, does not science become subjective rather than objective? How can we know objective reality? There is no end to these questions.

We will pre-empt this game by making an assertion: all relativistic effects *are real.* Just as real as anything else in physics.

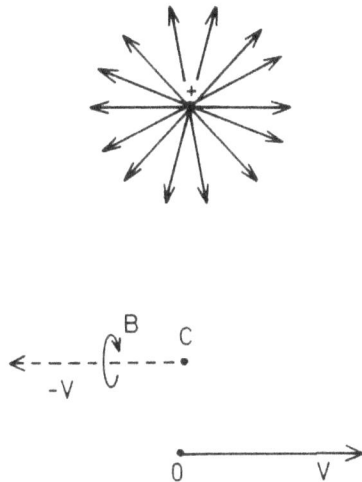

Fig. 5.16. If we (O) move past a charge at C, it is moving relative to us and thus produces a magnetic field.

Consider the following relativistic effect. If an electric charge is at rest (in a given frame of reference), it produces a static electric field, but no magnetic field. Now take another frame of reference, *moving* with respect to the charge. To measuring instruments in that frame, the charge is moving, and so produces a magnetic field. Its electric field will also appear modified.

Now we ask: is the magnetic field real? Of course it is; as real as any other magnetic field in nature. Still, there is a frame of reference in which it does not exist.

The argument here is no different from the one concerning length contraction, time dilation, relativistic mass increase and the rest.

The length contraction effect was in fact predicted before Einstein formulated his theory, and is generally known even now as the Lorentz contraction. But Lorentz predicted it as an effect resulting from Maxwell's electromagnetic laws. Recognizing matter as being made up of elementary negative and positive charges, he noted that when an object was in motion it would produce electric and magnetic fields *different* from those produced at rest. He went on to determine how these particles would interact with each other through these modified forces, and came to the conclusion that the result would be a contraction of the object by a factor $\sqrt{1 - v^2/c^2}$ along its direction of motion!

(Actually, his argument introduced some additional assumptions, but these fortunately did not alter the result.)

Lorentz's work was a technically brilliant exercise, far more complicated than Einstein's deduction from simple principles. But most interesting is the fact that those principles did not appear to play any role in the deduction.

In retrospect we can now see that, in a way completely hidden to Lorentz, they had nonetheless been implicit in his use of Maxwell's equations in the derivation. For we now know that Maxwell's theory was, rather miraculously, a relativistic theory (a fact of which neither Maxwell nor the other physicists of his time had the slightest notion).

In any case, Lorentz was certainly not using a relativistic framework of explanation. He was deducing a consequence of a well-established physical law.

It is interesting to note in passing that, in the course of his argument, Lorentz found it convenient to introduce something he called "local time", which was in fact what we would now call "proper time". What he did *not* do was provide us with a new interpretation of nature, a new set of basic principles which laid bare the essential simplicity of the new phenomena of space and time, while at the same time encompassing the whole structure of electromagnetic theory. Only when Einstein entered the scene was the great step made which carried us far beyond Lorentz.

The moral of the foregoing discussion is that the length contraction was, to Lorentz, a physical phenomenon following Maxwell's electromagnetic theory. That relativistic phenomena have a physical reality, and that the relativity concerns only the *interpretation* of that reality, is easily seen from the even simpler but quite dramatic argument which follows.

5.12.1. *From electricity to magnetism by change of reference frame*

A charge at rest produces an electric field which is spherically symmetrical; the lines of force radiate out from the charge (Fig. 5.16). It does not, of course, create a magnetic field.

Suppose that we move with a speed v relative to the charge in a given direction. The charge appears to be moving relative to *us*, with the same speed but in the opposite direction. We know, however, that moving charges create a magnetic field. A succession of electric charges moving in this way is in fact equivalent to an electric current. By Oersted's law, the current will produce a magnetic field whose lines of magnetic force circle around the direction of flow in closed circles as shown in Fig. 5.16. (If the flow of charge, i.e. the current, is steady, so will be the magnetic field produced. If there is only a single charge, the magnetic field will move with the charge.)

We can see in these facts evidence of a profound connection between electric and magnetic fields. To one observer, who is at rest relative to the charge, there is only an electric field. To another, moving relative to the first, a magnetic field will also appear. But the difference only depends on the point of view! What appears as a pure electric field to one observer will appear as a *combination* of electric and magnetic fields to another! And that magnetic field is, of course, perfectly real. It will, for instance, deflect the needle of a compass.

Another interpretation is: what appears as a simple manifesta-tion of the electric field law in one frame of reference is explainable only by also using Oersted's law in another. A profound connection between these laws is thus revealed.

Because of the symmetry between electric and magnetic fields revealed by Faraday in his law of induction, the above argument leads us to ask how the fields of a magnet appear to an observer moving with respect to it. If the magnet is at rest, it produces only a magnetic field. But what happens if it is moving, or if one is moving relative to it? If the field is between two poles of a magnet, the needle of a compass at rest will point along the lines of force (Fig. 5.17).

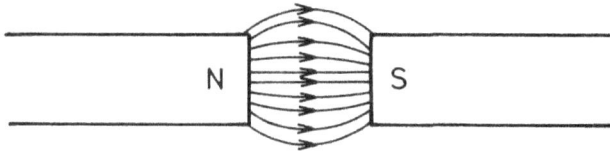

Fig. 5.17.

We might expect from analogy with the previous example that, in a frame of reference moving relative to the magnet, an *electric* field would appear. This could be tested by a charge, e.g. of an electron. If the electron is at rest relative to the magnet, the magnet would exert no force on it (we ignore that the electron is a tiny magnet. This, however, would produce only a minute effect).

Imagine that the charge moves relative to the magnet (or vice versa). We might consider, for example, a small loop of conduct-ing wire, which moves horizontally from back to front in Fig. 5.17, remaining always parallel to the pole faces. As it moves, the flux through it changes (or, as Faraday would say, it cuts through mag-netic lines of force). The effect would, of course, be exactly the same if the loop were kept fixed and the magnet moved horizontally backward relative to it.

In either case, by Faraday's law, a current, or a motion of charges, would be set up around the loop. This is a manifestation of the *electric* field associated with a changing magnetic flux. This electric field is not present in the situation of relative rest of magnet and electron. The situation is therefore analogous to the preceding one.

The electric field detected by the moving coil does not depend in any way on the existence of the loop of wire, but would also be felt by an electron, or a beam of electrons, passing between the magnetic poles. If the charge infringes on the magnetic lines of force from the rear, the magnetic flux would be *increasing*, tending to create a current or electron flow curving in a *clockwise* direction about the flux lines. As a result the electron would be deflected *downward*.

Another way of summing up the previous discussion is: a magnetic field observed in one frame of reference may be "transformed away", i.e. will not be detected, in another. Physical effects attributed to magnetic fields in one frame of reference must be attributed to electric fields in another, and vice versa.

5.13. The Twin Paradox[4]

We have no experience of motion at the speeds at which relativistic effects become significant, so our intuition fails us when dealing with relativistic phenomena. One way of expressing our inability to reason about them intuitively is to seek "paradoxes" (or more appropriately, *apparent* paradoxes) in which relativistic principles, applied to circumstances within our comprehension, lead to some sort of *reductio ad absurdum*. Sometimes it is only a matter of a conclusion which is strange and unexpected, in which case there is no paradox at all. Sometimes, by virtue of flawed reasoning, we think we detect logical contradictions which create the appearance of a paradox where none really exists.

---

[4] For a more detailed discussion, see the Appendix at the end of this chapter.

Of these supposed "paradoxes" the most famous is the so-called "twin paradox". It can be expressed in the following familiar and simple form: imagine two identical twins one of whom (B) is going on a long space voyage at speeds approaching the speed of light. The other (A) stays home. Now because B is moving very fast relative to A, B's clock, and at the same time his physiological processes, appear to be slower than A's. Thus when B returns home, he will be younger than A.

The so-called paradox is this: just as B is moving away from A, A is moving away from B. Since Einstein says that all frames of reference are equally valid, from B's viewpoint A appears to be aging more slowly. Can *both* be right?

For this argument to be valid, the two frames of reference must be equivalent and Galileian. As long as the frames are moving relative to each other with constant speed, it would appear to each of the twins that the other is aging more slowly than himself. This may appear strange, but it is not paradoxical unless they meet again. Only then will it be evident whether one or the other is older, or whether they are of the same age.

Of course they cannot meet if they continue to move with constant speed relative to each other. One or the other, if not both, would have to decelerate to a stop, and then accelerate backwards. Let that be the voyager B. However, this process of deceleration will have physical effects in him. Of course, he could say that A was accelerating relative to him, but the simple fact is that A feels no force of acceleration while B does — an important difference. If A remains in his Galileian frame, B cannot.

Let us spell out how the two twins keep track of each other. A, for instance, could send out a message at a given time, "tagged" for B with this information. For simplicity, suppose B's speed relative to A is $\frac{4}{5}c$. B receives constant signals from A specifying the passage of time by his (A's) clock. But due to the Doppler effect, the signals will come to B at a frequency

$$\sqrt{\frac{1 - \frac{4}{5}}{1 + \frac{4}{5}}} = \frac{1}{3}$$

as great as that specified by A as he sends them. Thus, when he receives the announcement from A that one year has passed, B will have seen *three* years pass. A will appear to be aging only $\frac{1}{3}$ (not $\sqrt{1 - \frac{16}{25}} = \frac{3}{5}$) as fast. The situation will be precisely reversed if B's frame is used.

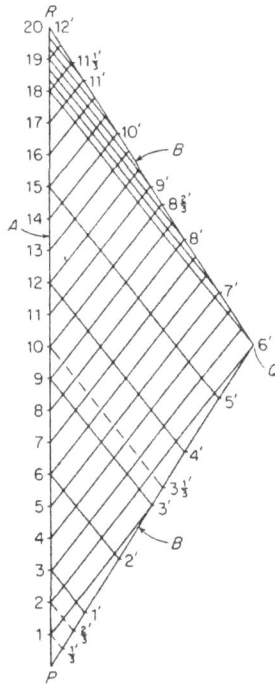

Fig. 5.18. How each of A and B can "survey" the motion of the other. (Reprinted with permission from *Introduction to the Special Theory of Relativity* by C. Kacser, Prentice Hall. Copyright 1967 by Claude Kacser.)

If B is going to a planetary system eight light years away, A, noting that B's speed is only 4/5 that of light will clearly expect B back in 20 years. But, being a logical man, he is aware that there is one flaw in their procedure of sending messages. That is, B will be further away at the end of a period than at the beginning. The end will thus appear delayed, so that the period will be perceived as longer than it really is (this in fact is the essence of the Doppler effect). So they decide beforehand to eliminate this factor by using a better system. When B receives his "end of the first year" message (at the end of his third year) he immediately signals this fact back to A. Since A knows that, due to the constantly increasing distance between them, B's clock will appear slowed down by a factor 3, he will not be surprised when B's acknowledgement reaches him at year 9.

A now makes a simple deduction. Since the "out" and "back" messages travelled the same distance, and always at the universal speed of light, he will deduce that B would have sent the return message halfway between his years 1 and 9; that is, during his year 5. But B's message will say that he was at year 3 at that time. So A will now deduce (not directly observe) that B is aging 3/5 as fast as he is, exactly as Einstein would predict. Since there is nothing in this argument that does not still hold if the positions of A and B are reversed, B will also deduce, with equally impeccable logic, that A is only aging 3/5 as fast as he is.

Going a stage further in the argument (made from A's viewpoint), A will deduce that after 10 of his years, B will have reached his destination, being only 6 years older than when he left. Since the return journey should take the same time, B should arrive back home only 12 years older, and find his twin A older by 20 years.

But wait: how can B go 16 light years at 4/5 the speed of light in only 12 years? Should it not be 20 years? All is well if B went a distance equal to $\frac{3}{5} \times 8 = 4.8$ light years, rather than 8. This is of course precisely the Lorentz contraction. The earth and the distant star have remained in place, but because of his speed, B sees the earth as only 4.8 light years away from the star.

Incidentally, A could, from the data at his disposal, verify how far away B must have been when his 6-year message was sent. The message which A sent at year 2 was answered in year 18, 16 years later. Since the trip to and fro took equal time, B must have been 8 light years away. The same method could equally well be used to verify his distance at each point of the trajectory, e.g.

– The message A sent at year 1 was received by B at year 3, a fact which could be signalled in the return message received at year 9. His distance at that time was $\frac{1}{2}(9-1) = 4$ light years.

– A message sent at year 5 was received by B at year 7; his reply was received at year $18\frac{1}{3}$. Thus, his distance at his year 7 was $\frac{1}{2}(18\frac{1}{3} - 5) = \frac{20}{3} = 6\frac{2}{3}$ light years. A then verified that B was at this point on his return journey home.

Let us recapitulate the essential point. Going purely by observation A "sees" B aging only one-third as fast as he is on the outward journey, but three times as fast on the inward one. He is able to deduce from their exchange of messages, however, that B is really aging 3/5 as fast as he is, a fact which can be verified when B arrives back home. Note clearly the difference between the physical reality and what is "seen". Most important of all to understand is that relativity is a theory of what really happens, though how that reality is described depends often in a very dramatic way on the frame of reference used for its description.

The difference between appearance and reality manifests itself, not only in the case of time dilation, but also in that of Lorentz contraction. An illustration will be briefly outlined. Let us consider a rigid wire bent at right angles at its centre, passing across our line of sight and moving in the direction of one of its arms. The following facts may be established:

(i) We do not "see" the arm contracted according to the Lorentz formula. Rather, it appears contracted less as it moves toward us, and more as it moves away.

In the situation shown in Fig. 5.19, we note first that all measurements of length made in the frame of fixed observer (O) in the direction of motion AB are Lorentz-contracted. In that frame, AB *is* shorter than AC. But the signal from the end B of the arm AB

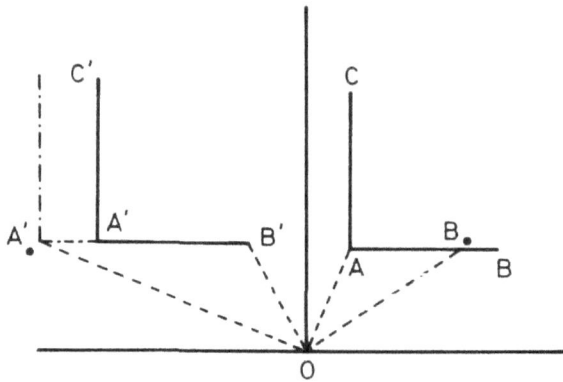

Fig. 5.19.

which arrives at O simultaneously with the signal from A will have been emitted at an earlier time, when B had not moved as far to the right. Thus, there will be an additional *apparent* shortening.

If, on the other hand, the structure is to the left of OA (position C′A′B′) it is the signal from the A end which must be emitted at an earlier time if it is to arrive simultaneously with that from the B end; thus, there will be an *apparent lengthening* of AB superimposed on the Lorentz contraction. Only when the centre of AB is opposite will the observer at O *see* the true Lorentz contraction.

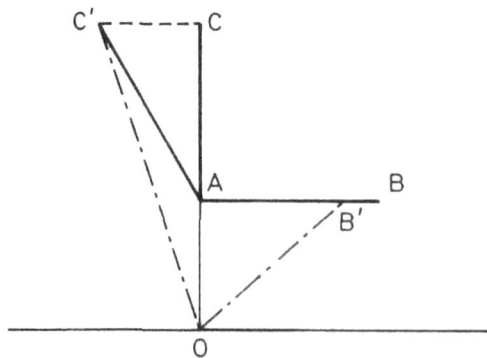

Fig. 5.20. Light from B′, C′ reaches us (at O) at the same time as that from A.

(ii) If the extremity of the perpendicular arm is the part of the structure furthest from us, that arm will appear to be rotated backwards relative to the direction of motion. Its back end will then be visible before its front end reaches its point of nearest approach to us! The reason for this is of course simple. Since the far end is further away, light from that end will take longer to reach us. Thus, coincident with light received from the *near* end we will receive light emitted *earlier* from the far end.

5.14. An Alternative Viewpoint on the Twin Paradox

The situation may be viewed in another way which casts the phenomenon in a different light. On the outward journey, we can look at matters from the viewpoint of twin B, who is in a frame of reference moving with speed $\frac{4}{5}c$ relative to the frame of A, and vice versa. Thus, to B it seems that A is aging more slowly. So, when B reaches his destination, after six B years, he finds that A has only aged $\frac{3}{5} \times 6 = 3.6$ years. That is, in his frame of reference, the event at time 3.6 years by A's clock is *simultaneous* with B's arrival at his destination at year 6 by his time.

The moment B begins his return journey, he will be in *another* inertial frame of reference, moving in the opposite direction to his rest frame on the outward journey, with the same speed. We can now ask: in this frame, what event at A, back on earth, is simultaneous with his arrival at his destination? Simultaneous events are events which lie along lines parallel to his space axis. But we can see from Fig. 5.21 that, since his time-axis, his own space-time trajectory, now slopes to the *left*, and his space axis must make the same angle with it as before, that space axis is as shown. It is the vertical reflection of that for his outgoing journey. He arrives at his destination at time 16.4 by A's clock. This is an example of time dilation; six of his years correspond to the lapse of $20 - 16.4 = 3.6$ of A's years. Thus we have accounted for 7.2 years of A's time.

What of the other 12.8? They must be attributed to the non-inertial part of B's journey, when he decelerated to rest at his desti-

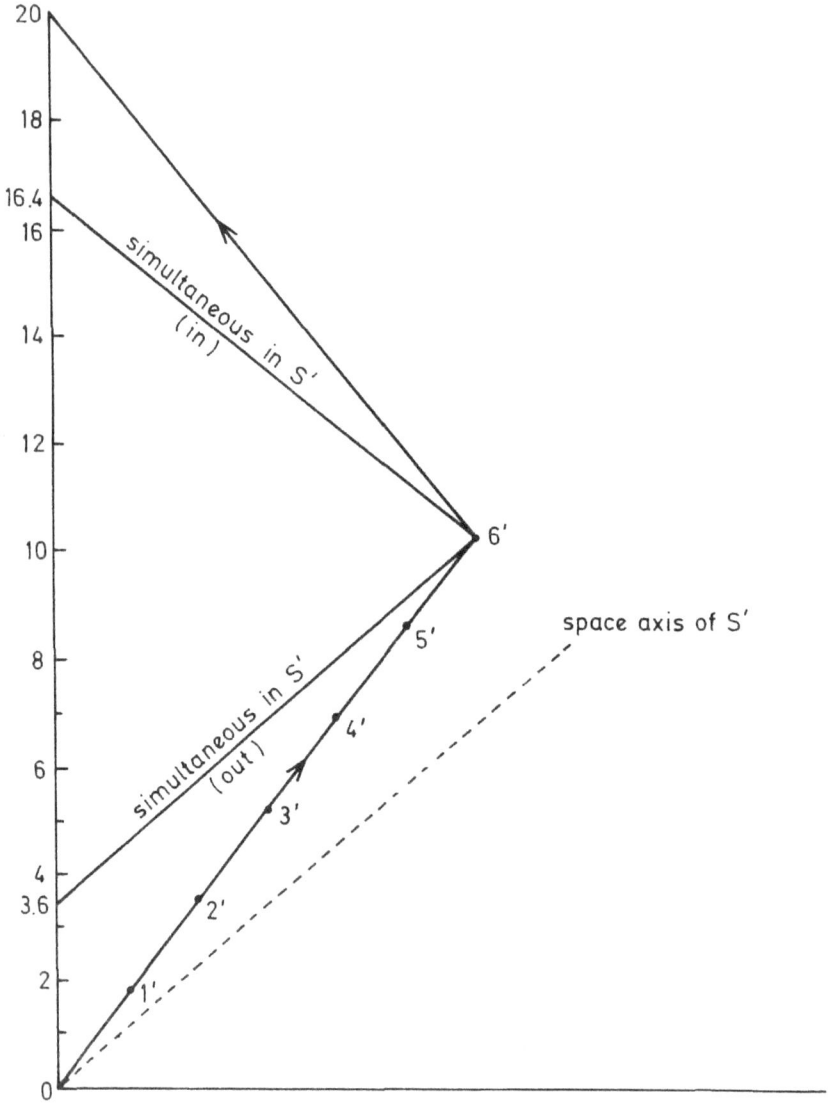

Fig. 5.21.

nation, and when he re-accelerated back to $\frac{4}{5}c$ for the journey home. As seen in Fig. 5.21, this is assumed to have happened instantaneously. This is an idealization. In reality, a finite time will have been involved (in his frame of reference). In any case, the switch from the outgoing inertial frame to the returning one must have taken 12.8 years of A's time. The interesting thing is that we do not have to know *how*; in fact, a detailed description in terms of the general theory of relativity has been worked out, yet the special theory is sufficient for a correct answer.

5.15. The Reality of Relativity – An Addendum

Relativity arose in answer to the conflicts between Newton's classical mechanics and Maxwell's electromagnetic theory. Arthur Koestler pointed out that the essence of scientific discovery is in the unification of what had previously been thought of as unrelated physical concepts and phenomena. R. P. Feynman defined the process as "taking the world from another point of view", of viewing reality in a way which reveals a new and broader set of relationships between natural phenomena. Discovering unity in what had previously been perceived as diversity and even contradiction, allows us to imagine new possibilities in nature, as well as to understand more profoundly previous knowledge.

The significance of a new theory is found in its generality and the depth of understanding it gives us. Relativity has provided us with new insights into phenomena ranging from the properties of fundamental particles of matter to the large-scale structure of the universe. It has profoundly changed our concept of space, time, energy and their relationships. Relativity is one of the most profoundly important theories in human history.

Yet its strangeness, its unfamiliarity, makes it abstract and esoteric. It is perhaps useful then to note that even the most practical and conservative elements in our society unconsciously subscribe to its truth. Great sums of public money are being poured into projects for the construction of ever more powerful particle accelerators. The design of these accelerators is based on our knowledge of physical law. If engineers designed such machines on the basis of pre-relativistic (Newtonian) mechanics, they simply would not work. Their construction is based on a very detailed application of relativistic principles, and they work *only* if these principles are correct. The voting by politicians all over the world for substantial sums of public money for their construction is nothing more, nor less, than a resounding vote of confidence in the principles of relativity. What greater testimony to the practicality of a supposedly abstract theory could one ask for?

Ronald Reagan has rather deprecatingly referred to Darwin's theory of evolution as "only a theory". There is no evidence in his words or actions of a disparaging attitude toward relativity, nor a suggestion that contradictory theories be given equal weight in the educational system. To give the word "theory" a pejorative connotation is to deny the essence of science.

Our entire experience is meaningless until it is correlated by theory. Theories are tenable only as long as they survive critical testing and observation. An intuitive opinion is not a theory. A theory must have precisely determinable consequences which are verifiable by experience. Relativity or evolution *are* theories rooted in experience. "Common sense" or "creationism" is the substitution of dogma for reason.

Our goal here has been to expose the underlying concepts of the theory of relativity, and their philosophical implications. It is important to understand, however, that it is the expression of these concepts in a precise and general form which makes it possible to explore all of its detailed consequences in specific situations. (Poincaré had some intimations of the concept, but presented no coherent new theory.) Where the physicist has an advantage over the non-physicist is in the technical exploitation of the theory. Where the non-physicist sometimes has the advantage, if and once he has grasped the concepts, is in appreciating the truly radical nature of the theory and its impact on our processes of thought. Too profound an absorption in technical detail may blunt the sense of wonder and reduce it all to banality.

5.16. Appendix: The Twin Paradox Elaborated

5.16.1. *How to use the space-time diagram*

1. Let us recall the following established facts:

(a) If B is leaving A with speed v, signals emitted from B at time interval T_0 will be received by A at interval

$$T = \sqrt{\frac{1 + v/c}{1 - v/c}} T_0 \ .$$

Example: if $v/c = 4/5$, $T = 3T_0$.

(b) If B approaches A with speed v, the corresponding formula will be

$$T = \sqrt{\frac{1 - v/c}{1 + v/c}} T_0 \ .$$

Thus, if $v/c = \frac{4}{5}$, $T = \frac{1}{3} T_0$.

2. Here is how the diagram is constructed:

The vertical line is the world-line of twin A, who stayed home (Fig. 5.18).

Suppose that his twin makes the journey to a star 8 light years away. If he travels at 4/5 the speed of light, the journey will, as A sees it, take *10 years*. So he will be *back* in 20 years (10 years out, 10 years back). We mark off A's time axes into 20 segments of a year.

Now what about the world-line of B? It makes a slope with the vertical time axis such that, for each unit of A's time (1 year) it moves 4/5 of a light year in the horizontal direction. Displacements in the horizontal direction represent displacements in space toward the distant star. In one year, B goes 4/5 of a light year, since he is travelling at 4/5 the speed of light.

How do we now scale off time intervals on B's line? What we know is this: if A sends out a signal after 1 year, B will receive it after 3 of his years ((a) on the previous page). So we draw a light line out from 1 on the vertical axis, and where it intersects B's light line, we mark 3 years (we'll designate this 3').

Similarly, if A sends B a signal at year 2, it will be received by B at his year 6, which we designate 6'.

If we do this we see that this point is just opposite A's year 10. That is, a line drawn from A's 10 to B's 6' represents a spatial displacement. These points correspond to the same A time.

And this agrees with time dilation, for the time dilation formula tells us that, according to A, B ages more slowly than he (A) does by the factor $\sqrt{1 - v^2/c^2} = \frac{3}{5}$ when $v/c = \frac{4}{5}$.

We also know how *far* away he is (again on A's distance scale). At year 10 he is at 8 light years, in other words, he has arrived at his destination.

He now turns back, heading for the reunion with his brother A at the latter's 20th year. The intermediate years in B's history can now be scaled off. He will arrive back at his year 12', eight years younger than his brother who stayed at home!

3. How each observer perceives reality

(a) In A's frame of reference:

A keeps track of what is happening to B in the following way: he sends out a signal toward B at some time t_0. It is arranged that as soon as B receives this signal he will immediately signal back his age. Suppose this signal reaches A at time t_1.

A now says: it took just as long for my signal to travel out as back, therefore B must have received it at a time half-way between times t_0 and t_1. Thus, he must have received the signal at my time $\frac{1}{2}(t_0 + t_1)$.

A also knows how far away he must have been when he received it. For the signal to go out and come back, it took $(t_1 - t_0)$ years. To go one way it took half that, i.e. $\frac{1}{2}(t_1 - t_0)$ years. Therefore, since the signal travelled at the speed of light, he must have been distant by $\frac{1}{2}(t_1 - t_0)$ light years.

Let us do an example and take a value from Table 5.1. Consider the signal sent out at A's year 1. It will arrive back at year 9, so A reasons that it must have reached B at time

$$\frac{1}{2}(1 + 9) = 5 \text{ years} .$$

B has however indicated that he received it at year 3, so he has aged only 3/5 as fast as A, exactly as time dilation says he should.

Table 5.1.

Signals sent from $A \to B \to A$

Leave A at t_1	Arrive B at	Back to A at t_2	B-event at A-time = $\frac{1}{2}(t_1+t_2)$	B-distance according to A = $\frac{1}{2}(t_2-t_1)$ light years
1	3$'$	9	5	4
2	6$'$	18	10	8 (max)
5	7$'$	$18\frac{1}{3}$	$11\frac{2}{3}$	$6\frac{2}{3}$
8	8$'$	$18\frac{2}{3}$	$13\frac{1}{3}$	$5\frac{1}{3}$
11	9$'$	19	15	4
14	10$'$	$19\frac{1}{3}$	$16\frac{2}{3}$	$2\frac{2}{3}$
17	11$'$	$19\frac{2}{3}$	$18\frac{1}{3}$	$1\frac{1}{3}$

This slower rate of aging persists at the same rate throughout the whole journey.

(b) In B's frame of reference (see Fig. 5.22):

We now have to change our viewpoint. Now B monitors A's aging by sending out light signals and receiving back signals stating A's age. We now note that, while he is on the outward journey, if he sends out signals at a certain interval, the interval at which they are received by A will be three times as great. Thus, a signal which he sends out at his year 1 will be received by A at his year 3. This is indicated on the diagram by the light line joining 1' to 3.

On the return journey the situation will be reversed. Now, signals which he sends out at 3 year intervals will be received by A at intervals of *one* year. This is because they are now *approaching* rather than receding from each other. We see this on the diagram by looking at the signal sent by B at his year 6' which reaches A at his year 18 and that sent by B at his year 9' which reaches A at his year 19.

If we now use the space-time diagram, we can follow how things look to B. It is shown in Table 5.2 and Fig. 5.23.

While the final situation — that A has aged by 20 years while B has only aged by 12 — is the same from the two viewpoints, the *way* in which B sees it is surprising. In the first $3\frac{1}{3}$ B years, A seems to be aging more slowly, in fact, only by 2 years, which is quite consistent with time dilation, since

$$\frac{2}{3\frac{1}{3}} = \frac{3}{5} \ .$$

A then seems to have receded $2\frac{2}{3}$ light years from B. But from that time on until the last $3\frac{1}{3}$ years, A appears to be aging 3 years for every B year elapsed. For all that time, B deduces that A is only $2\frac{2}{3}$ light years away. Then, in the last $3\frac{1}{3}$ years, A once more appears to be aging only 3/5 as fast as B.

So let us do the overall accounting: For $6\frac{2}{3}$ years of B's time (half at each end of the voyage) A only seems to age by 4 years.

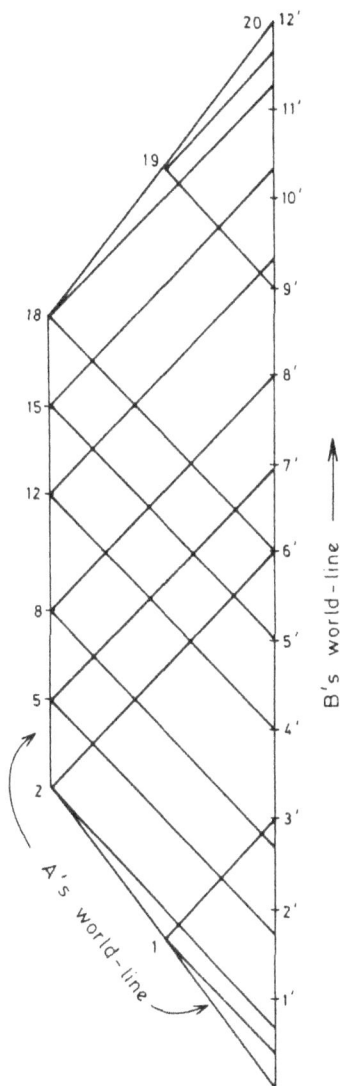

Fig. 5.22. How B sees it. (Redrawn with permission from *Introduction to the Special Theory of Relativity* by C. Kacser, Prentice Hall. Copyright 1967 by Claude Kacser.)

But for the remaining $5\frac{1}{3}$ B years — the interval in which he sends signals on his outward journey and gets responses on the inbound journey (during which time he has decelerated at his destination and re-accelerated back), A seems to have aged *three times* as fast as B, so that in a lapse of $5\frac{1}{3}$ B years A has aged

Table 5.2.

Signals sent from $B \to A \to B$

Leave B at t'_1	Arrive A at	Back to B at t'_2	A-event at B-time = $\frac{1}{2}(t'_1+t'_2)$	A-distance according to B = $\frac{1}{2}(t'_2-t'_1)$ light years
$\frac{1}{3}'$	1	$3'$	$1\frac{2}{3}$	$1\frac{1}{3}$
$\frac{2}{3}'$	2	$6'$	$3\frac{1}{3}$	$2\frac{2}{3}$
$1\frac{2}{3}'$	5	$7'$	$4\frac{1}{3}$	$2\frac{2}{3}$
$2\frac{2}{3}'$	8	$8'$	$5\frac{1}{3}$	$2\frac{2}{3}$
$4'$	12	$9\frac{1}{3}'$	$6\frac{2}{3}$	$2\frac{2}{3}$
$5'$	15	$10\frac{1}{3}'$	$7\frac{2}{3}$	$2\frac{2}{3}$
$6'$	18	$11\frac{1}{3}'$	$8\frac{2}{3}$	$2\frac{2}{3}$
$9'$	19	$11\frac{2}{3}'$	$10\frac{1}{3}$	$1\frac{1}{3}$
$12'$	20	$12'$	12	0

The results shown in this table are represented graphically in Fig. 5.23.

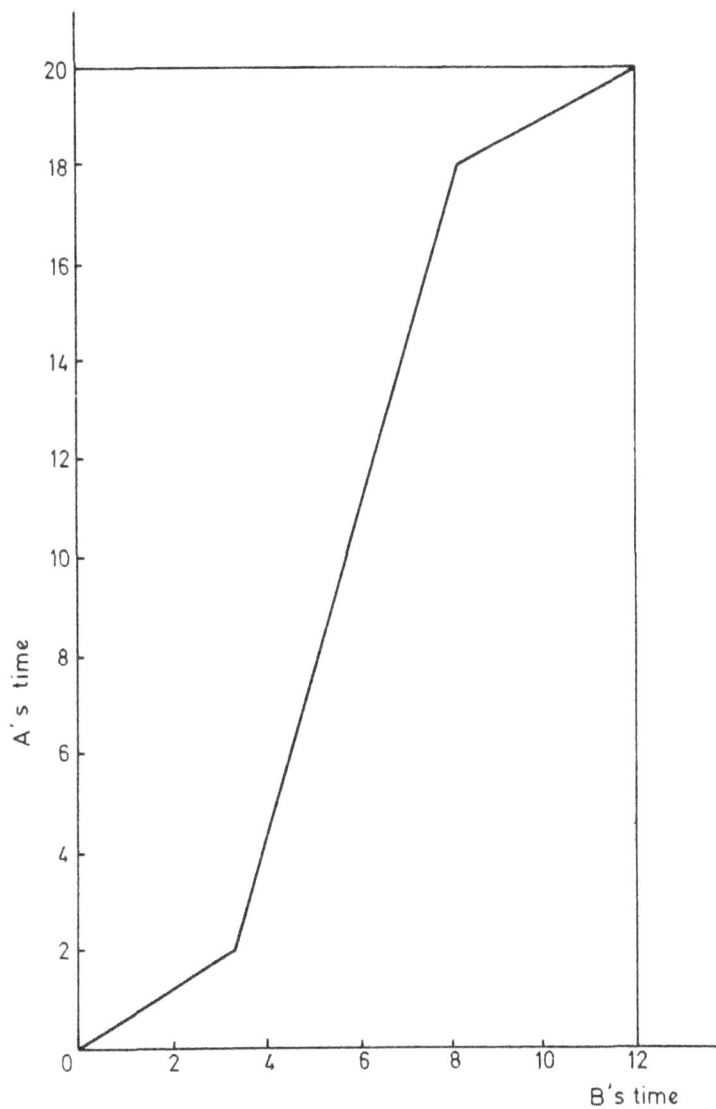

Fig. 5.23. How B "sees" the aging of A.

$$3 \times 5\frac{1}{3} = 16 \text{ years} .$$

A's total aging, then, as predicted, has been

$$4 + 16 = 20 \text{ years} .$$

We should not take too seriously the constant $2\frac{2}{3}$ light years distance which this calculation gives for A's distance from B during the latter's $5\frac{1}{3}$ middle years. After all, B will understand that in general the signal which he sent out and the one he received correspond to distances less than that of the star to which he travelled. He would therefore argue something like this: I sent out a signal at time $\frac{2}{3}'$ and it came back at 6'. A indicated that the signal reached him at his time 2, though mine was $\frac{1}{2}(6 + \frac{2}{3}) = 3\frac{1}{3}$. At that point I was at a distance of $2\frac{2}{3}$ light years. But I got to the star in 6 years. Therefore its distance must have been $\frac{6}{3\frac{1}{3}} \times 2\frac{2}{3}$ light years = 4.8 light years. Why not the 8 light years, which is the distance measured by A? Because of the Lorentz contraction due to B's motion *relative* to the star. The contraction is by a factor of 3/5, and $\frac{3}{5} \times 8 = 4.8$.

Chapter 6

THE GENERAL THEORY OF RELATIVITY

One can easily imagine that Einstein's "general theory" is more difficult to understand than the "special" one which we discussed in the previous chapter. Interestingly enough, this does not seem to be so, because in the process of generalization we are led naturally to gravitation, and then to the nature of force in general, and these are areas of which we have some intuitive experience. Further, the central idea, embodied in what we call the "Principle of Equivalence", is itself really a rather simple and intuitive one. It is only in following its consequences to their logical conclusions that we are led to new and wonderful discoveries.

What is the basic problem? Einstein already showed in the "special" theory that all Galileian frames of reference (i.e. frames of reference moving with *constant* speed relative to the "fixed stars" and to each other) were equally valid for the description of physical phenomena. The direction of Einstein's thought, however, was this: might it not be possible to cast the laws of physics in a form which was equally valid for *all* frames of reference; thus, for frames *accelerated* with respect to Galileian ones.

At first sight this idea seems ridiculous since it purports to express the equivalence of relatively accelerated frames, though acceleration seems to have *qualitative* manifestations. Donald Clayton in "The Dark Night Sky" recalls a ride at a funfair in which one stood against the walls of a large cylinder which was rotated on its axis at increasing speed until one was pressed firmly against its side, and

139

then the floor fell away. By what force was one held against the walls? By no force at all, according to the physics of the time; it was simply that, following Newton, objects wanted to continue to move in straight lines, which brought them into continuous collision with the restraining cylinder. It felt very much like a force, and was thus designated as a centrifugal force. Our physics teachers at that time, who should have known better because Einstein's general theory was by then almost two decades old, insisted it was only an imagined or "fictitious" force. They did not understand that the force was as real as that of gravity.

But back to Einstein, who realized that in Newton's equation of motion "force = mass × acceleration", it was purely a matter of convention whether one put something on one side of the equation or the other, provided the laws of algebra were respected. This is the Principle of Equivalence, on which the general theory of relativity is based. How, then, did Einstein go beyond what was already evident in Newton's equations? The great leap forward was the affirmation of the *generality* of the equivalence which, when translated into mathematics, led to general rules governing the description of the phenomena of nature. But as in other cases of the far-reaching and revolutionary consequences of simple but profound principles, the full implications of the Principle of Equivalence were not perceived all at once, but unfolded only gradually.

So much was gravitation at the centre of Einstein's thinking that the general theory of relativity has often been characterized as Einstein's Theory of Gravitation. Let us see how Einstein came to the conclusion that whether gravitation was a *real* or a *virtual* force depended only on the frame of reference which one used to describe a single (invariant) reality.

Consider the following imaginary situation. Out in space, away from the earth's gravitational field, Mr. X is in a closed spaceship. The ship is being accelerated at a rate of 9.8 m/s², which is the acceleration of free fall under gravity. The direction "up" for him is that of the ship's acceleration.

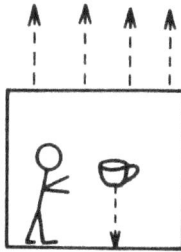

Fig. 6.1.

Suppose now that he jumps. The "floor" of the ship will come accelerating up toward him. For him, inside his little world, he will fall again toward the floor, just as he would in the gravitational field of the earth. Suppose he drops (i.e. lets go of) a teacup. The floor of the ship will accelerate upward toward the cup, which will finally crash on it. Mr. X will simply see the cup falling and breaking as it would have on earth under gravity. In fact, the mechanics of life in the earth's gravitational field will have been reproduced exactly. He can pour himself a drink, he can pick up a heavy object from the floor, he can sleep in his bed without fear of waking to find himself floating in space.

But we, on the outside of his ship, know that there is really no gravitational field at all. All that happens inside his ship is not due to gravity, but merely to kinematics. The important thing, however, is that the laws of physics inside his ship are exactly the same as in the static frame of reference on the earth's surface. We have made a transformation, so to speak, between frames of reference in relative acceleration, without any change in the laws! It appears that an important step toward Einstein's goal has been realized.

Now let us look at a somewhat different situation. Suppose the ship is not accelerated, but is falling freely in the earth's gravitational field. If inertial and gravitational mass are the same, the ship and everything in it will be falling with exactly the same acceleration. He, and everything else, will then float freely inside the ship. For him, gravity will have been abolished. This is, of course, a phenomenon with which we are quite familiar since the beginning of the era of space travel. In the spaceship, everything is weightless. In fact, dynamics inside the ship are those of an inertial frame, *despite* the presence of a gravitational field and forces of acceleration. One has cancelled out gravity over a region of space in which it is effectively constant. What has been realized, in fact, is a local inertial frame of reference. In this frame of reference, the special theory of relativity will hold, so we have made a link between the special and general theories.

We return, however, to the question, what does Einstein's contribution consist of, since the arguments we have just given seem based on familiar Newtonian mechanics? One answer to this question is the following: he was able to formulate a complete mathematical theory in terms of laws of physics which can be applied in *any* frame of reference, and which then fully incorporated the phenomenon of gravitation (as well as other "forces", including for example, the centrifugal force). His manner of doing this depended on a pre-existing mathematical theory, that of Riemannian geometry, or the geometry of "curved spaces". Such spaces have peculiar properties: "parallel" straight lines may ultimately meet, the sum of the angles in a triangle will not in general be 180°, and so forth. This is not a particularly esoteric or even unfamiliar situation. Imagine life on the surface of a sphere, our earth, for example. We must use our imaginations to visualize this *surface* as the totality of space, and that nothing can exist outside it. Mathematically, it is an example of a two-dimensional curved space; Einstein's universe is a four-dimensional equivalent. Its properties, different from those of the flat Euclidean geometry of secondary school mathematics include:

— It is possible to go south one kilometre, east or west one kilometre, and north one kilometre, and arrive back at the starting point (starting at the north pole).

— In the triangle traversed in this example, the sum of the interior angles is greater than 180°, since there are two right angles (turning east and turning north) and another angle (at the pole).

— Imagine two points on the equator, and go south from each (thus on parallel paths). If both continue always in the same direction, the paths ultimately meet (at the south pole).

— Imagine a circle of one kilometre radius with the centre at the north pole. Its circumference will be $2\pi = 6.28$ km to a very high degree of accuracy. Now imagine circles of greater and greater radii about the same point. Clearly, for such circles of sufficiently great radius, the circumference will be *less* than 2π times the radius. For instance, when the radius is one-quarter the circumference, the circle will be the equatorial circle; the ratio of circumference to radius has now been reduced from 2π (6.28) to four.

After that, as the radius increases, the circumference decreases — ultimately to zero.

— If you keep moving in the same direction, you will ultimately return to your starting point.

(In the above examples, I have talked of "north pole", "south pole"; north, east, south and west directions, and "equator". One should realize, however, that on a sphere everything is the same everywhere.)

The question now is: how do "curved spaces" enter the physical picture? What have they to do with gravitation? Under the action of gravity, and in the absence of all other forces, all objects launched with the same initial velocity follow the same path. This follows from Newton's equation of motion, force = mass × acceleration. The gravitational force on a body in a given gravitational field is proportional to its mass. Thus, provided that gravitational and inertial mass are identical (an essential requirement of Einstein's theory), the path of motion is independent of the body (i.e. of its mass). If the trajectory is the same for all bodies, it cannot in any way be characteristic of them. It seemed to Einstein natural to treat it as a property of the space through which the objects moved. If the path was curved, so was the space. At this point Einstein turned to Riemannian geometry as the mathematical framework for his theory. Since forces in themselves were to be abolished, being subsumed

into the geometry of space, particle motions were *free* motions, and
so must follow the geometry of space itself.

Thus it seems that observation of objects in "free fall" could be
used to demonstrate that geometry.

But this cannot be quite right, because if we throw objects with
different velocities from a given point, even if they are thrown in
the same direction, the trajectory depends on their initial velocity.
The greater the velocity, the flatter the trajectory and the greater
the distance of travel. Different curvatures are observed (Figs. 6.2
and 6.3).

Fig. 6.2.

Fig. 6.3.

We should be looking at the situation in terms not of space
exclusively, but of space-time. Suppose we draw the sort of space-
time diagram which we used in the special theory (Fig. 6.4). If we
suppress the vertical component of motion, the diagram looks like
this: the trajectory of an object thrown at low speed is a line making
a small slope with the x-axis (A). At greater speeds, the slope is

steeper (B). Thus, the curvatures we measure will be curvatures in different directions in *space-time*, which become well-defined. We can now add the third (vertical) dimension.

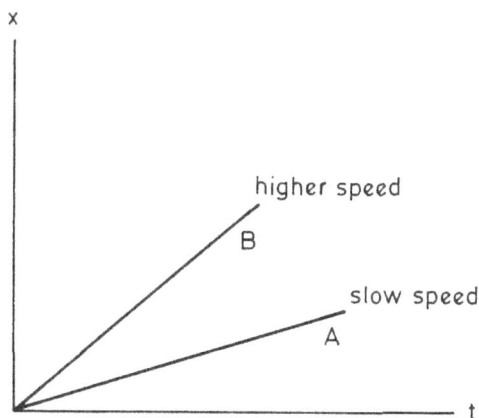

Fig. 6.4.

By launching the projectile from a point at different speeds, we can trace out the curvature of space-time in different directions around the point of projection P (Fig. 6.5).

Einstein found that the effects of "forces" were due to distortions in the geometry of space and time. Just as, on the surface of a large sphere, things seem flat over a sufficiently small distance, so, in Einstein's world, physics may seem Galileian locally. The special theory of relativity is valid over distances which are small compared with the *curvature* of space-time.

The best way to visualize this is to think of the freely-falling frame of reference in the earth's gravitational field. If an object of significant size is acted upon by that field, the pull of gravity on each point of the object is toward the earth's centre; thus, there will be a net sideways force of compression, which will be so minute as to be unmeasurable. Similarly, some points will be further from the centre

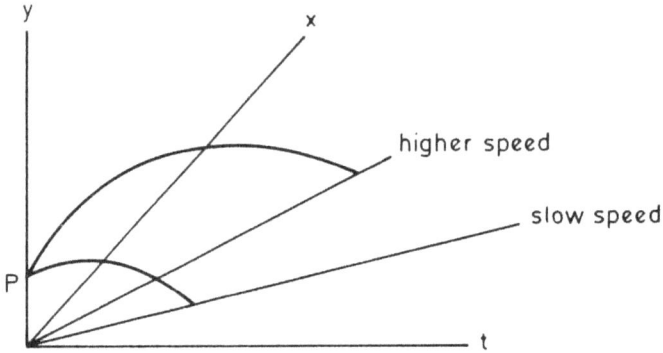

Fig. 6.5.

of the earth than others. Since gravitation follows an inverse square law with distance from the gravitating centre, the near parts of the object will be pulled a bit more strongly than the more distant ones. It is this effect of the gravitational pull of the moon on the earth which creates our tides. We thus call this sort of gravitational effect the *tidal* effect. Local inertial frames of reference can be defined only over regions small enough that these "tidal effects" are insignificant. But for our example of the effect of the earth's field we can see that the local inertial frame can be defined over distances small compared with the earth's size. If we have a laboratory whose greatest spatial dimension is 50 metres, for example, the relative importance of these effects will be $\frac{50 \text{ metres}}{\text{circumference of the earth}} = \frac{50}{6,500,000}$, (the circumference of the earth being about 6500 km). This ratio is then 7.7×10^{-6} or slightly less than eight parts in a million. To this accuracy, one can use a local inertial frame to describe the physical world in such a laboratory.

These remarks are important to note, for we shall use arguments to describe phenomena of general relativity which seem to be

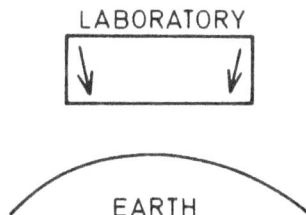

Fig. 6.6.

based on the use of inertial frames of reference. We can do this because Einstein's theory permits us to use *any* frame of reference to describe the physical world. Inertial frames then provide a familiar and convenient perspective for describing these phenomena.

We must, however, be careful, in making this sort of argument, that the inertial frame is sufficiently extensive to encompass our experiment. It is possible in most instances due to the *weakness* of gravitational forces. We shall see later, however, that on the astronomical scale, immensely *powerful* gravitational forces exist around neutron stars, galactic nuclei and black holes, which render such arguments impossible. Such exotic domains of the physical world therefore make severe demands on our imaginations (if we are not mathematicians), or on our mathematical skills if we are.

We may, then, make arguments based on the use of inertial frames. But we must remember that the general theory of relativity enables us to carry over these arguments into perfectly arbitrary frames, and is designed to describe physical phenomena which themselves are *independent* of the frames of reference used.

6.1. Early Tests of the Theory: Bending of Light in a Gravitational Field

Einstein put forward the general theory in 1917, when Germany was at war with Great Britain, France, Italy and finally, the United States. But war does not stop the spread of science, nor do scientists

feel themselves at war with their counterparts on the other side of
the battleground. In fact the theory created a great stir in the allied
countries. The end of the war in 1918 made scientific collaboration
between all the European countries possible again and it was an
English astronomer, Sir Arthur Eddington, who undertook the first
verification of this exciting new theory.

Einstein had predicted that rays of light passing close to the
limb of the sun would be bent slightly by the earth's gravitational
field. One way of describing this is to reason from the equivalence
of mass and energy, which implies that a beam of light (electromagnetic energy) must be pulled toward the sun in the same way as an
equivalent mass. Or, alternatively, one can say that the light beam
follows the curvature of space-time, so that the bending reflects the
curvature created by the sun's gravitational field.

Sir Arthur Eddington proposed to verify this prediction during
an eclipse of the sun which would be visible in South America in
1919, and led the expedition to make the necessary observations.

The effect is illustrated (greatly exaggerated and distorted in
scale) in Fig. 6.7. Light from a star at S is bent at the limb of
the sun and reaches the observer at O. If the sun were visible, it
would of course not be possible to see the star, but the observation
can be made when the sun is eclipsed and the surrounding sky is
dark. The light will arrive at the point of observation along the line
S'O, so that the star will appear to be at S'. The true position of
the star will of course be known from prior observation, so that its
apparent position makes possible the determination of the angle of
bending. The theory predicted that this bending should be 1'75" of
arc.[5] Since the second is 1/3600 of a degree, the effect is very small.
This smallness indicates the weakness of the gravitational field of the
sun. In fact, all early tests of relativity involved the measurement
of very small effects. Only recently have very strong gravitational

---

[5] Einstein first calculated it to be one half of this value. The next section, which
deals with the effect of gravitational fields on clocks, reveals the source of his
error. Of course, he later corrected it.

fields been found to exist. However, so far Einstein has always been proven right!

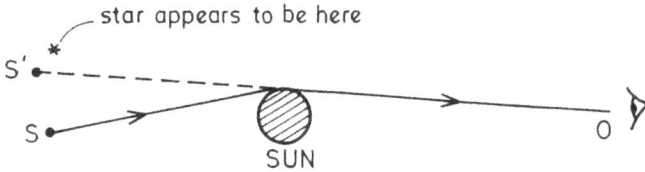

Fig. 6.7.

It is not within the scope of this book to reproduce the exact calculation of the bending of light, but a rough argument can be made which gives a good approximation to the correct value.

Firstly, we note that the energy of light comes in quanta, whose energy is hf, h being Planck's constant and f the frequency of the light. This represents an equivalent mass of hf/c^2, and the sun's gravitational force on it is

$$\frac{GMhf/c^2}{R^2},$$

where M is the mass of the sun and R its radius. Now the quantum is under a constantly varying force towards the sun's centre and is a maximum when it is just at the sun's radius. We shall simulate this by assuming that it acts over a length of path of $2R$ (R on each side of its closest approach), and that over this path the force has its maximum value. In reality, the force acts over a greater distance but gets constantly weaker, so we may hope that these two errors compensate each other to some extent.

The action of the force gives the photon (light quantum) a momentum *toward* the sun's centre (i.e. perpendicular to its direction of motion) of

$$(\text{Force}) \times (\text{time acting}) = \frac{GMhf/c^2}{R^2}\frac{2R}{c} = k\frac{hf}{c},$$

where

$$k = \frac{2GM}{c^2 R} \ .$$

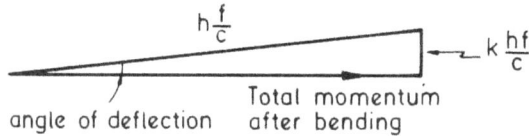

Fig. 6.8.

This corresponds to an angle of deflection of the fraction k of 360°, which by a simple arithmetical calculation is found to be about 0.87 seconds of arc. Though this is Einstein's original result, this precise agreement is fortuitous, especially in the light of the rather rough approximation we made to get it.

Before passing on to the question of the missing factor of two, let us note a very simple argument, based on the equivalence principle, which tells us that light *should* be bent in the sun's gravitational field. Imagine a closed laboratory out in free space, which is being accelerated relative to inertial frames at the rate of free fall in the gravitational field of the sun at its surface. Inside this laboratory, physics will be the same as in a laboratory experiencing the sun's gravitational field. The observer in the laboratory will have no way of knowing that he is not in fact in such a field so that, by the equivalence principle, everything will be exactly as though he were.

Let us now imagine that through a tiny hole in the shell of his laboratory, a narrow beam of light is allowed to enter in a direction perpendicular to that of his acceleration (Fig. 6.9). In the inertial frame of reference outside his enclosure, this beam of light will travel in a straight line. Inside the enclosure, on the other hand, its acceleration will cause the beam to be deflected toward the "floor". By the equivalence principle, it is exactly as though the light in the

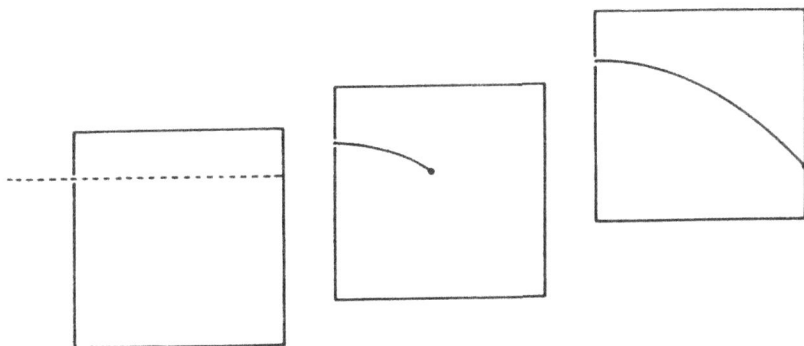

Fig. 6.9.

beam is "falling" under gravity; its path is bent as shown in the following diagram. From this we deduce that light will be bent in a gravitational field.

6.2. Gravitational Redshift

Light consists of photon quanta of energy hf where f is the frequency of the light. Due to the mass-energy equivalence, photons should be attracted by the gravitational force, just like any other particle.

Suppose a photon of a given frequency is emitted at the surface of a gravitating object, e.g. our sun. Its equivalent mass is hf/c^2, so that the gravitational force on it is $(GMhf/c^2)/R^2$, M being the sun's mass and R its radius. How much energy must be expended to displace it to "infinity", or, in other words, out of the sun's gravitational field? In Newtonian gravitation this amount is $(GMhf/c^2)/R$. The energy of the liberated photon is thus increased by

$$\frac{GMhf}{c^2R}.$$

The fractional decrease of its frequency in the sun's field must then be GM/c^2R, a small ratio which we came across in our discussion of the bending of light at the sun's limb. This decrease in frequency for ordinary light would be toward the red (low-frequency) end of the spectrum, so it is known as the *gravitational red-shift*.

While the above argument describes a valid phenomenon, not all the details are correct. The result is correct for the photon frequency at infinity, and it is very nearly right wherever the effect is small (and we have seen that for the sun, $GM/c^2 R$ *is* small). For collapsed objects, however, a mass comparable to the sun's may be compressed into a much smaller radius, increasingly so as we progress from white dwarf stars to pulsars and black holes. The correct formula for the gravitational redshift is then

$$f_\infty = f_R \sqrt{1 - \frac{2GM}{c^2 R}} \, ,$$

f_R being the frequency at distance R and f_∞ the frequency at infinity.

If we square this formula we get

$$f_\infty^2 = f_R^2 \left(1 - \frac{2GM}{c^2 R}\right) \, ;$$

whereas if $f_\infty = f_R \left(1 - \frac{GM}{c^2 R}\right)$,

$$f_\infty^2 = f_R^2 \left(1 - \frac{2GM}{c^2 R} + \left(\frac{GM}{c^2 R}\right)^2\right) \, .$$

If $\frac{GM}{c^2 R}$ is already small, the term $\left(\frac{GM}{c^2 R}\right)^2$ is even smaller. For the sun, $\frac{GM}{c^2 R}$ is about 2 one-millionths. The difference between the two formulae for the frequency shift cited here is about 2 parts in one million million — totally insignificant. In a pulsar, however, whose radius is perhaps 200,000 times smaller, the difference is quite significant.

What is most striking is that the frequency is red-shifted to *zero* at a radius of $2GM/c^2$ according to the correct formula, but only at half that value according to our approximate formula. What is the origin of this factor of 2? It is precisely the non-linearity of gravitational field as a function of mass mentioned earlier. Close to the surface of a strongly gravitating object the energy in the gravitational field produces a further gravitational field. Or, to put it more precisely, the total energy (rest-energy plus gravitational energy) is just enough to produce the gravitational energy itself. In the technical term of the theoretical physicist, the gravitational field is self-consistent.

A second argument for the gravitational redshift also makes use of the equivalence principle. It does not, however, make use of the quantum character of light.

Let us return to our accelerating spaceship, in which a desired gravitational field may be simulated. Suppose that a regular series of light pulses of frequency f_0 is emitted from its floor, and detected at its ceiling, whose height we shall call ℓ. When the first pulse is emitted, the ceiling of the spaceship is receding at a speed v; by the next pulse $1/f_0$ this speed has increased by g/f_0, where g is the acceleration of the ship. But to a first approximation, the time for the light signal to go from floor to ceiling is ℓ/c, and thus the extra distance the second pulse must travel is $g\ell/f_0 c$, and the extra time to make the journey $g\ell/f_0 c^2$. Thus whereas the time between the pulses at the floor was $1/f_0$, at the ceiling it is

$$\frac{1}{f_0} + \frac{g\ell}{f_0 c^2} = \frac{1}{f} .$$

The change in light frequency over a distance ℓ against the force of gravity is from f_0 at the bottom to

$$f = \frac{f_0}{\left(1 + \frac{g\ell}{c^2}\right)}$$

at the top. Similarly, if light is propagated downwards from the top, the frequency is *increased* by the factor $(1 + g\ell/c^2)$.

For the case of a gravitating body of mass M at a distance R, $g = GM/R^2$, so the factor of frequency increase is

$$1 + \frac{GM\ell}{c^2 R^2} .$$

We may now return to Einstein's error of a factor of two in the calculation of the bending of light at the limb of the sun. One can start by explaining the deflection by the action of the sun's gravitational field on the light (photon) energy. Detailed calculation shows that this gives exactly *half* the effect observed.

There is another effect due to the gravitational redshift: clocks placed in the sun's gravitational field seen by a distant observer seem to run more slowly than identical clocks in the observer's own rest-frame. Thus the frequency of the light in the local inertial frame at the sun's limb must be higher than that of its emission by the star, and its reception by us. This will result in a greater gravitational force than the one used in our calculation, and consequently a greater deflection. The light would seem to be propagated more slowly in the earth's gravitational field than in our inertial frame.[6] Detailed calculation shows that this temporal effect contributes the same amount to the deflection as that due to warping by the gravitational field. Only by taking account of this effect do we get agreement with the measured result.

6.3. Advance of the Perihelion of Mercury

This is the third phenomenon which verified Einstein's general theory. Up to the mid-sixties, the theory was accepted because it predicted *correctly* the departure from classical expectations in these three instances. This was not quite true, since Einstein's theory about the development of a cosmological theory of an expanding universe (first postulated by Hubble in the 1920's) is perhaps a more cogent argument than the three traditional ones. Not until the discovery in the 1960's of strongly collapsed stars – the pulsars or neutron stars – was it possible to put general relativity to the test in a domain where its predictions were radically and dramatically at variance with those of classical physics. For many years the theory remained unchanged and its impact on astronomy was marginal (save

---

[6] It should be recalled that, though the speed of light is a universal constant in all *inertial* frames, no such assumption was made about non-inertial ones!

perhaps in the realm of cosmology). This was followed by a revival, which continues to the present time. The theory now plays a central role in our understanding in all areas of astrophysics and astronomy. The general theory of relativity was some fifty years ahead of our ability to detect its manifestations in the physical world. Those fifty years were frustrating ones for relativists, but the following twenty have been full of excitement and activity.

Of the three early triumphs of the theory, the most subtle was undoubtedly the correct explanation of the small advance in the perihelion of the planet Mercury.

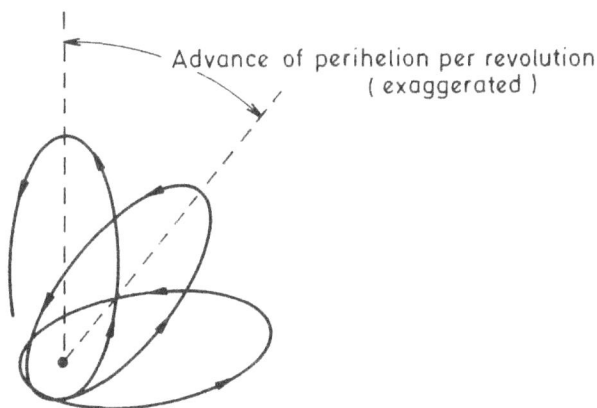

Fig. 6.10.

We can understand the phenomenon qualitatively by adopting arguments based on the special theory. The orbit of Mercury is slightly elliptical. It moves more rapidly near the aphelion (the point of *nearest* approach to the sun) than at the perihelion (its most distant point) (Fig. 6.10). In local inertial frames, clocks, and thus natural processes, are slowed down more in frames moving rapidly relative to the inertial frame of the fixed stars. Thus, the orbit of Mercury is slowed down more at aphelion, where the sun's gravitational force on it is strongest, than at perihelion, where it is weakest. The net effect is that the major axis of its elliptical orbit rotates slowly in the direction of the planet's rotation, so it arrives at the perihelion a bit later in each successive orbit. The effect is again very small; the rotation is about 43 seconds of arc per century (a second is 1/3600 of a degree). Of course, the orbit of Mercury is subject to perturbations from the other planets; it is therefore necessary to take into account these effects in order to assure that there is a residual anomaly. This, and the smallness of the effect, makes the verification of Einstein's prediction very delicate. The effect is, however, well outside the margin of observational error, and general relativity is in fact well vindicated.

It is worth noting the varying rate of natural clocks following Mercury in its orbit is *real* – the orbit of Mercury *is* physically modified. Thus, time dilation is again shown to be a real physical effect, and not simply an anomaly of observation.

6.4. General Relativity and Geometry

One of the fundamental postulates of Newtonian mechanics is that every object moves with constant velocity in a straight line unless compelled by external force to change that state. The geometry of the space in which Newtonian mechanics operates is that of Euclid. It permits another formulation of the postulate – that bodies not acted upon by force follow the shortest path between two points.

In the special theory of relativity, the Newtonian principle can be generalized as follows: bodies not acted upon by forces follow straight lines in space-time. If, for simplicity, we consider only one spatial dimension (as we did in giving a geometrical description of special relativity), space-time has two dimensions, a spatial one and a temporal one. The graphical representation of a freely moving object is a straight line (OA), as shown in Fig. 6.11, *whatever* the inertial frame of reference used to characterize it. The path of light (OL) is also a straight line, independent of the frame of reference.

How do we describe a body uniformly *accelerated* on a straight

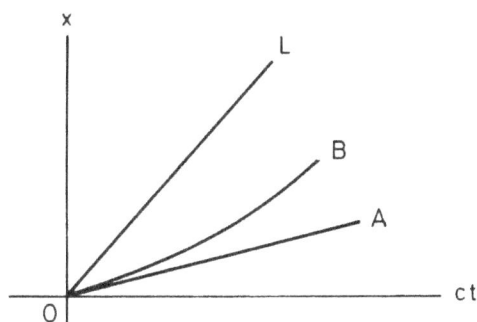

Fig. 6.11.

line? Its path would start out something like OB. In Newtonian mechanics it is a parabola. The path in space-time is no longer linear. From the equivalence principle, this could also be the path of an object in a uniform gravitational field. Gravitation curves the path. It appeared to Einstein that, if one wanted to make a theory which incorporated fields, or accelerated motion, one should again seek a law governing the paths of objects relative to arbitrary frames of reference, even if those frames of reference were *accelerated* relative to each other. This was facilitated by the knowledge of geometries *other* than that of Euclid, in which the shortest distance between two points is *not* the Euclidean distance between two points. A simple example is the geometry on the surface of a sphere. Here, the shortest distance between two points is a great circle, defined thus: imagine a plane through the centre of the sphere and the two points. The intersection between the plane and the sphere is a circle. Between any two points on this circle there is no route shorter than that defined by the circle. Thus, the great circle is the obvious generalization, in the geometry of our sphere, of the Euclidean straight line.

We have a problem here. These "straight lines" are in fact curved. Although geometry on the surface of a sphere is only two-dimensional, the curvature due to the lines (which are known as "geodesics") is a curvature in a three-dimensional Euclidean space which contains the two-dimensional spherical space. If we talk only in terms of the two-dimensional world which consists exclusively of the totality of points lying in the surface, we must find a way to define "curvature" which does not refer to anything *exterior* to the two-dimensional space. This is quite easy to do. We take lines starting out from a given point in two different directions, and continue along them indefinitely. We find that these lines, which initially diverge, ultimately meet at another point (the "antipodal" point). Between any point and its antipodal point, all great circles have the same length. These are characteristics of a curved space which are not shared by a Euclidean or *flat* space.

This is, however, a rather *special* curved space, since all points on it are equivalent. Its curvature is the same everywhere. To attain greater generality, we should find a method of measuring the curvature of the space *locally*, at an arbitrary point. This is not difficult to do (there are a number of ways, actually, of which we shall discuss only one). Suppose we take two adjacent points, and imagine the great circular path between them. From each we make a displacement of a given distance perpendicular to that path — "parallel" displacements. We then ask whether the parallel paths remain the same distance apart (the case for a flat space) or move closer (in which case we say the space has positive curvature). We may use the earth to illustrate the point. Take two adjacent points on the equator, then make equal but small displacements in a northerly direction; these are parallel displacements. The displaced points are closer together than the original ones. The relative magnitude of their approach is a measure of the curvature. (It should be noted that there is nothing special about choosing points on the equator, or northerly displacements. To the extent that the earth is a sphere, all positions and directions are equivalent.)

Einstein's program was to find the geometry of space-time such that the motions of objects under the action of force were geodesics (the equivalent of straight lines in a curved space-time). In this way forces were replaced by warps in the structure of space-time itself.[7]

Although Einstein was primarily concerned with gravitation, we shall see that his program can also be applied to other forces. Of particular interest is the case of *rotating* frames of reference, in which free-particle orbits become curved — a fact traditionally attributed to "centrifugal forces". Most of us were told in high-school physics courses that "centrifugal forces" were fictitious and were really only the result of looking at things in an incorrect way. The equivalence principle denies the distinction between "real" and "fictitious" forces, so it is quite legitimate to speak of a centrifugal force. It depends on your frame of reference, and Einstein tells us that all frames of reference are equally valid.

Let us recall our earlier discussion of the interpretation of motion in a gravitational field in terms of the curvature of space-time. If we throw a ball into the air with different speeds, we are throwing it in different directions in space-time. Thus, the curvature is different in different directions. This is not true for a spherical space, which has the same curvature everywhere and in all directions. If, on the other hand, we consider the space on the surface of a cylinder, or of a dirigible, we note that the curvature is quite different in different directions.

We must then look at all physical phenomena in this perspective. The motion of the moon around the earth reflects the curvature of space-time due to the gravitational field of the earth, that of the planets around the sun to the gravitational field of the sun, and so on.

---

[7] The essential idea of the theory, from a mathematical viewpoint, is to assume the curvature of space-time to be proportional at each point to the density of energy (energy per unit volume). This has the surprising and profound consequence that, contrary to Newton's postulates and our intuition, matter and energy do not exist in space and time, but themselves define and determine the very structure of space and time. This surely is one of the boldest ideas in the entire history of science!

Of course, we would expect these curvatures to be felt simultaneously. The moon curves the space around the earth; so does the
sun. Surprisingly, the moon's tidal effect is greater! Furthermore,
the amounts of curvature from the two sources are not additive, for
the energy of the combined gravitational fields of the two act as a
source of further gravitational fields. This is what makes relativistic
calculations so difficult. Fortunately, in practice, things are not so
complicated. The moon's gravitational force on the earth is about
200 times less than that of the sun, but its effect does make earth
and moon circle around each other, or, more precisely, around their
mutual centre of mass, which actually lies inside the earth. Both
make gentler and more leisurely circuits around the sun, reflecting
the sun's greater distance and the small curvature it creates at our
planet's great distance. The amount of gravitational energy involved
is so small (due to the weakness of the gravitational force as reflected
in the small value of the gravitational constant) that it has only a
miniscule effect on creating a "second order" gravitational field.

A simple model can be used to familiarize us with the theory of
motion, and the idea that motion follows geodesic paths in curved
spaces.[8] Imagine a flexible elastic membrane held at its perimeter so
that it is under uniform tension (Fig. 6.12).[9] Now imagine a heavy
ball placed on it. The membrane will become curved around the ball,
but the curvature will decrease as distance from the ball increases
— the membrane becoming ultimately almost flat (provided that it
is far enough from the edges of the membrane). Now take a very
light ball, whose effect on the membrane is slight, and project it
on a path which does not strike the heavy ball but passes close to
it. The path will in fact be a geodesic curve in the curved space of
the membrane's surface, that is, it will always follow the *shortest*
distance between any two points on its path. But this path will be

---

[8] We shall henceforth not distinguish between space and space-time, the latter
being merely a *space* with an additional (temporal) dimension.

[9] For an illustration of this model, see W. J. Kaufmann III, Black Holes and
Warped Space-time, pp. 76-77.

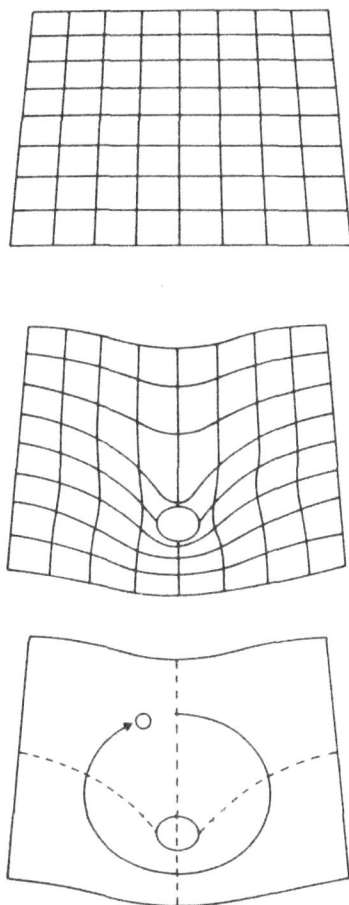

Fig. 6.12. A physical analog of the bending of space-time by the gravitational field of matter is this sort of elastic membrane. If a weight is placed on it, the surface is deformed. Other objects of negligible mass will move on this deformed surface. In particular, a small light ball could trace out a circular orbit, just as the earth moves about the sun.

deflected as the light ball passes close to the heavy one, not by virtue of a force between them in the usual sense, but because the motion of the light ball will be the free motion on the curved surface created by the heavy one.

Of course, we must remember that these simple images represent simplifications of reality. Real space-time has four dimensions, and we have no way to create mental images, much less models, in four dimensions. But two or three-dimensional images capture all the essential features of reality. One must imagine that the addition of an extra dimension or two does not alter the essentials; we simply have more of the same. Sometimes, in fact, a dimension (or two) become irrelevant, because nothing changes in them, and the interesting phenomena take place in the remaining dimensions (in free fall under gravity, for example, nothing happens in the two directions perpendicular to the direction of motion, so that in effect one has only one spatial and one temporal dimension).

6.5. Rotating Frames of Reference: Newton's Bucket and Mach's Principle

Relativity no longer allows us to retain Newton's "absolute space"; still, it does not permit us to get rid of absolutes. So, like Einstein, Newton evoked a "thought experiment" which seemed to him to be a persuasive argument in favour of absolute space, and, when approached from the perspective of relativity, still appears to lead us back to an absolute. The experiment involves what has come to be known as "Newton's bucket".

Imagine a bucket of water which is rotated on its axis. This rotation will in time, through friction, be communicated to the water in it. "Centrifugal force" then causes the water to move toward the sides of the bucket. The amount of this displacement, which causes the water level to be higher at the side than at the centre, will be determined by the balance between centrifugal force and gravity; the surface of the water will be concave as shown in Fig. 6.13.

However, ultimately, there will be no relative motion between bucket and water; the latter will rotate at the same rate as the bucket.

Now suppose that the bucket is suddenly stopped. The water will continue to rotate, and there will be a relative motion of water

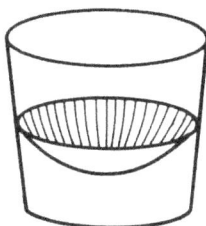

Fig. 6.13.

and bucket, though the surface of the water will at least initially keep the same form.

What then determines the shape of the water's surface? It cannot be the relative motion of water and bucket, since the shape can be the same both when the water is rotating relative to the bucket and when it is not. It must be then, Newton argued, that rotation, which causes the water to rise, is relative to something absolute; in fact, to his "absolute space".

This seems to be a failure of relativity. If the water is at rest when the bucket rotates, the water surface will be flat; if the water rotates while the bucket is at rest, it will be concave.

Once again, contrary to the contention of Feynman's "cocktail party philosophers", the statement "everything is relative" does not seem true.

The question is, can we define rotation without a reference frame which is not rotating? Is this reference frame then not absolute?

Before the time of relativity, this problem had bothered the philosopher-physicist Ernst Mach. He posed the question, what distinguishes inertial frames of reference? He perceived that their distinguishing feature was that they were at rest in, or on uniform motion relative to, the fixed stars, the firmament — the totality of matter in the universe.

But Mach, unlike Newton, did not interpret this as a manifestation of absolute motion or absolute space. Rather, he contended that

if the centrifugal forces on the water in Newton's bucket were due to
rotation relative to the firmament, it must be that these forces were
in fact created by all the matter in the universe, out to its furthest
reaches.

This proposition, which is known as Mach's Principle, was in
harmony with Einstein's own method of thinking, and Einstein
avowed that he was, in the evolution of the theory of relativity,
much influenced by Mach. Unfortunately for Einstein, however, he
later realized that Mach's forces could not be built into his theory,
and in the course of time talked less and less about Mach and his
principle. The situation has not changed. The principle, to which
Mach was never able to give a mathematical expression, still does not
appear to follow from, and is certainly not explicitly incorporated
in, Einstein's theory.

On the other hand the observation that distinguishable physical
phenomena were associated with rotation relative to the fixed stars,
which had led Mach to his principle, is in no way inconsistent with
Einstein's general theory.

6.6. Motion in Rotating Frames of Reference

Imagine trying to throw a ball to someone on the opposite side
of a merry-go-round. If you threw it *toward* him, he would no longer
be there when it arrived at the place where he *had* been. Mount a
movie camera, fixed to the merry-go-round at its centre, and arrange
its aperture so that it can film only what is happening *on* the merry-
go-round, but not any of its surroundings. The film will show the
ball following a curved path.

Some years ago, in a program of a television series on science
(The Nature of Things) which exists to this day, professors David
Hume and Donald Ivey of the University of Toronto used a variation
of this experiment. Instead of a ball, they used a puck floating freely
(i.e. without friction) on an air table, the whole set on a rotating
turntable. Hume would give the puck a gentle push toward Ivey, and
it would proceed to curve away from him. The secret was that the

movie camera, mounted above the turntable and looking downward at it, was fixed to it so that it shared the rotation. If one blocked out the surroundings, no evidence of rotation could be seen; the camera was stationary in the rotating frame. The strange motion of the puck, as illustrated in Figs. 6.14a–d, would, according to Newton's law, be explainable only by the invocation of forces.

If, on the other hand, the camera is fixed to a stationary object outside the turntable, the puck will be seen to move in a straight line. This is because the camera is now at rest in an inertial frame of reference. Thus, there is now neither friction nor any other observable force acting on the puck. So, according to Newton, it must move in a straight line. Any objects at rest in the rotating frame (such as, for instance, Professors Hume and Ivey) will, of course, be seen to be rotating.

Figure 6.14 shows orbits of the puck under different conditions. Suppose the air table has radius R and the speed of the puck is v.

(a) The time needed to travel the diameter of the table $4R/v$ equals half the period of rotation T of the turntable.

The speed of the puck is $8R/T$.

(b) Speed is $2R/T$.

(c) Speed is R/T.

(d) Speed is $R/2T$.

The "forces" we invoke depend on the speed of the puck (like the force exerted on a charged particle by a magnetic field).

Is the observed behaviour due to forces or due to the acceleration (rotation) of the turntable? For Einstein, the two explanations are equivalent. If we insist that it is due to rotation, because we can *see* the turntable rotate but can see nothing acting on the puck, we are insisting that our (rest) frame of reference is the "correct" one. The stars and galaxies in the night sky vindicate us. But the ghost of Mach might whisper that perhaps it is precisely the effect of these fixed "stars" that exert the forces causing the strange motion of our puck.

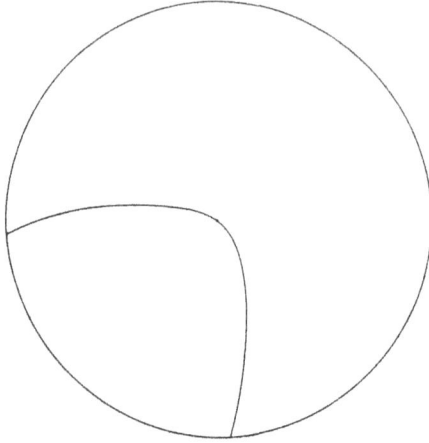

Radial velocity $V = 8R/T$ $T = $ rotating time

$R = $ radius of turntable

Fig. 6.14a.

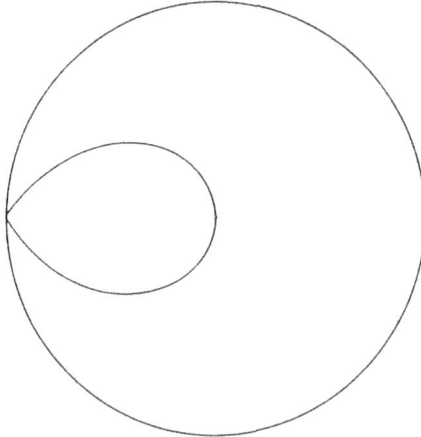

Radial velocity $V = 2R/T$

Fig. 6.14b.

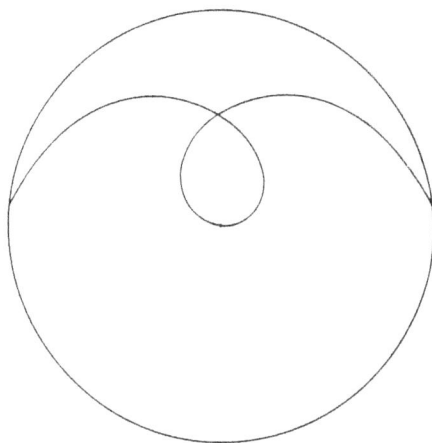

Radial velocity $V = R/T$

Fig. 6.14c.

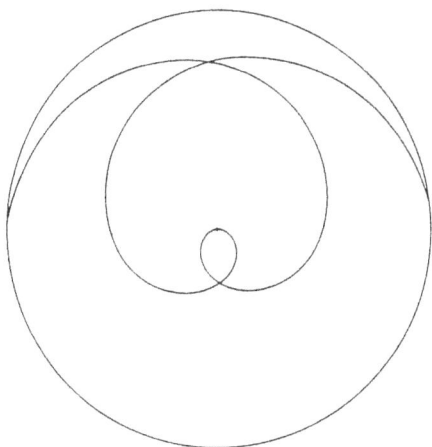

Radial velocity $V = R/2T$

Fig. 6.14d.

6.7. Rotating Systems Seen from an Inertial Frame of Reference

Let us look at our rotating turntable as a physical object observed from an inertial frame of reference, relative to which the centre of the turntable is at rest. "Centrifugal forces" will now be experienced by everything at rest on the turntable. Suppose that we have a monochromatic light source attached at a distance R from the centre of the turntable, whose period of revolution is T. The speed of the light source is $2\pi R/T$. According to the special theory, we expect the frequency of the light received in the inertial frame to have its frequency red-shifted by a factor $\sqrt{1 - (2\pi R/cT)^2}$. Such an experiment was in fact carried out by Hay, Schiffer, Cranshaw and Egelstaff in 1959, and had the anticipated result.

Note, however, that another argument also leads to this result. Now the explanation depends on the general theory, and is explained in a manner similar to that used for the gravitational redshift. In that case we considered a light-quantum (photon) losing energy, and thus frequency, pulling itself up out of the gravitational field. In the present case, the argument is the same, except that now the field of centrifugal force must be overcome. The phenomenon is most easily envisaged if we imagine detecting the light at the centre of the turntable, which is at rest in the inertial frame.

When I first met the theory of relativity, another problem bothered me. Suppose one imagined a larger and larger turntable, rotated at constantly increasing speed. Finally, at a sufficiently great distance R, $2\pi R/T$ would become equal to the speed of light c, and the circumference at that distance would have shrunk to zero, while all frequencies, and hence photon energies, would have decreased to zero, so that nothing more could be seen. And beyond that distance, what? Presumably all matter at this radius would have infinite inertia, so that nature would not permit us to attain such a rate of rotation, and we would have arrived at a contradiction.

But let us not go to such an extreme limit for the moment, and consider only length contraction. The result of the shrinking of lengths in the direction of motion, but not those measured radially outward, would change the ratio $\frac{\text{circumference}}{\text{radius}}$ of the circle in which a point moved. This ratio would no longer be 2π, but $2\pi\sqrt{1 - \left(\frac{2\pi R}{cT}\right)^2}$. The geometry of the world of the turntable would have changed. If one rashly pushed again to the limit $2\pi R/T =$ the speed of light, the ratio would have gone to zero.

Shocking though this may seem at first sight, on second thought we realize that we have already seen this situation in the geometry on the surface of a sphere! If we start making circles of increasing radius around the north pole (or any other point), the space seems quite flat at first and the ratio is 2π. But by the time the radius is one-quarter the circumference of the sphere, we are at the equator and the ratio has become $\frac{2\pi 4}{0.5\pi R} = 4$. This is a large decrease from $2\pi = 6.284$. If we continue, the circumference starts to *decrease* with increasing radius, until, for a radius of R, we reach the south pole and the circumference has shrunk to zero (Fig. 6.15). Quite aside from dynamical considerations, relativistic mass increase, materials strong enough to resist centrifugal forces, and the like, relativity presents a mathematical solution in the form of a finite spatially-closed, curved sub-universe which represents a theoretical limit to reality.

6.8. Postscripts on Gravitational Redshift and Advance of Perihelion

Several more recent experiments have been done to confirm the gravitational redshift:

1. Imagine that a radar beam is emitted and reflected from a planet on the other side of the sun, the beam passing close to the surface of the sun. When passing the sun, its speed is reduced due to the sun's gravitational field. This results in a delay in the reception of the echo relative to what might be expected. Such an experiment has been done and agrees precisely with the prediction of relativity.

2. A clock carried in a jet airplane at high altitude runs faster than on the earth's surface. It is easily verified that the gravitational

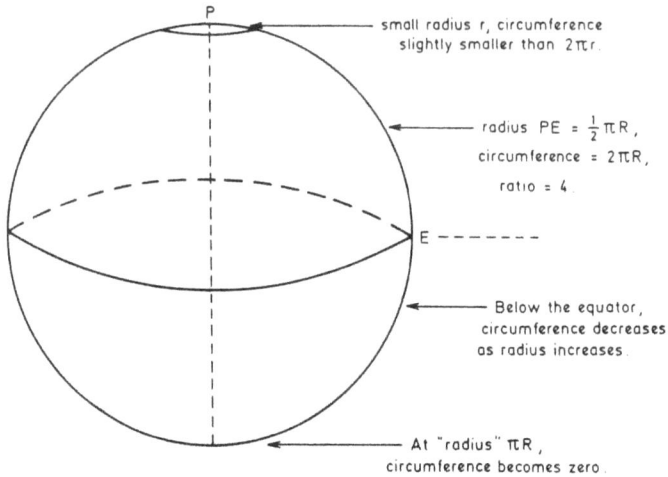

Fig. 6.15.

effect on frequency is much greater in this case than the redshift due to the plane's motion.

3. Since white dwarf stars are suns whose mass is packed into dimensions comparable to that of the earth, the gravitational field at their surface, and the gravitational redshift of light coming from them may be as much as a hundred times greater than those of the sun. An observation made on one such dwarf showed a redshift of about 3 parts in one hundred thousand. For the sun the comparable figure is about 1.4×10^{-6}, i.e. some twenty times less (Trumpler, 1955). Seven years later, Brault measured the redshift from the sun, and the measurements agreed well with the theory.

4. One can imagine that the above measurements, since they concern small effects, are difficult and delicate. The most delicate one of all was conducted by Pound and Rebka at Harvard, who measured the change in frequency of light between the top and bottom of a building (22 m) in Cambridge, Massachusetts. This was possible using the Mossbauer effect, which states that the absorption of

a photon of light at a precise spectral frequency of an atom of a material does *not* cause the atom to recoil, since the momentum of the photon is absorbed, not by a single atom, but by the whole crystal. If the absorbing atom did recoil, the energy of the photon and thus its frequency would be diminished. The exclusion of this effect made possible the detection of the much smaller gravitational shift. The experiment was done using a 14.4 keV gamma-ray emitted from the Fe^{57} nucleus, absorbed by a target of the same material. The very small mismatch between the frequency of the photon at its emission and that with which it arrived at the detector resulted in a small but measurable decrease in absorption by the detector. The theoretically predicted fractional shift of frequency of 2.5×10^{-15} was verified to within the limits of experimental error.

Turning next to the advance of the perihelion, a measurement made in 1979 on a double star system showed an effect many times greater than that for the orbit of Mercury.

The system was a strange one. One star in the system was a pulsar flashing with a period of 0.059 second. Its companion was invisible. What we know about it can only be deduced from its effect on its visible partner. This companion is probably also a pulsar, but with its axis tilted in such a way that we do not see its light, since pulsars radiate fairly narrow beams, somewhat like giant flashlights.

One can deduce the orbital period of the visible member from observation of the periodic variations of the Doppler effect as it circles its companion. Both stars will in fact orbit about their centre of mass. Kepler's law, which says that the cube of the size of the orbit is proportional to the square of its period, then permits us to estimate the orbital dimensions.

The system makes 1100 orbits per year, and the orbit is three times as eccentric as that of Mercury, which greatly enhances the effect, since the stronger gravitational forces and the consequent slowing of clocks which that entails are large. The greater eccentricity means a greater difference in this effect at perihelion and aphelion. Thus, there will be a greater advance per orbit. This, coupled with the large number of orbits per year, gives an advance of periaston of 4.2 degrees per year, as opposed to 0.43 second per year for Mercury.

Of course, such a large effect increases the accuracy of the measurement, and gives a correspondingly stronger confirmation of the theory.

Chapter 7

COSMOLOGY

From the time of Kepler, Galileo and Newton until the twentieth century, the astronomical universe was generally regarded as static; the earth appeared to move against an unchanging background of "stars" which stretched away into the depths of space. Newton himself thought that the universe must be infinite in extent. For this conclusion he had a simple argument: if the universe were finite, it would have a centre, toward which all matter would be pulled by the gravitational force of all other matter. Thus, a finite universe would be doomed to collapse. Since this appeared contrary to the constancy which was observed, Newton argued that the universe must be infinite, in which case there would be no centre toward which it could contract. Inevitable fluctuations in density would, however, according to Newton, cause collapse of matter toward local centres of higher than average density; this, he argued, was how the stars were formed.

The general theory of relativity gave a new dimension to the problem; since matter caused a curvature of space, the distribution of stars should introduce a global curvature; thus, a new cosmology was implied.

Prior to the renaissance of relativity stimulated by astrophysical discoveries in the 1960's (quasars in 1963, pulsars in 1967), experimental evidence for the validity of the general theory of relativity depended on the following three very weak effects: the bending of starlight near the limb of the sun, the redshift of radiation emitted from astronomical bodies and the advance of the perihelion of Mercury. There was a fourth piece of evidence, however, more impressive than any of the other three — the existence of a cosmological theory capable of describing the expansion of the universe which had been demonstrated by Edwin Hubble in the mid 1920's.

The development of a cosmology was a natural outgrowth of Einstein's theory. From the beginning this theory was developed as a theory of gravitation; its mathematical expression was in field equations which can be considered as the gravitational analog of Maxwell's equations for electromagnetism. It is true that Einstein gave them a geometrical interpretation, and considered the fields to be related to the local deformation of space-time. While there are compelling reasons for this interpretation, one does not have to adopt it; the equations can be treated simply as field equations. But fields are not *in* the universe; they *are* the universe, and for Einstein this was no less true of electromagnetic fields than of gravitational ones.

What then are the sources of these fields? The source of gravitation is matter or, more generally, energy. Einstein's equations therefore expressed the components of the field in terms of quantities describing the distribution of energy. In technical mathematical terminology, they describe the gravitational tensor in terms of the energy-momentum tensor. A tensor is a set of quantities obeying certain rules of transformation in going from one frame of reference to another. Thus, the gravitational fields characterizing the universe, or its geometry, could be determined if one knew the distribution of its mass-energy. It is important to note that this geometry includes time, for Einstein's universe is four-dimensional, and, as noted in our description of the special theory, phenomena which are viewed as spatial in one frame of reference may appear as temporal in another.

The key to Einsteinian cosmology is what is known as the cosmological principle. According to this principle, the universe is, on a sufficiently coarse scale, homogeneous and isotropic. It certainly does not appear so to the earthbound astronomer when he surveys the solar system or our galaxy. About 100 billion galaxies are estimated in the universe; however, might they not be considered mere specks of dust? If so, is the universe a sort of uniform dust cloud? It appears not, because groups of galaxies tend to form clusters which are held together by their mutual gravitational attraction. At what scale then is there uniformity? It appears only to be true for volumes of linear dimensions of some 10^8 light years. We now know the size of the universe to be 10–20 billion light years, i.e. a hundred times bigger than this in linear dimension and a million times in volume. The universe appears more or less uniform on this scale. It is not beyond the bounds of possibility that significant inhomogeneities might exist at still larger scales. Perhaps it will ultimately be necessary to take account of such features as an integral part of a cosmological model. For the present, however, we shall take the density of the universe to be homogeneous at least on the scale of galactic superclusters.

The geometry of the universe should, then, also be homogeneous on that scale. On smaller scales, however, it will have local deviations of its curvature on the galactic scale, and smaller, more local ones on the scales of star clusters and individual stars. Energy in the form of radiation should also be expected to be inhomogeneous on these scales. The theory of cosmology then describes the structure of the universe only on the large scale.

All of the above must be read as an exercise in hindsight. When the first relativistic cosmologies were being proposed, the very notion of galaxies had not yet emerged. Astronomers observed, however, that the universe appeared to be statistically much the same wherever one looked.

Another concept which survived from ancient times concerned the apparent unchangeability of the firmament. Astronomers from time immemorial had charted the motion of the same stars at the same relative positions in the sky; this was one of the most significant proofs of the local motions of earth and planets. Even in

Einstein's early years, the heavenly landscape seemed fixed and constant.

The new relativistic cosmology, however, did not accommodate itself to this constancy. Einstein's equations only allowed universes undergoing expansion or contraction. This disturbed Einstein, and led him to modify his equations to include a "cosmological term" designed to restore stability. Intuitively it would seem that an expanding universe should be decelerated by the mutual gravitational attraction of its parts. If the speed of expansion were great enough, it might expand forever, though always at a constantly decreasing rate. If the expansion were slower it might ultimately stop and the universe might start to recontract; such contraction would then be expected to continue at an accelerating rate. Since neither of these alternatives seemed palatable, something additional was needed to give the cosmos its presumed stability. This was the "cosmological term". It represented an antigravitational, repulsive force.

By 1922, Alexandre Friedmann had given a complete description of expanding and contracting models of the universe, and determined the criterion for distinguishing the different scenarios. At almost the same time, V. M. Slipher was publishing data showing the predominance of redshifts in the spectra of astronomical objects. Two years later, Hubble was able to establish the distances of spiral nebulae, and thus to recognize their true character as "island universes" consisting of innumerable stars or galaxies. Observation and theory came together in the work of G. Lemaitre in 1927, which extended that of Friedmann and tied it to the ongoing work at the Mount Wilson Observatory where Hubble was refining his studies on the redshift. By 1929, he was able to publish strong evidence of a linear relationship between the distances of galaxies and their redshifts. Einstein had been wrong in trying to stabilize the universe, which was in fact expanding. His "cosmological term" had been a mistake based on an erroneous prejudice, a fact which he was quick to acknowledge.

7.1. Measurement of Astronomical Distances

The key to Hubble's discovery was the development of methods for measuring the distances of distant astronomical objects. There exist a variety of different methods, applicable to different ranges of distance. Since some of the methods are indirect, and are based on plausible but sometimes uncertain assumptions, one builds a sort of chain of techniques which are linked at those regions where they overlap so that, as we move to greater distances, each new step is checked by the consistency of its results with those of previously established ones used for nearer objects. In this process one may verify Hubble's hypothesis linking velocity to redshift. To the extent that this is verified, we may then use the redshift itself as a distance measure. For the most distant objects that we can observe, we must usually rely on this method alone. The problem is that we do not know at what distance the Hubble law may become invalid.

For sufficiently near objects, distances may be measured by the simple technique of parallax. This is illustrated in Fig. 7.1.

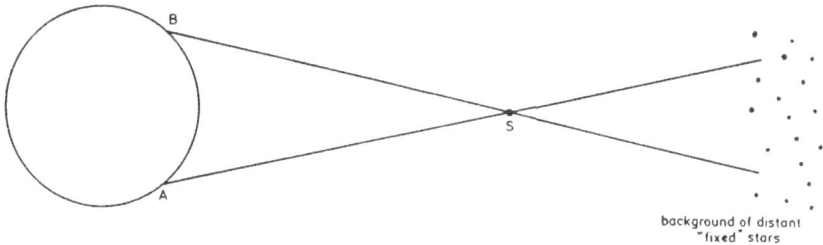

background of distant
"fixed" stars

Fig. 7.1.

Two observatories A and B on earth observe an object S. Seen against the background of distant fixed stars, S will appear to the two observatories to be at different points in the sky. From its apparent angular displacement, its distance is easily deduced. This method is the basis of a common astronomer's unit of distance, the parsec. It is the distance at which the parallax angle is one arc second from a baseline equal to the mean distance of the sun. In more familiar units, it is about $3\frac{1}{4}$ light years. If the baseline is the diameter of the earth, the distance of objects with the same angular separation is only about one ten-thousandth as great. The distance of planets can easily be determined by simultaneous measurement at two observatories; if, one the other hand, measurements are made six months apart from any observatory on earth, so that the baseline is the diameter of the earth's orbit, distances may be measured out to about 100 light years. Using the motion of the sun over several years, this distance can be extended to over 100 parsecs.

All the same, these are very small distances on the astronomical scale. To penetrate further into space, we need another method, which can also be used for stars within the limit of parallax measurement so that this new method may be calibrated. It came from researches of Henrietta Leavitt, who studied certain types of stars called Cepheid variables. Their luminosity varies periodically by as much as 20% (Fig. 7.2).

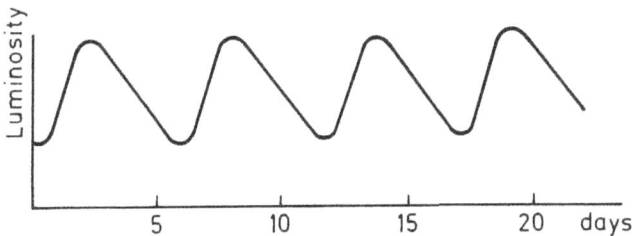

Fig. 7.2.

If the distance to a star is known (e.g. by parallax), its intrinsic brightness may be deduced from the amount of light reaching us. Imagine light from a source P falling on an area A at a distance d (Fig. 7.3). The same amount of light would fall on an area $4A$ at a distance twice as great $(2d)$; only a quarter as much would fall on the same area. The apparent brightness at twice the distance is thus only one-quarter as great. Conversely, for the same amount of light

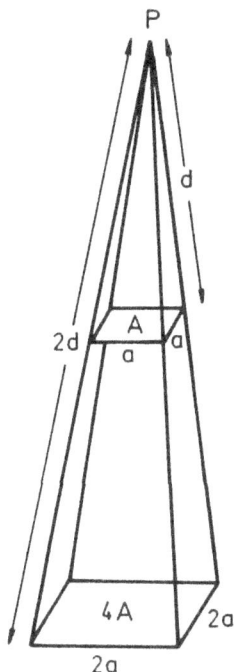

Fig. 7.3.

per unit area to arrive at distance $2d$, the source would have to be four times as bright.

In 1912, Leavitt discovered the relationship between the intrinsic brightness of a star and the period of its oscillations (see Fig. 7.4). For stars too distant for the parallax method to be applicable, the relation could be used in reverse; from the period of oscillation the intrinsic luminosity could be surmised. From the knowledge of apparent luminosity (the amount of light reaching us) and the intrinsic luminosity, the distance may be deduced. That near Cepheids and more distant ones had in fact the same properties could be verified by observations of their temperatures and the characteristic spectra of the light emitted.

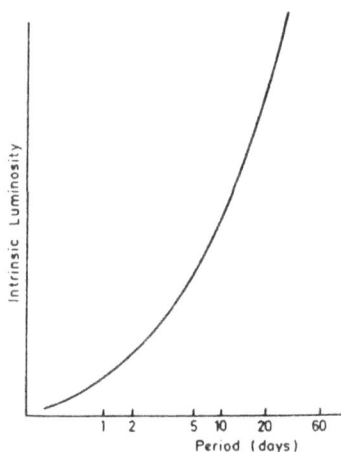

Fig. 7.4.

The Cepheid variable scale may easily be extended to other galaxies, so long as individual Cepheids in them may be identified. The practical limit for the use of this method is some ten million light years.

It is impossible to recognize individual objects inside more remote galaxies, though it may be possible to distinguish their form, that is, whether they have a spiral, a globular, or an irregular character. If clusters of galaxies are studied whose distances are known because they are within the range in which the Cepheid variable method can be used, a high degree of constancy is found in the intrinsic luminosity of their brightest galaxies. If we assume this feature to persist for more remote galaxies of the same type (spirals, in particular), it is possible to use their apparent luminosity as an index of distance. This index is not extremely precise, since the intrinsic brightness of the brightest galaxy may vary by as much as a factor of two, so it is again desirable to have a cross-check. At this point, if the distance inferred by the Doppler redshift gives the same value as the "brightest galaxy" method, confidence in the result is enhanced.

In the whole hierarchy of methods for estimating distances, the key is to find some sort of "fingerprint" of the object which establishes its relationship to a nearer object whose intrinsic brightness has already been determined by another method. There are evident hazards in the approach — are we, for instance, justified in the assumption that, by and large, the character of galaxies does not change as we go to greater and greater distances? Greater distance implies earlier time; might not galaxies have changed their properties by evolution over the age of the universe? If we go back far enough, this is almost certain to have been the case. The same consideration applies to the redshift method. The rate of expansion of the universe may not have been the same over cosmic time. In fact, we try to distinguish between different possible cosmological models precisely because they make different predictions concerning this history. The element of uncertainty becomes greater for the part of the universe which is most remote from us.

Indices other than those we have mentioned have been used to strengthen our confidence in our estimates of distance scales. At one stage, the realization that there were two different sorts of Cepheid variables with different period-luminosity relationships brought about quite abruptly a major revision of the whole astronomical distance scale; distance estimates of all but the closest objects were revised upward by more than a factor of two! Inherent uncertainties in astronomical measurement appear to be inescapable. The slow process wrought by improved observational technique, the accumulation of cross-checks, and better theoretical understanding serve somewhat to alleviate uncertainty about the reliability of our estimates of the distance of remote objects, but can never entirely dissipate it.

In any case, starting from data concerning the distances of astronomical objects over the whole range of current observation, it is in principle possible to retrace cosmic history; thus, to test the consistency of what we know about the evolution of the cosmos with models based on the theory of relativity.

7.2. Looking Backward: The "Big Bang"

Einsteinian cosmological theories led to one inescapable con-clusion: that a stable universe was possible only if one postulated a cosmic repulsive force such as that characterized by Einstein's "cosmological constant". Einstein had withdrawn this idea in the light of Hubble's discoveries, characterizing it as "the worst mis-take of my life". If the structure of the universe was determined by gravitation alone, dynamical evolution was inevitable. Hubble's observation of the recession of galaxies was perfectly consistent with the various relativistic models proposed by Lemaitre, de Sitter and especially Friedmann. But these models allowed for a variety of possibilities about the ultimate fate of the universe, as we shall see subsequently.

There was less ambiguity if one extrapolated backward rather than forward. Friedmann showed that the expanding universe must have originated from a singularity, an instant at which all matter was concentrated at a single point in space and time. A simple argument demonstrates how this can be deduced.

Recall that, according to Hubble, the velocity of recession of a distant galaxy is proportional to its distance from us; at a distance R the recession velocity is HR, H being known as "Hubble's constant". Consider now two objects A and B, B being twice as far from us as A (Fig. 7.5). It is therefore receding twice as fast, so that it will continue to remain twice as far indefinitely. Similarly, arguing backward in time, B will also in the past have been twice as far from us as A.

It follows that A and B, and we, must at the beginning have been together, along with all the rest of the universe. The universe, then, must have had a beginning at some definite time (measured backwards from the present). What happened at that time? The whole content of the universe must have come into existence and exploded "outward", though exactly what is meant by outward is not very clear in this context. Alternatively, we might say that space expanded. At any rate, the process by which it all started is

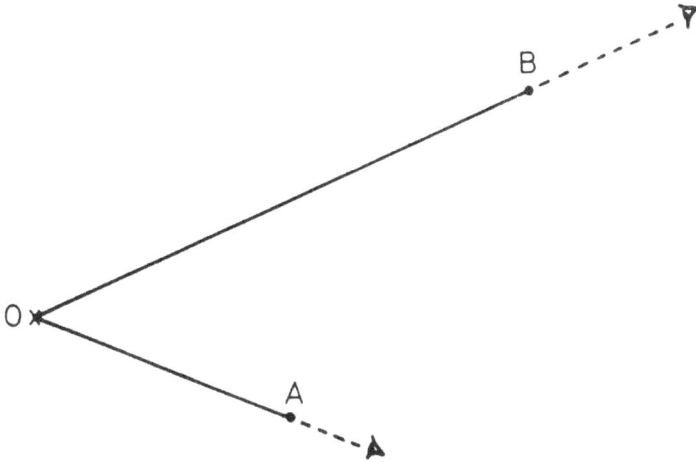

Fig. 7.5. The dotted lines in the diagram show the increase in distance of A and B over the next time interval.

generally referred to as the "Big Bang" for its apparent similarity to matter flying outward from an explosion.

This picture, of a universe starting from a big bang and expanding, is a difficult one to grasp, so let us examine it carefully. In the first place, when we say that everything started at a point, we do not mean some point in space. We mean, rather, that all space was a point; so that all matter, or energy, was compacted together at infinite density. But this must be interpreted in the light of the principle that it is matter or, more generally, energy which defines space.

Consider next the observation that every distant object in the universe is receding from us. This must not be taken to mean that we have any special status in the universe. A more general way to state the situation is to say that all objects are receding from any point in space, or that space is expanding in the sense that all distances between objects in it are increasing in time in the same ratio! A simple analogy, but one which must be used with caution, is that of the expansion of the surface of a balloon; astronomical bodies are like points painted on its surface. If it is blown up at a uniform rate, all points recede from each other at a uniform rate.

We must, however, be aware of all the implications of the analogy. The surface of a balloon is two-dimensional, but the balloon itself exists in a three-dimensional space. There is a point inside which is at its (three-dimensional) centre, but this point is not in the two-dimensional space of the balloon's surface. It is the surface of the balloon which is the (two-dimensional) analogy of the three-dimensional space of our universe. There is, then, no centre from which everything is receding.

What, then, of Einstein's four-dimensional space-time? Within its framework, the big bang has indeed something of the character of a unique point from which the whole universe evolves.

We can extend our picture of the two-dimensional universe of the surface of the balloon to three dimensions, maintaining the hypothesis of statistical homogeneity. We can imagine a cake in which raisins are uniformly distributed and which is rising (expanding) uniformly. All the raisins in this cake will then be receding from each other at a rate proportional to their mutual distances. But this "universe" has an edge, being embedded, not in a space of more than three dimensions, but in an infinite three-dimensional space. Thus, our image is that of a bounded universe; points on the boundary are not the same as those inside. The real universe must be unbounded. It must be an infinite cake. But here analogies fail us, because our universe may be finite.

These observations lead us back to an argument of Newton, who faced the same difficulty. He imagined a universe with a fixed density of matter, acting under its internal gravitational forces. If it were finite, he contended, it would have a centre, and everything would be pulled together toward that centre, to which it would ultimately collapse. Newton concluded that the universe must be infinite, since there would then be no point to which it could collapse. The flaw in the argument is Newton's assumption that space is absolute, with the universe inside. For Einstein, space was defined by the distribution of energy; the greater the energy density, the greater the local curvature of space. But a space which is curved in the same manner everywhere cannot be infinite, but must close on itself without having any boundaries. Space-time may be infinite in the time dimension, but this is another question to which we shall address ourselves subsequently.

All in all, then, the surface of an expanding sphere, where the surface is the whole universe, is the best and simplest analog for Einstein's universe.

Remember, too, that for Einstein time also was not absolute, but was defined by the evolution of matter and energy in the universe. Students frequently ask the question: what was there before the big bang? The answer was given long ago by St. Augustine, in slightly different terms: before the creation, he said, there was nothing and therefore there was no time. The answer is perfectly consistent with Einstein. Time began when the universe was created, time being a property of the world, a manifestation of its evolution.

So much for the past. Let us now turn our attention to the future and the ultimate fate of our universe.

7.3. The Future of the Universe

While we can retrace most of the history of the universe with reasonable confidence, we are much less certain of its future. Alexandre Friedmann's theoretical models allowed three possibilities. In one case the universe will continue to expand forever, with a constantly diminishing density of matter and energy and decreasing temperature. On the other, gravity will ultimately bring the expansion to a stop, and the universe will recontract toward the "big crunch" in which all matter will disappear into the singularity from which it was born. Or finally, with exquisite fine tuning, the expansion might ultimately grind toward a halt, taking an infinite time to do so, but without ever recontracting.

Which of these possibilities is realized depends only on the density of the matter-energy of the universe. If this density is high enough, its internal gravitational forces will be sufficient to halt the expansion; if not, expansion will continue forever.

Remembering our analogy with the expansion of the surface of a balloon, we can see that a simple way to characterize the expansion is by a "scale parameter" which characterizes changes in the distances of different astronomical objects. Everything is presently receding from everything else. In a given time, then, all distances will change in the same proportion p. This is our scale parameter. According to Hubble's law, this parameter varies linearly with time; Hubble's constant is the fractional rate of change in all distances. But our intuition tells us that, in the long term, this cannot be precisely correct, since gravitational forces must "decelerate" the expansion. This deceleration is characterized by a new parameter (the deceleration parameter) which defines the rate at which the expansion is slowing down. This parameter is defined as q_0. Clearly its value will increase with the density of matter in the universe (matter encompasses both matter and energy). The critical value of q_0 is $1/2$. Below this value there will be recontraction and a closed universe, above it expansion ad infinitum and an open one. ("Closed" and "open" refer to space-time; in one case both space and time coordinates in the universe always remain finite, in the other both can go to infinity.)

Figure 7.6 shows the various possibilities. The scale of the universe is plotted vertically and time horizontally. The vertical line through N corresponds to our present time; the distance ON corresponds to about 20 billion years. The curve marked $q_0 > 1/2$ represents the history of a closed universe. The area above the curve $q_0 = 1/2$ corresponds to open universes. $q_0 = 0$ corresponds to an empty universe (zero density).

A simple mathematical calculation will enable us to specify the critical density above which the universe will re-collapse. Imagine a shell of matter at a distance R. A simple Newtonian argument says that this matter will be pulled toward us by the gravitational force of all the matter inside the shell. This matter occupies a volume

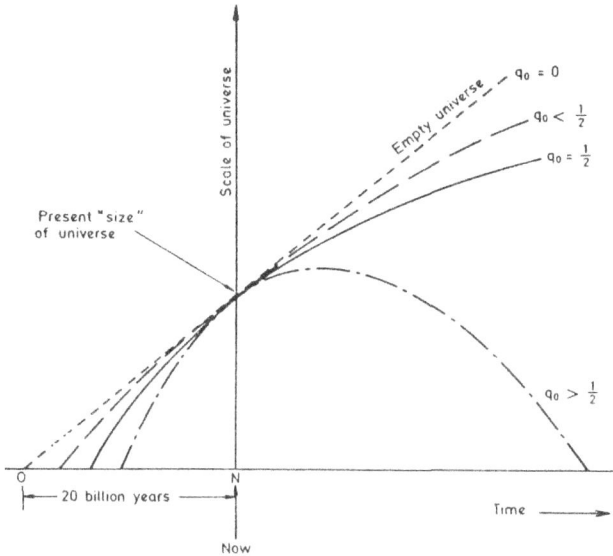

Fig. 7.6. The history of the universe.

of $4\pi R^3/3$ and has a mass of $4\pi R^3 \rho/3$, ρ being the matter density, assumed uniform everywhere. The gravitational potential energy per unit mass of the material in the shell is then

$$-4\pi R^3 G/R \ ,$$

G being the gravitational constant. The kinetic energy, again per unit mass, is $v^2/2$ where $v = HR$ is the velocity of recession (Hubble's law). So the total energy per unit mass is

$$H^2 R^2/2 - 4\pi R^2 \rho G = R^2 [H^2/2 - 4\pi G\rho/3] \ .$$

If the quantity in brackets is positive, the expansion will continue forever; if negative, the expansion will ultimately stop and recontraction will set in. The critical density is then $3H^2/8\pi G$ which, for the range of possible values of H, gives $0.5 - 2 \times 10^{-23}$ g/m^3.

Before commenting on this number, let us note that no good justification appears for the way we have done the above calculation; if there were, we would scarcely need the general theory of relativity in cosmology! Fortunately, the answer we have obtained is precisely that given by the most refined theory.

All the same, there is a plausible rationalization for what we have done. At 100 million light years, the recessional velocity is only about 0.5% the velocity of light, so that the use of the non-relativistic formula for kinetic energy may be justified. By the same token, the gravitational field at the surface of a sphere of radius R, due to the matter inside it, is not very strong, since the black hole radius of the mass of matter in it is only a thousand or so light years compared with its extent of 100 million light years. Therefore, the gravitational theory of Newton should suffice.

Finally, is 100 million light years sufficient to give a reliable sampling of the mean density of the universe? The mass involved is about that of 10^{15} suns, or some 10,000 galaxies. We might then expect our calculation to be reasonably valid.

The ultimate fate of the universe can in principle be decided if we can determine with reasonable accuracy the energy density of its matter. Unfortunately, this is a very difficult task. If we look at the galaxies and their stars, we can roughly estimate the amount of energy represented. It is, by a large factor (ten at least), insufficient to cause a re-collapse of the universe.

For the past twenty years or so, we have become more aware that much of the mass of the universe may be unseen. As we have discovered the "hidden" energy stored in the universe, it appears more likely that the universal density may attain the critical value. For the present, too many uncertainties exist. Each new discovery pushes us closer. Let us look at some of the factors in the equation.

7.4. Forms of the "Missing Mass"

Kepler showed us that a simple law governed the motion of planets circling around the sun: the squares of their periods are proportional to the cubes of the dimensions of their orbits. This may be expressed in the equation $T^2 = ba^3$, where b is a constant. The speed of the planet in the orbit is

$$v = 2\pi a/T$$

so that $T = 2\pi a/v$. Putting this in Kepler's law we obtain the relation

$$(2\pi a)^2/v^2 = ba^3 .$$

It states that the square of the velocity in the orbit is inversely proportional to its radius. The outer planets move more slowly than the inner ones. In an article in "The Universe of Galaxies" (Freeman, 1984), Vera Rubin reports on studies of stellar motions in spiral galaxies as determined by Doppler shifts. She finds that in many such galaxies the velocities rise rapidly at first as one leaves the centre of the galaxy, and then become fairly constant up to distances well beyond the visible limits of the galaxy (Fig. 7.7).

Matter in the galaxy does not become more tenuous beyond those visible limits, it simply becomes invisible! The galaxy is more extensive and more massive than it appears.

This is an important factor when one tries to estimate the total density of the universe. It is the existence of the "missing mass", the difference between the mass of what we can see and what is necessary to explain the motions of this visible matter.

There seems to be a large amount of "missing mass" within many galaxies.

The situation is repeated on the level of clusters of galaxies. Once again, if we measure the motions of individual galaxies within clusters, we find that they are usually moving much faster than they should if they were acted upon only by their mutual gravitational interaction. This again bespeaks the presence of large amounts of invisible matter. It may be of the order of ten to a hundred times the mass we would attribute to the visible stars.

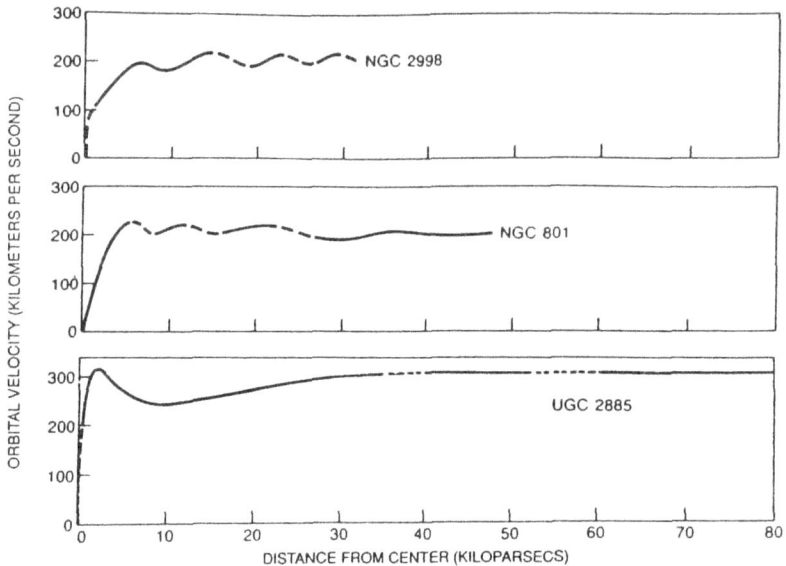

Fig. 7.7. Rotation curves show orbital velocities of three Sc galaxies from the centre outward. Galaxies increase in luminosity from top to bottom. With increasing luminosity galaxies are larger, orbital velocities are higher and velocity gradients near the galactic centre are steeper. (From *Dark Matter in the Spiral Galazies* by Vera C. Rubin, *Scientific American*, June 1983, p. 101. Copyright ⓒ1983 by Scientific American, Inc.)

But there may still be other manifestations of missing mass. One has been revealed in recent work in the field of X-ray astronomy. One way of locating precisely the source of cosmic X-rays is to observe the change of X-ray intensity as a galaxy passes behind the moon. It should thus be possible to determine how much X-radiation is emitted by that galaxy. Such measurements were made several years ago for two distant galaxies in the constellation Aries, known as Abell 339 and Abell 401. As the moon passed in front of first one galaxy and then the other, a continuous plot could be made of the X-ray strength. What was found was that much of the X-radiation was coming, not from the galaxies alone, but from a vast invisible region between them. The source of this radiation, invisible to the optical telescope, must be at very high temperatures.

More recently, a team of astronomers from Cornell University, using the 300-metre telescope at Arecibo, Puerto Rico, discovered, quite accidentally, a huge cloud of gaseous hydrogen floating in the space between galaxies. It is larger than our galaxy, some 300,000 light years across, and has the mass of a billion suns. But once again, its Doppler shift reveals that the surface of the cloud is moving at a speed which can only be explained by the existence of a mass within the cloud 100 times larger than the mass of the cloud itself, once again, a huge "missing mass".

In the last few years a diffuse background of X-rays and gamma-rays has been observed, corresponding to a gas at a temperature of some 500 million degrees, which is isotropic and thus intergalactic. At the same time, the Einstein X-ray Observatory revealed that individual quasars at large distances are powerful X-ray sources; in fact, they account for a large proportion of the total X-ray intensity detected by the Observatory. That this radiation is isotropic indicates that the same is true of the distribution of the quasars themselves, though this could not be verified in itself.

These discoveries, which suggest the existence of large amounts of invisible mass both within and between galaxies, are not sufficient to permit us to conclude that the mean mass density of the universe is sufficient to assure its ultimate recontraction. It simply suffices to throw the question into doubt. More indirect indications, such as the study of the Doppler shifts of the most distant galaxies or of the proportions of hydrogen and helium in the observable universe, must also be taken into account. But beyond all this is another consideration which may in the final analysis be determinant; it is the question of the possibility that the neutrino has a small but not negligible mass. Recent experiments suggest that it may have. If this is so, it is possible that the mass of these neutrinos may be the largest single contribution to the total mass density. Particle theory tells us that there should be, in the universe, between 10^8 and 10^9

neutrinos for every nucleon. If mc^2 for the neutrino is as much as 20 electron volts, compared with a bit less than 10^9 ev for a proton, they may well be the dominant factor in the universe.

Recent developments in early-universe cosmology, and specifically the inflationary model, require a universe more or less perched on the boundary between the two limiting regimes, a flat universe. This evidence is not convincing, but it seems consistent with most of the observational evidence presently available. For now, the resolution of the problem is still very much in doubt.

Many physicists favour, on grounds that are more emotional than rational, one or the other resolution. The widely-proclaimed objectivity of scientists probably is not, and possibly should not be, so overwhelming as is sometimes assumed.

7.5. The Microwave Cosmic Background Radiation

While we have dealt with the relativistic and cosmological aspects of the Big Bang model of the universe, other consequences are also of primary importance. These concern the detailed structure and content of the cosmos. On the one hand, there are the problems of the formation of galaxies; on the other, questions concerning the nature of matter (or, more broadly, energy) at the various stages of cosmic evolution. Modern nuclear and particle physics has much to tell us about these questions. The central factor is the variation of the temperature of matter and radiation as the universe has expanded. The clue to their variation is that the wavelengths of photons increase in precise proportion to the expansion parameter (a conclusion suggested by the "surface of the expanding balloon" model). Since the energy of radiation varies in this way, so does its temperature. Furthermore, early in cosmic history, matter and radiation must have been in equilibrium, since most matter would have existed in ionic (charged) form, so that it interacted strongly with the electromagnetic field.

At the very beginning, say, one hundredth of a second after the "creation", when the temperature of the universe was about 100 billion degrees Kelvin, the universe would have been filled with a "soup" of matter and radiation whose contents (quanta) were in continuous collision with each other. Thus, everything was in thermal equilibrium. (This, of course, raises interesting questions about the application of the second law of thermodynamics, since an equilibrium state is a state of maximum entropy, or disorder; yet the order which characterizes its present state, in the form of stars, galaxies and the rest, must have evolved from it!)

In this primeval soup, every sort of known particle, along with equal, or almost equal, amounts of the corresponding antiparticles, must have been present. "Particles" of zero rest-mass, or particles whose rest-energies were far below the mean thermal energy, such as photons, neutrinos, and electrons would have been most plentiful. There would have been only about 10^{-9} as many protons or neutrons as of these light particles.

Under very hot early conditions, particles and antiparticles of the different species would have been made in equal numbers. They may, however, almost immediately re-annihilate each other to create electromagnetic radiation. In any case, though, a surplus of neutrons and protons over their antiparticles must remain. Whence this bias in what should apparently be a symmetrical situation? Why does our world consist of protons, neutrons and electrons rather than anti-neutrons, anti-protons and positrons? Each seems *à priori* equally plausible. Only two answers are possible, neither of which provides an explanation: either the imbalance was there from the beginning, or the bias is in the laws of physics as nature has given them to us.

Not everything in the world is explainable. No question can be answered except in terms of something else. But we may then always ask "why?" of the something else. The best we can do is to find that the answer to one question also implies the answer to another. This is the goal of physics. If one asks questions like: what caused the "Big Bang"? or, what happened before the Big Bang? or, why are the laws of physics this rather than something else? one must respond: on what basis am I asked to answer this question?

One cannot create an explanation from nothing. Answers to such questions can only be sought in hypotheses which are outside the context of the questions themselves; otherwise all answers would be circular: A is true because of B and B is true because of A.

If, therefore, you thought that all questions can be answered by science, you are wrong. Life always contains an edge of mystery; logic requires this. Questions arising within a given system of thought cannot all be answered within that system.

Of course, questions may be related to each other; since the universe must above all be electrically neutral (a conclusion which may be deduced from observation and physical law), a bias toward protons rather than antiprotons must be accompanied by a similar bias toward electrons rather than positrons. Thus, two questions become one.

To the question: is the asymmetry in the initial conditions or in the laws of nature, we shall not propose an answer here; the issue is at this point not resolved beyond a doubt. It is enough to say that theories can be, and have been, conceived which provide for such an asymmetry (but no answer to the question: why is the asymmetry in the direction in which we find it and not in the opposite one!). The question is most properly resolved by the testing of the theory which contains the element of bias.

Let us continue our sketch of the chronology of the early universe. The next milestone comes at about one minute after the beginning. Now, due to expansion, neutrinos travel farther between collisions with other particles; furthermore, their interaction with electrons and nucleons is extremely weak. Thus, they no longer maintain an equilibrium with other matter, and go their own way, essentially cooling, just as photons would, to temperatures inversely proportional to the expansion parameter. From here on their numbers will not change, but due to cooling their contribution to the total energy of the universe will decrease. If, as has been thought until recently, the neutrinos truly have zero rest-mass, that energy will decrease indefinitely with the temperature. Current evidence, not yet conclusive as of this writing, suggests that they may have a very small rest-mass, say, 15–20 electron volts. If this is so, the energy will diminish to their rest-energy when the temperature decreases below about 200,000 K, that is, when the age of the universe reaches between 250 and 300 years. Their energy will not appreciably decrease beyond that point.

Up to now, neutrons and protons have been present, but they have not been able to form heavier nuclei. A stumbling block is the low binding energy of deuterium, whose formation must precede that of still heavier nuclei, e.g. He^3, tritium (H^3), and then He^4, which is very much more stable. The formation of deuterium depends on the relative densities of photons and nucleons, since helium formation is more likely the greater the nucleon density. For a ratio of 10^9 photons per nucleon, nucleosynthesis begins at a temperature less than a billion (10^9) degrees. This event occurs three to four minutes after the Big Bang. Very rapidly, most free neutrons will have participated in the formation of helium; nuclear matter is thus in the form of hydrogen or helium, in the proportion of approximately 75:25. It is from these materials that the first stars will be formed.

At this point the main components of the universe are gamma-rays, hydrogen and helium nuclei, neutrinos and antineutrinos, and electrons. As the universe expands the temperature drops proportionately. Charged matter and photons remain in equilibrium at this temperature; the universe is opaque to light because of strong particle interaction with the electromagnetic field. Most of the energy of both electrons and protons is rest-energy; thermal (kinetic) energy is relatively low. The energy of light quanta, however, continually decreases with expansion. Because there are more photons than particles, at first the photons account for the greater part of the energy; thus, we refer to this epoch as radiation-dominated. After a few hundred thousand years however, the radiation had sufficiently cooled that it was matter which accounted for most of the energy (remember, however, the uncertain role of neutrinos).

At about the same period, something even more important happened: the mean photon energy became too low to prevent the combination of electrons and protons to form hydrogen atoms or to excite them once formed. Thus, the mechanism for maintaining equilibrium between photons and matter disappeared, and the universe became transparent to photons. We are aware that astronomers view objects billions of light years away; thus, the light that we "see" consists of photons which have travelled freely through the cosmos. It is clear, however, that this was not always possible.

Photons, out of thermal equilibrium with the material world, continue to cool due to expansion. Matter cools independently, at its own rate. The numbers of material particles and of photons in the universe are "frozen" at the values which they had at the time of decoupling, at some 10^8–10^9 times as many photons as material particles. That "archaeological" radiation should still be with us.

Something of this sort was envisaged by George Gamow and his collaborators in 1949! In speculating about the consequences of the "Big Bang" model, they showed that, to explain the present abundances of the light elements, it was necessary to assume the existence of about 10^9 photons for every nuclear particle. Since they could estimate the present density of nuclear matter, they could then deduce the density of photons. But further than that, pursuing a line of reasoning similar to the one we used above, they predicted the temperature of these photons to be about 5 degrees Kelvin.

It is quite astonishing that such a precise and dramatic prediction did not arouse much excitement. In fact, the paper was almost totally ignored. So much so that when this radiation was finally observed in 1965, it was not immediately recognized.

Like many other important discoveries in science, this one was quite accidental. A. Penzias and R. W. Wilson of Bell Laboratories were attempting to measure the intensity of radio waves emitted by the gas surrounding our galaxy. They found that their measurements were bedevilled by the presence of unexpected microwave noise, which they succeeded in identifying as having a wavelength of 7.35 cm. Their efforts to locate the source of the noise led to a strange conclusion, it appeared to be coming equally from all directions. In discussing their problems with a colleague at M.I.T. it was pointed out to them that Robert Dicke at Princeton and his collaborators, including James Peebles, were building an antenna precisely to look for the radiation predicted by Gamow, at a wavelength of 3.2 cm. As a result the groups met in early 1965 and agreed to publish simultaneous papers. That of Penzias and Wilson dealt with their observations, and was entitled "A Measurement of Excess Antenna Temperature at 4080 MHz". Dicke, Peebles, Roll and Wilkinson explained that the 3K radiation had its origin in the "Big Bang". Penzias and Wilson won the Nobel prize for their work in 1978; probably their paper had the least glamorous title and one that gave the least indication of the rationale for its choice, of any that ever won the prize.

Its importance was that, properly interpreted, it provided very strong evidence, not only for the general idea, but also for the detailed theory underlying relativistic cosmology.

It should be noted, however, that Penzias and Wilson's measurements were only made at one wavelength (or frequency). But the theory made a more detailed prediction: that the radiation should have a black-body distribution corresponding to a temperature of about 3 K. This distribution would have been established when the universe was perhaps a thousand times smaller than at present. At that time its temperature was perhaps 3000 K, since the radiation would only have cooled due to expansion. If the radiation could be shown to be distributed over a black-body spectrum, the interpretation would be even more firmly established. There was a problem, however. As long as measurements were made at the earth's surface, i.e. under the blanket of the earth's atmosphere, an important part of the black-body spectrum would be blanked out by atmospheric absorption. Radiation at wavelengths larger than that of the peak of the black-body distribution penetrates the atmosphere quite easily, but that on the high frequency side is strongly absorbed. This difficulty can be circumvented to a degree by using high-flying balloons (30–45 km), or, even better, by observing from a space-based observatory. In this way, accurate verification of the theory should soon be possible. Results obtained by balloon observations in 1979 are depicted in Fig. 7.8. The solid curve represents the theoretical black-body curve for a temperature of 2.96 K. Observational results, taken over a wide region, provide quite a strong confirmation of the predictions from the Big Bang model.

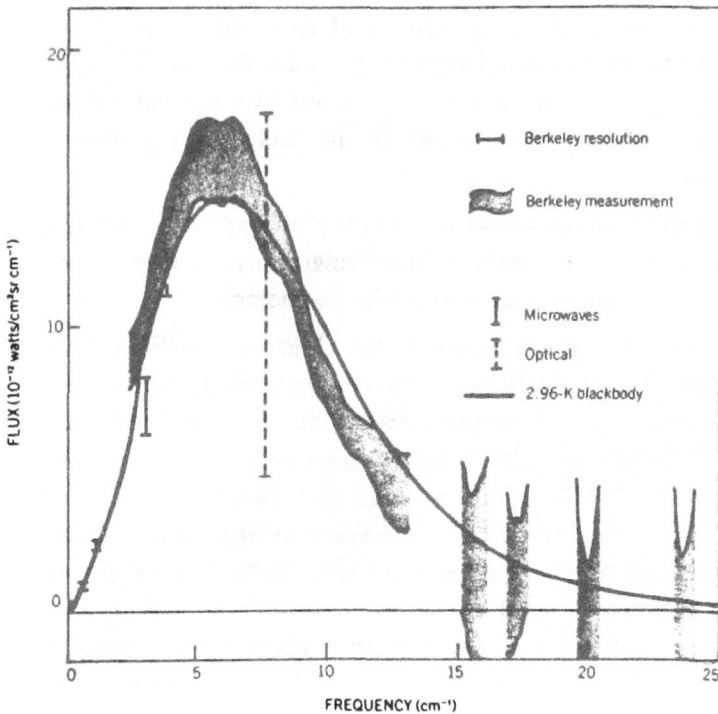

Fig. 7.8. Spectrum of cosmic background radiation. The solid curve is the spectrum of a 2.96-K blackbody. The shaded area is the rms sum of all Berkeley experimental errors with $\pm 1\sigma$ error limits. The gaps are left at 14, 16, 18 and 23 cm^{-1} because of strong atmospheric absorption there. (From *Physics Today*, June 1979, p. 17 by D. P. Woody and P. L. Richards, with permission from the authors and the American Institute of Physics.)

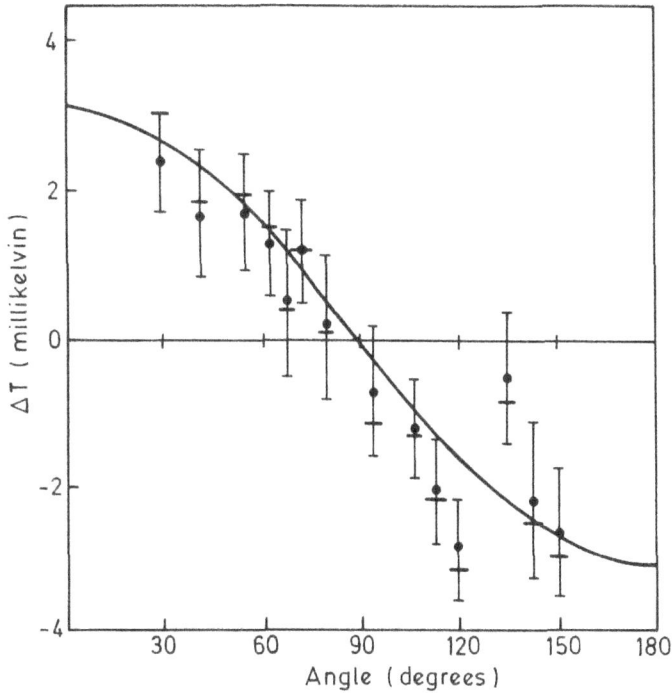

Fig. 7.9. (From *Physics Today*, Jan. 1978, p. 19 by G. F. Smoot et al. With permission from the authors and the American Institute of Physics.)

As for the directional distribution of the radiation, a remarkable degree of isotropy is observed; it is verified to one part in a thousand. The deviation from isotropy is, however, measurable and systematic.

Figure 7.9 shows the variation in the radiation "temperature" with direction. The direction represented is that of maximum anisotropy. The scale is in thousandths of a degree.

Why is there anisotropy? The answer lies in the motion of the earth relative to the firmament. The earth moves around the sun, the sun moves in the galaxy, the galaxy moves in the cluster of galaxies, the cluster of galaxies moves against the background of fixed stars. These observations lead us to conclude that the earth is moving, over all, with a velocity of about 400 km/s in the direction of the constellation Leo. (It is more dramatic if one translates the result into km/h – nearly 1.5 million; however, this is still only of the order of one thousandth of the velocity of light.)

But there appears to be another difficulty. Is not this an absolute motion of the earth, and does it not contradict the special theory of relativity? Or, to put it differently, does it not signify the existence of a sort of aether after all?

Two comments help put this question into perspective. One is that the frame of reference of the fixed stars has a privileged role, in the sense that it provides the basis for defining inertial frames of reference, which constitute the very foundation of the special theory. The other is that the universe of relativistic cosmology is statistically homogeneous, but it is not locally so. If the curvature of the universe is uniform on the cosmic scale, it has nevertheless local inhomogeneities. All the motions we have mentioned, that of the earth around the sun, of the sun in the galaxy, of the galaxy relative to the fixed stars, are manifestations of these local inhomogeneities, the curvature caused by the gravitational field of the sun, by the gravitational field of the galaxy, or that of the local cluster of galaxies, and these are small fluctuations on the global cosmic curvature.

Whether, then, we do or do not wish to talk of a "new aether", it has nothing to do with the aether of the 19th century physicists. It is in no sense a substance permeating space. Hence, the idea is neither fruitful nor properly representative of reality. Rather, we should learn that it is rash to extrapolate to the whole universe the conceptual framework which properly describes the interaction of its parts.

7.6. The "Paradox" of the Dark Night Sky

Scientists often claim that the darkness of the sky at night is itself a proof of the expansion of the universe. To understand this, we have to go back in history to an observation made by the German astronomer Olbers in 1823. Actually, the question goes back further than that, to Halley (1720) and Loys de Cheseaux (1744), but Olbers is the one to whom history, somewhat unjustly, awards the credit.

The argument starts from Newton's concept of the universe. He observed that the stars were uniformly distributed over an infinite static universe. In such a universe, if we look out from our vantage point into space in any direction, we should see a star. If we imagine successive "shells" of space of equal thickness surrounding us, these shells should contain volumes of space which increase as the squares of their distances, and the number of stars in them would increase in the same way. But conversely, the amount of light we would receive from each star would decrease with the square of the distance. Thus, every succeeding shell would illuminate the sky equally. Since the number of shells was infinite, the sky, even at night, would be brilliantly light.

But that, said de Cheseaux, was not right, since the nearer stars would block out the light from the more distant ones; this, to him, resolved the paradox. Subsequent astronomers rejected his argument, however, on the grounds that the energy of the "hidden" stars would be absorbed by the nearer visible ones, heating them and causing them to radiate more intensely, so that their luminous energy would reach us all the same. Note the assumption in this argument: the stars would come into equilibrium with each other, so that each would radiate as much energy as it absorbed. If the duration, as well as the size of the universe were infinite, this assumption would be valid. Modern cosmology states, however, that the universe has existed for "only" 10–20 billion years. How long, on the other hand, would it take to reach equilibrium? A simple calculation gives 10^{24} years! So the answer to de Cheseaux' argument is not valid, though this does not make the argument correct. The universe is not in equilibrium at the temperature of the surface of stars; if it were, the night sky would certainly be bright.

A solution is to be found in the observation that if the whole energy density of the universe were converted into radiation in equilibrium, this radiation would have a temperature of only about 20 K. Could it be that there is just not enough energy in our finite universe to make the sky bright?

We have to consider another complication. Stars do not shine forever, but have "luminous lifetimes" which vary according to their size. Of course, the death of old stars is marked by the spewing out of large quantities of stellar matter which will form the stuff of our new stars, so we might better speak of the luminous lifetime of stellar matter, which is limited by the ultimate conversion of all stellar material into iron. A reasonable estimate of the luminous lifetime is something like 10^{10} years. Because this is short compared to the equilibration time by some 14 orders of magnitude, there is far too little time to heat the universe up to equilibrium at the temperature of starlight. According to conventional wisdom, the darkness of the sky was proof of the expansion of the universe; however, we conclude that the night sky is expected to be dark independently of expansion. The confirmation of expansion must then to be found in other, more definitive evidence.

Chapter 8

THE EVOLUTION OF STARS

Traditional astronomers believed the heavens to be unchanging.
However, we know that over the longer scale of cosmic evolution radical change takes place. The astrophysical world has had a turbulent
history. From some primeval elemental matter and some debris from
earlier generations of stars, new stars are formed, and ultimately age
and die. Stars agglomerate into galaxies held together by their mutual gravitational forces, while galaxies similarly form clusters, all in
a manner which seems to deny the second law of thermodynamics.
In the earlier stages of the universe at least order appears to have
evolved somewhere out of cosmic disorder. We do not as yet fully
understand these processes, yet our increased understanding of natural law enables us to speculate on their history, to draw conclusions
from these speculations, and to subject these conclusions to the test
of observation.

The origin of stars, for example, can be traced to dark clouds
of molecular hydrogen within our galaxy, and thus, presumably, in
others. This conclusion is reached from spectroscopic observation.
Gravitation within these clouds then tends to collapse them. But to
collapse they must not be too hot; too much kinetic energy mitigates
against cohesion. So too does turbulence, but that turbulence may
dissipate itself in shock waves. Shock waves from supernovae, or even
cosmic rays, may play a role in star formation. We have some clues,
but not enough to provide definite answers. At some stage of the
process they are thought to be protostars, balls of gas characterized

by strong winds, but not yet hot enough from collapse to generate the nuclear reactions which power full-fledged suns. Violent and complex processes are constantly in action.

Leaving the mysteries of star formation aside, we can study fully-formed stars, which finally settle into more or less long periods of stable energy production. Our sun will serve as an example. It is made up primarily of hydrogen and helium, the former being more plentiful, and traces of other heavier elements. The process of energy production is that of the nuclear burning of hydrogen into helium. This is not burning in the usual chemical sense, in which energy is released in the binding of atoms into molecules; rather it is the nuclei themselves that fuse. The energies of nuclear binding may be a million times greater than those of chemical binding. Nuclei are positively charged and so are subject to strong mutual electric repulsion as they approach each other. Nuclear forces are of very short range and do not come into play until the interacting nuclei have penetrated the coulomb barrier. This requires collisions at high energy. How high this energy must be depends on the charges of the interacting nuclei. It increases very rapidly as the product of the nuclear charges of the particles increases.

In the ultimate stage of star formation, the mass of matter which will constitute the star is pulled together by its own gravitational forces. As the matter falls inward under these forces, the nuclei gain kinetic energy — the star heats up — and ultimately the nuclear velocities become sufficient to overcome the repulsive electric barrier. The nuclei then collide, react exothermically and emit the energy released by their fusion. The situation stabilizes when the radiation pressure created by the burning at the centre balances the pull of gravity. This is a controlled thermonuclear reaction, in which just enough heat is produced to keep the reaction going. The reaction is now similar to that of a thermonuclear bomb.

Although the scale of energy is vastly different, the process has much in common with the process of ordinary chemical burning. In a forest, oxygen molecules of the air collide constantly with the molecules of the wood in the trees, but they do not combine in significant amounts. The wood does not oxidize. If one heats the molecules sufficiently, as may happen for instance when lightning strikes a tree, large numbers of molecules may combine with the release of enough energy to increase the frequency of collision of adjacent molecules, causing them in turn to fuse, releasing still more energy. This is a self-sustaining chain reaction, but this time it is a chemical one.

In both burning processes, energy is lost to the surroundings. In chemical fires it is mostly infrared and optical radiation, though the surrounding air molecules are also heated (accelerated). In nuclear reactions, energy is carried away in very high energy electromagnetic radiation (gamma-rays), and by the by-products of burning, in the form of high energy electrons (beta-rays), energetic neutrons, etc. These processes will be discussed more fully in a later chapter.

Since the matter at the centre of the star has "fallen" furthest, the central core is the hottest part of the star. The temperature decreases with distance from the centre. It is a mere 6000 K at the surface. It is interesting, however, that the sun's gaseous envelope, the corona, which can be seen during eclipses, is at a much higher temperature — as much as a million degrees. How can this be? Sunspot activity and the giant flares seen during eclipses provide the clue. They are related to the intense magnetic activity welling up from within the sun, which create the equivalent of great particle accelerators on the surface, capable of exciting hydrogen nuclei up to 100 million electron volts. This is more or less the energy produced in a modest cyclotron.

But this is incidental. The main point about the energy production of the stars is that they operate through a remarkably efficient feedback system which maintains the thermal stability of the star through much of its lifetime. The high temperatures at the centre of the star create intense radiative pressure (remember that light quanta, and even unquantized light, carries momentum proportional to its energy). This outward pressure ultimately checks the gravi-

tational infall. Suppose there were a sudden rise in temperature
in the stellar core. The radiation pressure would cause the star to
expand, which would decrease the temperature, thus restoring equilibrium. If, however, the star begins to contract, the added gravitational energy increases its temperature, thus producing an outward
compensating pressure.

It is difficult to imagine the scale of the phenomena involved.
The radiation pressure at the centre of the sun is of the order of a
billion times the air pressure at sea level on earth. The energy which
the sun radiates is the equivalent of that of a thousand nuclear bombs
per second. This rate of energy output has been continuing for some
5 billion years, and should continue for as many more.

Various interesting consequences flow from the picture we have
traced. Imagine a star with a mass twice that of our sun, for example,
Sirius. Its mass is 2.2 times the sun's mass. Its gravitational field
will be correspondingly stronger, and a greater radiation pressure
at the core will be necessary to resist it. Heavier stars burn at
higher temperatures than light ones. Consequently, the temperature
throughout, and thus the luminosity, will also be greater. Sirius is
in fact 21 times as bright as the sun. The luminosity varies as the
fourth power of the surface temperature.

A star burning at a higher temperature burns its fuel faster,
and thus will have a shorter lifetime. This is due to the fact that the
higher temperature will augment its ability to burn heavier elements;
all stages of their evolution will be speeded up.

What happens when the hydrogen in the core of the star is exhausted? The outward radiation pressure will diminish, but though
the infall following its collapse will cause a dramatic rise in the core
temperature, no thermonuclear reaction will be sustained in the core
and a countervailing radiation pressure will not be created until the
temperature rises enough to ignite a reaction other than the hydrogen one. Before this happens the sphere of burning hydrogen spreads
outwards, causing the star to expand. It may swell to many times its
previous size and is then characterized as a "red giant". An example

is the giant star Betelgeuse in the constellation Orion. It is so large that if its centre coincided with that of our sun, its surface would extend out well beyond Mars!

An implosion in the core will cause its temperature to rise until it reaches 100 million degrees, when a new reaction becomes possible, in which helium burns to produce carbon. This could come about by a series of reactions involving absorption of protons. It does not work, however, because no nucleus of five nucleons exists with sufficiently low energy. Nuclear helium, like atomic helium, for reasons which will become evident when we study the states of nuclei, have a special stability; adding one more nuclear particle requires a particularly large additional energy. The formation of the heavier elements therefore requires a different mechanism.

There is less energy in two separate helium nuclei than in a nucleus in which they were combined. In terms of energy, a reaction in which three helium nuclei can combine to form carbon-12 is possible, but this seems unlikely. The probability that three helium nuclei will be found close enough to fuse is much lower than that for two; thus were it not for a fortuitous accident, carbon and the elements beyond it would never have been formed, and our world as we know it, including the phenomenon of life, would never have come into existence. The conditions for the existence of life would simply never have appeared in the universe! But heavier elements do exist, and from this fact Fred Hoyle drew a bold conclusion about nuclear physics. He argued that nuclear forces, which are very complicated, might be such that if a system of three carbon nuclei did by chance come together, they might stick together for a long time before separating. This would require the existence of a very long-lived excited state at the energy at which they came together. Such a state might in fact last long enough for helium nuclei to fuse into a carbon-12 nucleus before the helium nuclei could separate again. Hoyle could predict exactly what energy this excited state would need for this to happen. It is prohibitively difficult using the laws of nuclear interaction to verify the existence of such a state. On the face of it, it seems highly unlikely. There was certainly no simple way to see it; rather, it appeared that it would have to happen as a sort of accidental consequence of the application of empirical laws whose fundamental basis was not too well understood. That is to say, those laws of interaction were themselves rather complicated manifestations of simpler and more fundamental ones.

Such states as Hoyle invoked are known as resonance states. There are no simple relations giving the energies of resonant states. It appears to be some sort of miraculous accident that the universe is as complex as it is; that this degree of complexity makes possible the existence of life in all its diversity. This is but one example of a proposition of Richard Feynman: when one knows only the basic laws of nature one does not really know much of significance of the real world. It is a good illustration of the limitations of reductionism.

In summary, Hoyle argued that since the universe contained the known heavy elements, there must be a resonant state of the C^{12} nucleus at 7.82 million electron volts. Later experiments in nuclear physics laboratories directly verified this.

Let us now return to the evolution of stars. We have shown that the production of carbon becomes possible at sufficiently high energies.

When the temperature of the core rises sufficiently due to its gravitational collapse, a new mechanism of burning takes over. At the same time, the burning of hydrogen will take place in a surrounding shell, where the temperature will not be so high. The outer mantle of the core will now be enormously expanded, making a red giant of the star.

Let us now go to a step further. Ultimately, the carbon in the core will itself be exhausted, having burned to produce still heavier elements. The heavier the elements, and thus the greater their charge, the higher the temperature necessary to make them participate in further thermonuclear reactions. The exhaustion of carbon at the core will give rise to further infall of matter surrounding it, which will raise the temperature, ultimately triggering further reactions involving elements heavier than carbon.

A cautionary remark is in order. Lighter stars do not burn as hot as heavier ones. Thus, stars that are too light will not be able to sustain all stages in the sequence represented by the successive burning of ever heavier elements. Stars that are substantially lighter than our sun do not have enough gravitational energy to carry on the burning processes past hydrogen. Most stars end their lives as white dwarfs. These white dwarf stars usually have a core of carbon, surrounded by shells of helium and then, perhaps, of hydrogen. It is possible that the hydrogen may be evaporated off because of insufficient gravitational pull.

For heavy stars, the core temperature always rises with the burning of heavier and heavier elements. At any particular stage the heaviest elements will be burning at the centre; proceeding outwards reactions will take place in a succession of shells in which each successive one will involve the burning of lighter elements and lower temperatures. Thus, in different shells of the star various reactions will be taking place simultaneously; reactions of the heavier elements near the centre of the star, and those of lighter ones further away.

The process cannot proceed indefinitely because beyond iron, energy is necessary to fuse nuclei. The forces holding nuclei together decrease until in the region around uranium they are not able to hold the nuclei together in a stable configuration. These heavy nuclei spontaneously become radioactive and ultimately split into two comparable parts. This is known as the process of fission. Iron is found at the watershed; all lighter elements can liberate energy by fusion, while the heavier ones become increasingly less stable.

What happens when the nuclear fuel is exhausted to the point where the fire goes out and the temperature is no longer high enough to sustain it? Chandrasekhar was the first to propose a solution to this problem. Clearly, the star will tend to collapse under its own gravitation, but how far? What will finally check the collapse?

The answer lies with quantum mechanics. It concerns the "**uncertainty principle**", which says that particles can be localized within a spatial dimension Δx only by giving them a spread of momentum $\Delta p = \hbar/\Delta x$ and by virtue of Pauli's exclusion principle, which says that only one electron can occupy a given state. As the star becomes smaller the volume available to each electron is reduced, which forces the electrons into higher momentum, and thus higher energy states. There may come a point at which the cost in energy of this process will exceed the gravitational energy reduction associated with gravitational collapse, so that the collapse will be checked. The energy increase involved in compressing the electrons, due to quantum laws, is responsible for what is called the "quantum degeneracy pressure"; when this pressure balances the inward force of gravity, collapse is checked. What remains is known as a "white dwarf star". Such stars were in fact known in Chandrasekhar's time.

What are the conditions for the formation of a white dwarf? Can electron degeneracy pressure always check gravitational collapse? If the radius of the star is designated as R, the distance within which an electron is confined is proportional to R so its momentum is proportional to $\hbar/R$ and its energy to $\hbar^2/2mR^2$.

But the gravitational energy only varies as $1/R$ (and is of course negative). Thus, at large R the gravitational force will dominate, and collapse will take place. At small enough R, it appears that kinetic energy will dominate, and collapse will be avoided. But this argument has a flaw, in that we have used the classical expression for energy in terms of momentum. If the infalling velocities approach the speed of light, the energy becomes proportional to the momentum and thus also to $1/R$. Which term will be the bigger? The gravitational force depends on the mass of the star, while the electron degeneracy pressure does not. Thus, for a star of large enough mass, gravitation will always dominate, and collapse cannot be halted.

It was Chandrasekhar again who first analyzed this problem in detail, and he found that the star would stabilize at the white dwarf stage if its mass was less than 1.4 solar masses. This is now known as the "Chandrasekhar limit". Above it, the star would collapse beyond the white dwarf stage; this process would be sudden and violent and entail the explosive ejection of a large amount of matter. Our sun is below the limit, but even stars two or three times as massive may eject enough material on collapse to reduce the mass of the remnant below it to produce a white dwarf. The material ejected constitutes a nebula.

A white dwarf is a very dense object; so much so that the material of a normal star is compressed into a volume comparable to that of the earth. The sun, with a radius of about 700,000 km would, if collapsed to a white dwarf, be squeezed down into a volume with a 15,000 km radius. Whereas the sun has a density about 1.5 times that of water, the density of the white dwarf would be about 50 tonnes per cubic centimetre.

One should not make the mistake of thinking that a star too heavy to stabilize as a white dwarf will pass through that stage before further collapse; it will in fact collapse directly into something still smaller and more compact.

What mechanism may halt further collapse? Nucleon (i.e. neutron and proton) degeneracy pressure is the obvious one. The greater mass of these particles ensures that the degeneracy pressure is sufficient to check collapse only at much higher densities.

Could the electrostatic repulsion of protons do the trick? Stellar matter is, after all, electrically neutral, so electric forces cannot be a factor, except in the following sense. Continuing collapse forces both electrons and protons to higher densities, though the electron pressure is stronger. It becomes so strong that finally electrons and protons combine to form neutrons. Since neutrons are neutral, collapse is now limited by their degeneracy pressure. What remains is a very dense agglomeration of neutrons — a "neutron star" — an object whose radius is 5–10 kilometres and whose density is more than 100 million tonnes per cubic centimetre!

By the same argument that we used for the white dwarf, we know that, if the mass is large enough, even neutron degeneracy pressure can be overcome. In this case, nothing remains that can prevent collapse from continuing indefinitely. All matter is compressed into an object of infinitesimal size — a black hole. Neutrons become relativistic at energies some 2000 times higher than electrons; all the same, if the mass is great enough, the star cannot be stabilized as a neutron star.

Again, collapse causes violent ejection of enormous quantities of matter — possibly a large proportion of the total mass of the star. Only if the residue is heavy enough to overcome neutron degeneracy pressure will this ultimate collapse come about. Since some stars have masses more than 50 times that of the sun, some such black holes must exist. The critical mass for black hole production would be about 2.5 solar masses.

The foregoing picture was fully developed just before the Second World War by J. Robert Oppenheimer and two of his students. Oppenheimer, who later became director of the American nuclear bomb project, assigned the problem of the collapse of a star to the neutron star stage to George Volkoff. Volkoff was a Canadian immigrant from Russia, who later became a professor of physics at the University of British Columbia. During the war he worked on the Canadian nuclear project; afterwards, he returned to U.B.C. and became the first chairman of the department and then dean of science. Another student, Hartland Snyder, was given the problem of the ultimate collapse to singularity, the infinitely condensed object which we now know as a black hole.

At that time these problems were radical theoretical speculations with no known manifestations in the physical world. The solution to Einstein's equations describing a point mass, as developed by Schwarzschild, contained a mathematical singularity (a point or region at which a mathematical quantity becomes infinitely large) at a certain radius much smaller than that of real astronomical objects of the same mass. This singularity was quickly realized to be a mathematical artifact which could be transformed away by an appropriate change of coordinates, in the sense that there was no sudden physical "jolt" at that radius; an object could pass through it in a smooth and continuous way. However, it was curious that it marked the boundary of a region inside which the roles of space and time were mathematically reversed. Most physicists did not feel comfortable about this, and some doubt was expressed as to whether Einstein's equations actually described physical reality.

In any case, the work of Oppenheimer and his students was largely ignored; it was generally assumed that the ultimate fate of all stars, including our sun, was to become white dwarfs which would ultimately burn out.

The situation changed radically in late 1967. Jocelyn Bell, working for her doctorate under the supervision of Anthony Hewish at Cambridge, was observing radio signals from outer space with the aid of a vast array of antennas stretched out over a field. The project was to look for radio evidence of quasars, which had first been discovered several years previously.

What Jocelyn Bell found was something quite different — very regular and rapid pulses of unknown origin. The pulses from the first such object detected came at intervals of 1.3373109 seconds! That this can be cited to seven decimal places testifies to the extremely precise regularity of the signals.

The fact that the signals came from a fairly precise point in the sky signified that they were coming either from a stationary or a very distant source. The first wild guess was that we had finally discovered an alien civilization trying to make contact with other inhabitants of the universe. What made this implausible, however, was the very low informational content of the signal. In any case, by the end of the year three more such pulsating sources or pulsars were found, coming from quite different directions. Soon it became evident that what had been found, in fact, was a new sort of astronomical object.

It was clear that it could not be a white dwarf. No mechanism could be conceived by which a white dwarf could pulsate that fast. Various other hypotheses were also unsatisfactory. It was Tom Gold

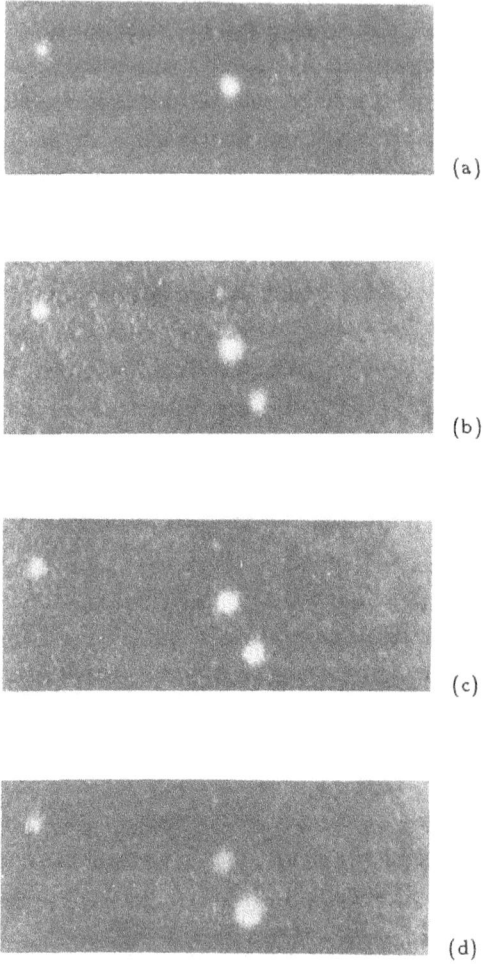

Fig. 8.1. Successive pictures of a pulsar (the lower object) through half a cycle. (Lick Observatory photographs.)

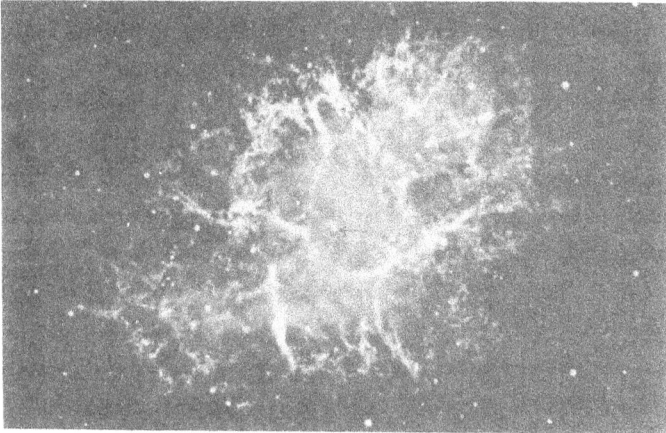

Fig. 8.2. Crab nebula (Arrow points to the pulsar) (Lick Observatory photograph).

at Cornell who put forward the idea that the pulsar in the Crab nebula was a rotating neutron star. The fact that it was in the centre of the Crab nebula was, to him, significant. When a star underwent collapse to a neutron star, it was to be expected that the "rebound" from the violent implosion involved would blow off a large fraction of the star's mass. In fact, ancient Chinese records related the appearance in the sky, in 1054, of an extremely bright "guest star", which gradually dimmed over the subsequent days. This is known to modern astronomers as a "supernova". In the same place in the sky we nowadays see the nebula known as the Crab (shown in Fig. 8.2). The Crab has been shown to be expanding at a rate of some thousand kilometres per second and now measures about 6 light years across. This is quite consistent with the assumption that the Crab nebula and the object noted in the Chinese records are in fact one and the same.

8.1. Pulsars (Neutron Stars)

What might be the properties of a neutron star?

Firstly, its density would be very nearly that of the matter in an atomic nucleus, that is, about 1.5×10^{14} g/cc — over 100 million tonnes per cubic centimetre (compared with a tonne or two per cubic centimetre in a white dwarf). A star must have a mass at least 40% greater than that of the sun to collapse into a neutron star. From this we can estimate its size; its radius is between 15–20 kilometres. The surface velocity at the "equator" of the pulsar is therefore 3 or 4 thousand km/s, which will clearly produce enormous strains in its interior. These can be sustained only by an equally enormous rigidity.

How did the rapid rotational velocity come about? The answer lies in the law of conservation of angular momentum. The part of the star's mass which is found in the pulsar will have decreased in volume during collapse by a factor of 10,000 or so. Since the angular momentum is proportional to $(\text{radius})^2/(\text{period})$, the rotational period after collapse will be something like 10^8 (a hundred million) times greater after.

Despite the fact that the rotation provides an extraordinarily accurate clock, the rotational rate does slow down very gradually — at the rate of 38 nanoseconds (38×10^{-9} second) or less than one part in a million million per day. There is, nevertheless, still a very large rate of loss of energy. Gold calculated this loss and found it to be just about the right amount to account for the luminosity of the nebula (or supernova remnant). The connection between the pulsar and the nebula thus seemed well established, and a coherent picture of the collapse and explosive release of the nebular material, as well as the evolution of the subsequent system, emerged.

8.2. The Crab Nebula and its Pulsar

Another important issue must be addressed: given that the pulsar rotates at a given rate, what is the origin of the periodic pulsation (presumably at the same period)? The source of the answer lies in its strong magnetic field.

The magnetic field of our sun is very weak (about 1 gauss).

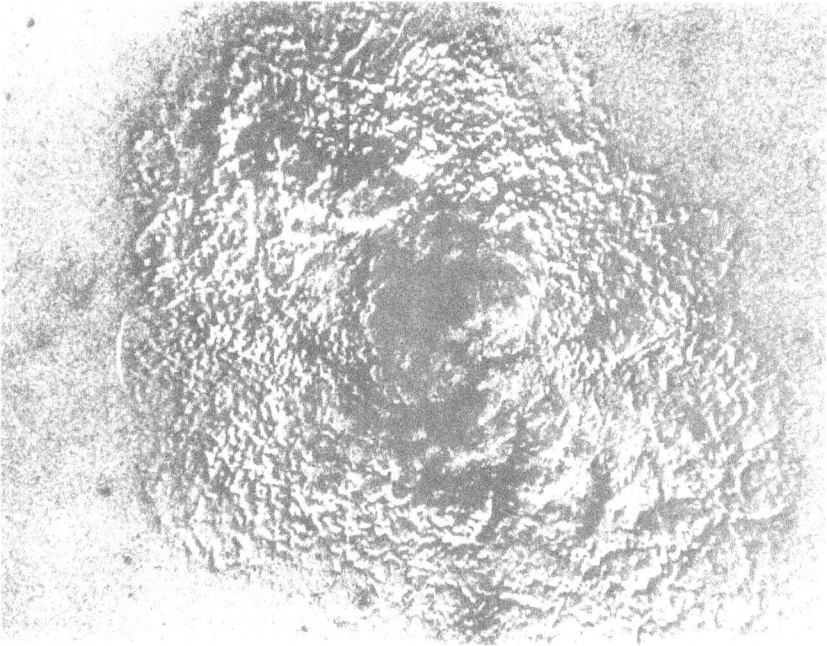

Fig. 8.3. This photograph was taken by superposing a *negative* of the Crab over a positive image taken 14 years earlier. In the absence of expansion, nothing should be seen. White regions are those which were not luminous at an earlier time and became so later. (Palomar Observatory photograph.)

Other stars however, might have fields ten or a hundred times stronger. The collapse of the star has radical consequences, which are best understood by using Faraday's image of lines of force. In the original state, the density of lines of force is small. We consider only that part of the star which is not ejected — the pulsar. When this material condenses to a radius of 10^{-4} or 10^{-5} of its original value, the lines of force are squeezed correspondingly, and this implies that the pulsar may have a magnetic field 10^8 to 10^{10} times that of the parent star. This compression of the lines of force is shown schematically in Fig. 8.4.

As seen in Fig. 8.4, the magnetic field becomes extremely intense at the two magnetic poles. Charged particles move along these lines

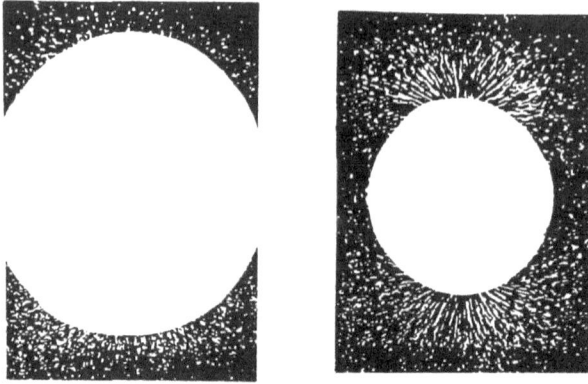

Fig. 8.4. Compression of lines of magnetic force under contraction.

of force, and also circulate around them.

When the field is very strong, these "orbits" are of extremely small radius; as a result the particles seem to follow the lines of force, and are attached to them (Fig. 8.5).

Suppose that the magnetic axis is not in the same direction as the axis of rotation. Intense beams of these particles will then be carried, at extremely high speeds, in circular orbits. According to theory, a very intense radiation will be emitted, predominantly in narrow beams about the magnetic axis. These will behave very much like searchlight beams. Their luminosity will be largely concentrated within 10 degrees of the magnetic axis. The source of the periodicity is then easily understood. Rotation of the pulsar causes the beam to sweep through space; if we are within the range of the beam we see two flashes per rotation. If we are outside, we will not see the pulsar at all. If the edge of the beam catches us, we will see a very brief flash; if we are near its centre, the flash will be longer.

The above model appears to suggest that pulsars should be associated with supernova remnants (SNRs) and vice versa. In fact, a rather small number of such associations have been observed. The reason for this is nevertheless fairly simply understood. As the rem-

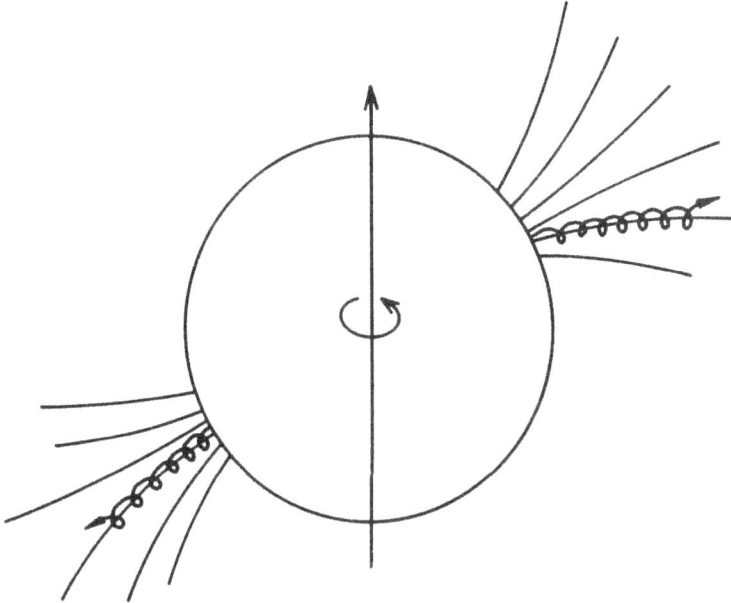

Fig. 8.5.

nants expand, they cool and become more diffuse. Thus, their visible lifetimes are expected to be only something of the order of 10–20 thousand years. From the rate of energy loss of pulsars, we estimate that they should radiate for one or two million years. Therefore, only a small percentage of pulsars should be surrounded by visible remnants. By the same argument, SNRs should accompany only the pulsars of a short period, since these are in general the youngest ones.

The pulsar creation rate in our galaxy has been estimated to be one every 5–30 years. This is consistent with our knowledge of the density and death rates of stars of mass greater than 2.5 solar masses. If a pulsar is created in our galaxy every 20 years, and their average age is a million years, our galaxy should have about 50,000. The number which can be detected is, however, much smaller, because of the beam-like character of their light emission and because many are hidden by other matter in the galaxy.

It should be noted in passing that pulsar radiation extends over the electromagnetic spectrum from the long wavelength (radio) end to the X-ray region. Gamma radiation is also present, though it gradually peters out for very high energies.

May 10, 1940   January 2, 1941

Fig. 8.6. (Palomar Observatory photographs.)

Because of the infrequent appearance of supernovae in our galaxy, attention has been turned in recent years to the search for them in other galaxies. Figure 8.6 shows an example in a galaxy in Coma Berenices. No pulsar has been detected in it, nor would one be expected to be seen at that distance.

We are left now with one more possibility for the ultimate fate of stars, a fate reserved to those stars which, even after the ejection of matter during collapse, are left with more than 2.5 times the solar mass. In such cases, further collapse should be sudden and complete, and leave behind a black hole. Since a substantial number of stars exist with masses 50 to a hundred times that of the sun, the search for such objects should be fruitful.

This is not the only context in which black holes should appear in nature. Before pursuing the question further, we shall consider some general properties of black holes.

Chapter 9

BLACK HOLES

Pierre-Simon Laplace, in 1798, gave intimations of a strange sort of object which we would today call the black hole. He wrote:

"A luminous star, of the same density as the earth, and whose diameter should be 250 times larger than that of the sun, would not, in consequence of its attraction, allow any of its rays to arrive at us; it is therefore possible that the largest luminous bodies in the universe may, through this cause, be invisible."

His insight was in seeing that gravitational fields could be so strong that nothing, even light, could escape from them. So far as the nature of the objects necessary to produce this effect is concerned, however, he got it quite wrong. What Laplace could not know was that such a mammoth object could not resist total collapse under its own gravitation. It would have the mass of nearly 80 million suns, whereas a mass of three suns is already enough to create a black hole. But he was correct in that if an object of a mass comparable to that which he envisaged could exist, and were confined within its prescribed dimensions, it would indeed have the properties of a black hole.

Let us recall the basis for the prediction of black holes. We saw that the gravitational redshift was given by the formula:

$$f = f_0 [1 - 2GM/c^2 R]^{1/2} ,$$

where f_0 is the light frequency emitted from the surface of a gravitat-ing object (with mass M and radius R), and f is the light frequency received at a distance where the gravitational field is negligible. The decrease of the frequency can be understood on the basis of Planck's relation between energy and frequency: $E = hf$. Light loses energy in pulling itself out of the gravitational field; it therefore suffers a proportional decrease in frequency. In the limit when $R = 2GM/c^2$, the frequency is shifted to zero, and no light energy escapes.

What is the value of this quantity $2GM/c^2$ for various values of the mass? (We shall call it the "black hole radius", though we must remember that it is not the radius of any material object, but only of the region from which light may not escape.)

Table 9.1.

Mass	Scale	"Radius"
6.7×10^8 tonnes	Mountain	10^{-13} cm
6.7×10^{13} tonnes		10^{-8} cm
6×10^{21} tonnes	Earth	1 cm
2×10^{27} tonnes	Sun	3 km
10^{37} tonnes	Galaxy	6 light days
10^{50} tonnes	Universe	16×10^9 light years

Let us look next at the strength of the gravitational force at the black hole radius (we shall call this the "event horizon"). We shall express it in terms of g, the gravitational force per unit mass at the earth's surface. There is a simple formula for this quantity: $GM/[2GM/c^2]^2 = c^2/2R$, where R is the black hole radius. This is inversely proportional to the mass, because the black hole radius is proportional to the mass. Thus, the smaller the black hole, the stronger the gravitational force at its surface is.

Physics: Imagination & Reality

Table 9.2.

Mass	"g"
10^9 tonnes	$6 \times 10^{30} g$
$M_\odot$	$3 \times 10^{12} g$
$10^{11} M_\odot$	$30 g$

$M_\odot$ is the mass of the sun.

It is interesting here to learn that, for black holes on a galactic scale, the gravitational forces are reduced more or less to the scale of our familiar world.

More interesting than simply gravitational fields, however, are differential or tidal effects. Thus, if we were near a black hole of a stellar mass (i.e. near its event horizon) which we approached feet down, we would feel a huge difference between the force on our feet and that on our head, large enough to pull us apart. We would also feel another effect: since everything would be being pulled toward the *black hole singularity itself*, the forces on our opposite sides would not be in the same direction; each has an inward component. For this reason gravitation would be squeezing us, as though we were being pushed down a funnel (Fig. 9.1). The two effects involve comparable "tidal" forces; lengthwise, about 2 million tonnes/cm^2 and sideways, 5 or 6 times less.

An object falling radially into a spherical black hole undergoes no discontinuity on crossing the event horizon. The time it takes to arrive at the singularity is fairly easy to calculate; it is $2R/3c$, where R is the event horizon radius. For a black hole of solar mass, that is a bit less than one ten-thousandth of a second. For one having the mass of a galaxy equal to that of 100 billion suns, the time is about 7.5 days; evidently, if one found oneself inside such an object, one could cruise around a while before the final crunch! Finally, it is amusing to see what comes out if one takes a mass equal to that of the universe. Of course this mass is not very well

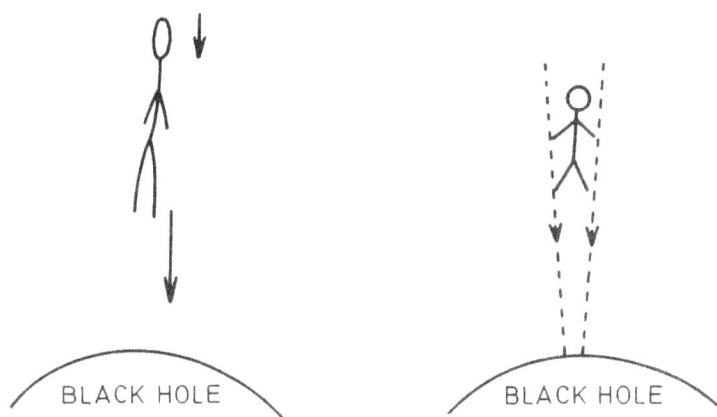

Fig. 9.1.

known; taking 10^{56} g, we arrive at about 10 billion light years. At first sight this seems quite extraordinary, since this is so close to the age of the universe, to the best of our present knowledge. On second thought, it is perhaps less surprising; since we believe that the universe was created in a singularity about that long ago, we can regard its subsequent evolution as the time-reversed process of the collapse to one. Such a way of reasoning enables us to answer another question – how can we talk about our universe in terms of a black hole, since it *has* no central singularity? To this we can now reply — ah, yes, but there will be one by the time we get there. We, and all else, will be crushed in it simultaneously!

Does it make any sense to think of the universe as a black hole? Not really, since we have defined a black hole as a closed-off region embedded in space, whereas our universe has been defined as being space. Nonetheless, one cannot avoid feeling that there is a message here, though we do not understand the language in which it is written.

9.1. Other Features of Black Holes

Stephen Hawking has proven a very interesting and general theorem about black holes, which has been given a whimsical name:

the "black hole has no hair" theorem. This theorem states that the black hole, considered as an object embedded in our universe, is completely characterized by at most three numbers, which specify respectively its mass (or energy), its angular momentum and its charge. In other words it can interact with anything else in the universe only through these three attributes; it exerts a gravitational attraction through its mass; it can exchange angular momentum, and it can exert electromagnetic influences. But it has no structure, no more detailed characteristics capable of manifesting themselves in the outside world.

At least in the astronomical sphere, the possibility of charge does not seem to be very interesting. From the very nature of charge and electricity, overall electrical neutrality is, in the long run, assured for all stable macroscopic objects. The existence of angular momentum, that is to say, of rotating black holes, is much more interesting. In fact, it is hard to imagine black holes without angular momentum.

Whereas a non-rotating (spherical) black hole is characterized by one critical surface, from the inside of which nothing can escape, a rotating black hole has several, as shown in Fig. 9.2. Let us for the moment, however, ignore the inner ones. Nothing may escape from the spherical surface between regions 1 and 2. It is an analogue of the "black hole radius" for spherical black holes. Its radius is diminished, however, by the presence of angular momentum having the value

$$r_s = R + [R^2 - (J/mc)^2]^{1/2} \ ,$$

where $R = 2Gm/c^2$ and J is the angular momentum of rotation. The region 1, between the foregoing surface and the elliptical one which forms its outer boundary, is known as the ergosphere. The outer surface is known as the "infinite redshift surface"; light from within may not escape to large distances ("infinity"), so that distant external observers cannot see anything that goes on inside. This does not mean that nothing can escape from the ergosphere; in fact, both light and mechanical objects may escape from it, while remaining gravitationally bound to the singularity. In fact, this possibility of escape provides a mechanism for extracting energy from a black hole. The process by which it does so was proposed by Penrose. The mechanism is this: A particle splits into two parts, one of which is ejected in a direction opposite to the direction of rotation of the hole, the other is ejected in the same direction. The first loses energy and is captured, resulting in a decrease in the angular momentum of the hole; the other has its energy increased, and escapes.

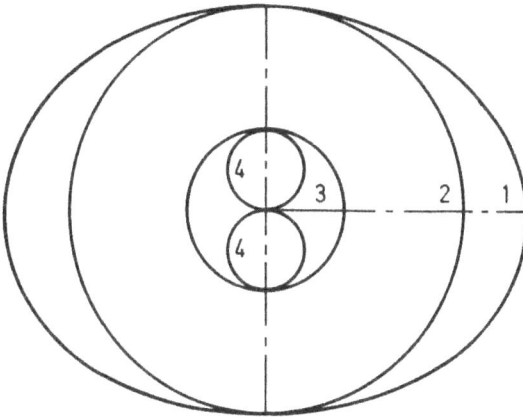

Fig. 9.2.

A second interesting property of the ergosphere is that anything within it is forced to rotate with the black hole itself; it may not come to rest, however much energy may be expended in the attempt. Even outside the ergosphere, the hole's gravitational field pulls objects around in its direction of rotation, but here, by expenditure of enough energy, the pull may be resisted and the object may circle in the opposite direction.

The inner surfaces are more curious. Region 3 is like an inner ergosphere; in it, all objects are again carried around with the angular motion of the hole itself, but are not drawn in inexorably to the singularity. But most extraordinary of all, in the regions marked 4, everything is again free to move in any manner whatsoever!

Of course one must be careful about these inner regions. We must remember that the gravitational field of the rotating black hole is that of a mass singularity, and not of any other mass distribution. Thus, the remarks about the inner critical surfaces make sense only if the moving objects have negligible gravitational fields. Physicists refer to them as test particles.

9.2. Another Stephen Hawking Theorem

No physical process can decrease the surface area of a black hole. Thus, if two black holes were to amalgamate, the surface area of the resulting hole would have to be greater than the sum of the areas of the original ones. (By "surface area of a black hole", we mean the surface area of the event horizon — that surface from the inside of which nothing can escape.)

This theorem is really more general than it first appears. No special assumption is made about the detailed nature of the process; for example, radiation might be emitted in the process of combination. Consider the case of two spherical (Schwarzschild) black holes of masses M_1 and M_2. Their surface areas are respectively $4\pi[2GM_1/c^2]^2$ and $4\pi[2GM_2/c^2]^2$; that of a black hole containing all the energy of the two components is $4\pi[2G(M_1 + M_2)/c^2]^2$. This is clearly bigger than the sum of the other two areas. But what the theorem really states is that there is an upper limit to the amount of energy which can be extracted by the process of amalgamation. The minimum mass that the residual black hole can have is M_0 where M_0^2 must, by the area theorem, be bigger than $(M_1^2 + M_2^2)$. Thus, the minimum value M_0 can have is $[M_1^2 + M_2^2]^{1/2}$, and the greatest amount of energy which can be extracted is $M_1 + M_2 - [M_1^2 + M_2^2]^{1/2}$. As a fraction of the total energy of the system, this is

$$f = 1 - [M_1^2 + M_2^2]^{1/2}/[M_1 + M_2] \ .$$

The second term has its minimum value when $M_1 = M_2$, in which case $f = 1 - (1/2)^{1/2} = 1 - 0.707 = 0.293$. The maximum percentage of the energy which can be extracted is 29.3%.

The "area theorem" played a very important role historically in the formulation of black hole thermodynamics. We shall discuss this in the last chapter. Since the basic law of thermodynamics is that, in the interaction of two systems, the entropy never decreases, Jacob Bekenstein thought that there might be a connection between the entropy and the surface area of a black hole. Subsequently, using more fundamental arguments, the entropy was found to be proportional to the surface area. Thus, the extension of the second law of thermodynamics to include black holes necessitated the area theorem.

9.3. Black Holes and the Quantum Theory

In classical, non-quantum relativity, the black hole is characterized by a mathematical singularity: all matter is concentrated at a geometrical point. But quantum mechanics does not permit this; the uncertainty principle tells us there is a minimum dimension into which one can squeeze a finite amount of energy.

What is required is that the black hole radius $2GM/c^2$ be large enough to accommodate the spreading of the energy inside it due to the uncertainty principle. By the uncertainty principle that spread cannot be less than $\hbar/2p$, where p is the momentum to be confined. Under relativistic conditions the energy-momentum relationship is $p = E/c = Mc$, so we must have

$$2GM/c^2 > \hbar/2Mc \; ,$$

from which we conclude that M must be greater than $\frac{1}{2}[\hbar c/G]^{1/2}$. We will designate $\frac{1}{2}[\hbar c/G]^{1/2}$ as the "Planck mass". Its significance is that, for objects below this mass, classical relativity will not suffice; one must take account of the quantum nature of matter. Above it, the black hole will be large enough that the quantum uncertainty can be accommodated within the event horizon, where everything is hidden from the view of the outside world.

If a quantum black hole has a minimum mass, then there is also a minimum length, which is the radius of its event horizon.

Putting the Planck mass $\frac{1}{2}(\hbar c/G)^{1/2}$ into the formula for the black hole radius, viz. $2GM/c^2$, we get the "Planck length"

$$L_{\text{Planck}} = (G\hbar/c^3)^{1/2} \ .$$

The significance of this length is the following: at distances smaller than this from a particle, which would be represented classically by a point singularity, quantum effects cannot be ignored, i.e. the "particle" must not be represented by a point at all, and the "singularity" presumably does not exist.

Finally, one can also specify a minimum time interval over which quantum effects may be ignored. This may be obtained in two ways. On the one hand, the quantum uncertainty principle (see Chap. 13) says that the minimum time for defining an energy E is $\hbar/2E$, which gives a time

$$T_{\text{Planck}} = (G\hbar/c^5)^{1/2} \ .$$

Another interpretation of the same time interval is that it is the time taken by light to travel the Planck distance. Again, the Planck time is the minimum time interval over which quantum effects are not significant.

Let us note the numerical values of the various critical parameters:

Planck mass $= 2.18 \times 10^{-5}$ g
Planck energy $= c^2 \times M_{\text{Planck}} = 1.96 \times 10^{16}$ ergs
$\qquad\qquad$ or 1.22×10^{19} GeV (billion electron volts)
Planck length $= 1.62 \times 10^{-33}$ cm
Planck time $= 5.39 \times 10^{-44}$ s

One of the consequences of the existence of the Planck mass is that the fundamental particles of physics cannot be black holes; their masses are too low.

All above considerations have a negative character. What we have done is to establish the limits of our capacities. In fact, there is no quantum theory of gravity, despite many years of efforts to find one. We therefore cannot say anything convincing about what goes on below the "quantum limits" that we have defined. We only know that non-quantum relativity is of no use to us there.

So much for black holes as mathematical objects. A more important question remains: given that black holes are theoretically possible within the framework of the general theory of relativity, do they in fact exist in the real world of astrophysics?

Chapter 10

BLACK HOLES IN ASTROPHYSICS

10.1. Black Holes in Stellar Collapse

A star whose core remains too massive to be stable as a neutron star, even after it has lost a substantial part of its outer layers in the violent implosion involved in collapse, seems to have no other possible fate than to finish its life as a black hole. We know that very heavy stars do exist, so such objects should exist. But how might one go about searching for them? By definition, black holes themselves are invisible; all we can hope to see, are the effects of their gravitational fields. These effects will be most evident when those fields are strong, i.e. close to the event horizon. But we know that a large proportion of stars, perhaps as many as half, have companions; they exist in double, and occasionally even multiple star systems. It is logical, therefore, to look for black holes in such systems.

If a double star has an invisible member (which *may* be a black hole), what will be observed is the strange motion of its visible companion, a motion in a closed orbit, with no direct indication of its cause. The first such object identified was a star named Cygnus X1. A supergiant blue star, HDE 226868, showed periodic Doppler shifts in its spectrum. A star circling around another object will be moving toward us part of the time and away from us part of the time. These changes take place in a regular and periodic way. The period of HDE 226868 was 5.6 days, and it emitted very intense X-rays, as well as radio waves. Its distance has been estimated as at least 2.5 kiloparsecs (8150 light years); from this, its intrinsic luminosity can be estimated. Knowledge of its intrinsic luminosity and of the nature of its spectrum enables us to deduce that it is a very large, hot star. By virtue of the relation between mass and luminosity (remember that massive stars burn hotter than lighter ones), its mass has been estimated as about 30 times that of the sun.

Finally, then, the mass of the invisible companion in the double-star system can be estimated to be about six solar masses (see Appendix). Thus it cannot be a pulsar; it must be a black hole.

Other data give more detailed clues as to what is going on in this system. It is a strong X-ray source. What is the mechanism for its X-ray production? Certain Doppler-shifted emission lines indicated that gas was flowing from the visible star to its invisible companion. Finally, the X-ray output has been observed to be flickering with variations as short as a millisecond, indicating, by an argument which we shall develop below, that the source must be as small as a light-millisecond or less. This is about 300 km. The black hole radius for a mass six times that of the sun is 17 km. The material which is emitting the radiation is outside that radius.

The radius of the large star might be of the order of 2 million km, while the distance between the stars would be about 37 million km if the orbit were circular (remember that the distance of the earth from the sun is about 150 million km!). The outer envelope of the star will be gaseous and at high temperatures; it will be acted on by very strong tidal forces due to the black hole.

The currently favoured model, then, which provides a satisfactory explanation of these observations, is that gas streams from the supergiant star. This gas is concentrated near the plane of the orbit, forming an "accretion disk" spiralling around the black hole. If the matter in the accretion disk consisted of macroscopic particles rather than gas, these particles would be expected to orbit around the black hole, and would not be drawn into it. We noted, however, in the Appendix, that the velocity of orbiting particles was inversely proportional to the square root of their distance from the attracting object; thus, the matter further away is moving more slowly than

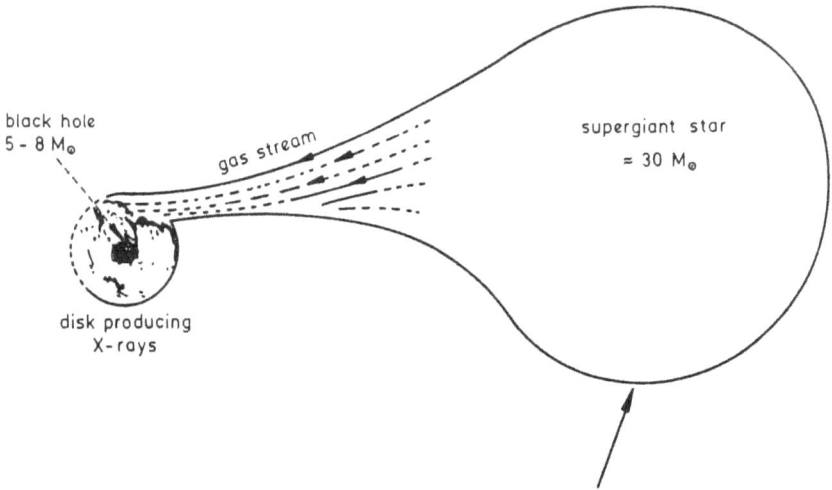

black hole
5 - 8 M$_\odot$

gas stream

supergiant star

$\approx$ 30 M$_\odot$

disk producing
X-rays

Fig. 10.1.

that further in. In Fig. 10.2, we portray schematically the orbit-ing matter in the accretion disk. If we focus on adjoining shells of matter, that in the outer shell is moving more slowly than that in the inner one. But let us now take into account that the matter is gaseous; this implies that the individual molecules of the disk are moving randomly at high speed while, in bulk, they sweep around more slowly in orbits (in a windstorm, the speeds of random mo-tion of the oxygen and nitrogen molecules of the air will be moving at several thousand km/h, while their "transport" velocity in the direction of the wind would be less than 100 km/h in strong wind).

As a result of these rapid random molecular motions, there are constant molecular collisions. In particular, collisions between molecules in adjoining shells will create a viscosity between them, which will tend to slow down the faster molecules close in, and speed up the slower ones further out. Thus a loss of energy takes place as the molecules spiral in, and this creates a continuous flow into the black hole. At the event horizon, the molecular speeds will approach

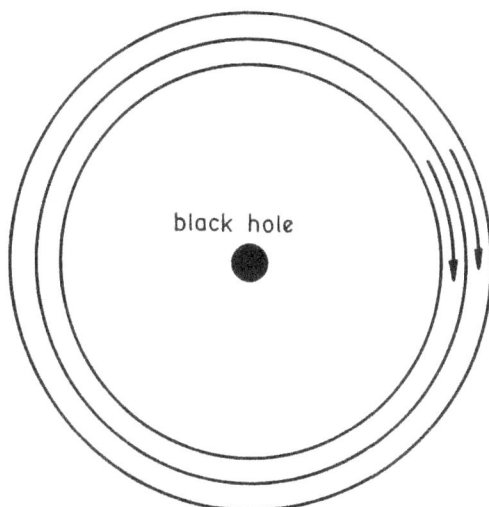

Fig. 10.2. Random collisions due to thermal motions cause inner rings to lose energy at the expense of outer ones, and material spirals in!

the speed of light. At these high energies the gas will be completely ionized, and will emit very intense high-frequency radiation (X-rays).

The study of such systems, which may be characterized as "mass-transfer binaries", offers the best prospects for discovery of further stellar-collapsed black holes.

10.2. Quasars and Exploding Galaxies

We have all seen photographs of spiral galaxies; several are shown here for illustration (Fig. 10.3). M74 shows itself almost face on; M31 in Andromeda almost sideways. We use M81 to give a rough idea of where we, and our sun, are in our own galaxy which, as far as we know, is quite similar to M81 (the position is at the intersection of the cross-lines, in a spiral arm at about 10 o'clock from the centre of the galaxy).

Physics: Imagination & Reality

(a) Galaxy M 74 (Palomar Observatory photograph).

(b) The Andromeda Galaxy M31 (Lick Observatory photograph).

(c) Galaxy in Ursa Major, M 81 (Palomar Observatory photograph).

Fig. 10.3.

The distances and redshifts of these galaxies are the following:

galaxy	distance (light years)	redshift
M74	20 million	0.001
M31	2.2 million	0.00011
M81	7 million	0.00035

These are relatively close galaxies with small redshifts.

Now look at three other objects, which do not show much structure. The first of these, known as Cygnus A, was discovered in the 1950's by radio astronomers. What is startling about it is its redshift, which is 0.057. This places it at a distance of about a billion light years. More astonishing still is that despite its great distance, it is one of the strongest radio sources in the sky. Its radioluminosity is about 100 million times that of the near galaxy M31. It is quite an extraordinary object!

In 1963, Maarten Schmidt and his collaborator Jesse Greenstein studied an object known as 3C273, which had already been identified by Alan Sandage as a new sort of strange object which came to be known as a quasar. 3C273 had at first been assumed to be a star, though its spectrum was not recognizable. On closer observation, however, he found that a thin jet protruded from it. A similar phenomenon had been observed in the case of M87, a galaxy which has a very strong radio source. It therefore appeared that 3C273 might also be not a star, but a galaxy, in which case it would be at a much greater distance than had been assumed. Schmidt measured the spectrum and found that it "made no sense". He looked at it in frustration for a long time before he was able to recognize four familiar hydrogen lines, shifted by an unbelievable 16%. For this implied that it was at a distance of 3 billion light years! He explained his discovery to Greenstein, who wondered whether this could be the answer to another puzzle, that of the quasar 3C48, which also had an incomprehensible spectrum. Looking more closely, the same lines were found, this time with a still larger redshift of 37%, which put it at some 5 billion light years away.

Cygnus A, 3C273 and 3C48 are shown in Figs. 10.4 (a), (b) and (c) respectively. The astronomical world was stunned. Never before had such huge redshifts been measured; never had one seen objects at such incredible distances. But most astonishing of all were its implications, for if these quasars were at such enormous distances, they must be radiating almost inconceivable amounts of energy to be seen at all. Their energy output would have to be at least that of a thousand billion suns, or ten galaxies. Subsequently, quasars have been found which radiate, in X-rays alone, 10^{47} egs/s, as much as 10^{14} suns radiate at all energies, that is, they have the brightness of more than a thousand galaxies. The total mass energy of the sun, its Mc^2, is about 2×10^{54} ergs; if this much energy were radiated in a year it would be equivalent to 6×10^{46} ergs/s. Thus, the energy output may be equivalent to the total energy of several suns per year.

But bigger shocks were still to come, for it was soon discovered that there were quasars which fluctuated in brightness over periods of weeks, or days, and sometimes even hours. This implied that these very distant objects, with their enormous luminosities, must in addition be very small; in fact their dimensions must be measured in light-weeks, or light-days, or even light-hours. One had started with the idea that they were like galaxies, but galaxies were normally a hundred thousand light-years across, five million times bigger than a quasar whose dimensions were a light-week. But what conceivable mechanism could produce the energy output of a thousand galaxies in a space millions of times smaller than a galaxy? Many astronomers felt that the very foundations of physics had been shaken. Might we be on the threshold of a startling new revolution in physics?

Before addressing this question, let us see why luminosity fluctuations of quasars over short periods of time imply that they have small dimensions.

In Fig. 10.5 we schematically show the crux of the argument. The sphere represents a luminous object. Suppose that its radius is a light week. Imagine that for one day its total light emission

(a) Cygnus A (Palomar Observatory photograph).

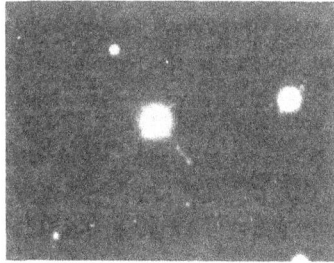

(b) The Quasar 3C 273 (National Optical Astronomy Observatories).

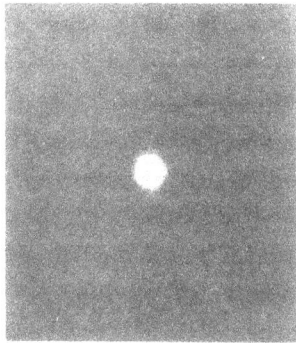

(c) The Quasar 3C 48 (Palomar Observatory photograph).

Fig. 10.4.

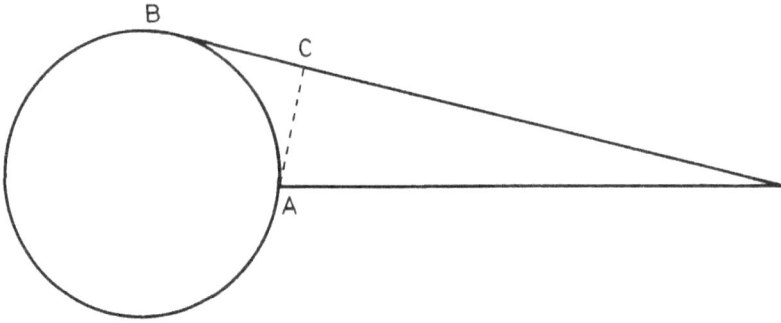

Fig. 10.5.

increases by 20%, and that the light coming from point A arrives
at our telescope at noon on a Saturday which is the first day of the
month. However, we are then receiving the greater light intensity
only from that point. Since it will take light a week to go from B to
C, the end of the flash from B will reach us only 8 days later, that is,
at noon on Sunday the eighth. At noon on Sunday the second, the
light emission at A will revert to normal, but from then until a week
later we will be receiving increased light intensity from one seventh of
the emitting region. In short, the increase in luminosity which took
place over a day at its source will only reach us over eight days, and
will on the average be less than 3% rather than 20% because at any
given time we will only be seeing the increase from one-seventh the
distance BC. Thus, the observation of the total extra light emitted
is spread over 8 days, at an intensity which builds up over the first
day, to nearly 1/7 of all the extra intensity, then diminishes as we
get light from areas which are more and more oblique to us. Finally,
even this diminished intensity dies out more rapidly on the last day,
as even the emitting area which we "see" decreases linearly.

Figure 10.6 shows the observed amplitude of a periodic variation
of extra luminosity as a function of that at the point of emission.

It shows the effect of "averaging" due to a variable time delay
of signals from different parts of the emitting object as a function of
the ratio of its size to the period of fluctuation. When the size in

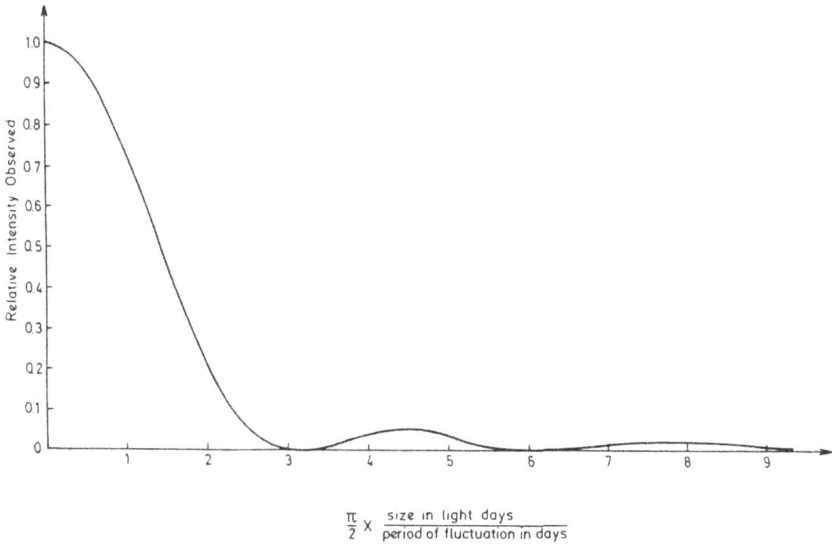

Fig. 10.6. If the size of the emitting region were much bigger than the period of fluctuation, we would only receive a very weak signal.

light days becomes twice the period of fluctuation in days, evidence of the fluctuation has effectively disappeared. Of course, the situation we describe here is purely illustrative; the variation over time of the fluctuations in emission intensity may take any form whatsoever. Qualitatively, however, the effect will be similar; the extra luminous intensity that we see at any time is diminished because the light emitted in a short period (in our example, a day) is observed only over a longer period, the time it takes light to travel across the dimensions of the emitting object. The fluctuation in observed luminosity will thus be "averaged out".

The following analogy may make the phenomenon easier to understand. The speed of sound in air is about 1130 ft/s (340 m/s). Suppose that the musicians of a symphony orchestra were spread out randomly over an area of several thousand feet, and we listened from a distance (we will assume that the musicians are kept in time by electromagnetic communication with the conductor). We would receive the sound emitted by the nearest members of the orchestra at one time, and the sound emitted by the most distant ones several seconds later, since these sounds take longer to reach us. Thus the musical message would be badly blurred; it would sound more like noise than music. Suppose that the musicians were cued to move toward a central point while continuing to play. As they did so, the musical sound would become more and more coherent; it would again become music. (This effect presented a very real problem at a special concert in honour of the 800th anniversary of Notre Dame de Paris a few years ago; the program featured simultaneous participation of two organs at different places in the cathedral, a symphony orchestra and massed choirs. Coordination for recording was achieved electronically.)

Let us now summarize our observations on quasars. First, their large redshifts prove that they are at enormous distances from us. Secondly, the brightness of the light signal which reaches us from these great distances means that they must have an intrinsic brightness equivalent to as much as that of thousands of galaxies. Finally, the variation in light output in short periods of time (from hours to weeks) tells us that these extraordinarily luminous objects are astonishingly small; light days or weeks, compared with 100,000 light years for typical galaxies.

Theorists were presented with a problem which seemed insoluble in their framework. It took them nearly 15 years to come up with a plausible explanation.

10.3. Unravelling the Puzzle of the Quasars

The size attributable to quasars was clearly so small as to suggest a black hole. As we showed earlier, a black hole with the mass of a galaxy of 10^{11} suns would have an event horizon of a radius of about 10 light days, which is comparable to the dimensions of a typical quasar. Could the quasar be a collapsed galaxy, or at least a collapsed galactic core? As noted in our earlier discussion of "missing mass", many galaxies appeared to have cores of extremely high density, so that the hypothesis that they might collapse to form massive black holes is not outrageous. But of course there is a problem with this notion: how could there be such stupendous energy output from a black hole, or rather, from its immediate neighbourhood? Only at a conference in the summer of 1977 did a consensus begin to emerge in favour of a model that had been the focus of attention of several groups, whose calculations had shown that it could account for the energy production of quasars.

Imagine a supermassive rotating black hole (of a billion or so solar masses) surrounded by an accretion disc of rotating gas (Fig. 10.7). Close to the black hole the particles of gas will be moving at a rate which is a significant fraction of the speed of light; the gas will be very hot and, as a consequence, ionized. It will produce an enormously strong magnetic field perpendicular to the plane of the disc. When jets are associated with quasars, they may be attributed to explosive processes very close to the black hole which send matter flying into space; according to theory the ejected matter should be strongly concentrated in this direction.

One interesting feature of this model is that as the black hole "swallows up" the surrounding gas, it increases in mass. But in our discussion of black holes we saw that an increase in mass implied a decrease in the gravitational force per unit mass at the event horizon surface $[GM/(2GM/c^2)^2 = c^4/4GM]$. For the black hole to swallow a whole star, it would not give rise to the emission of much energy. Imagine, however, a star near the black hole, and orbiting about it (the density of stars in the core of galaxies is very high). The black hole will exert very strong tidal forces on it, pulling off the star's outer layers. This will create a gas whose absorption will be accompanied by the emission of extremely intense radiation. But as the black hole grows, its tidal force will diminish, and one would expect it to dim.

The tidal force depends on the difference of the gravitational

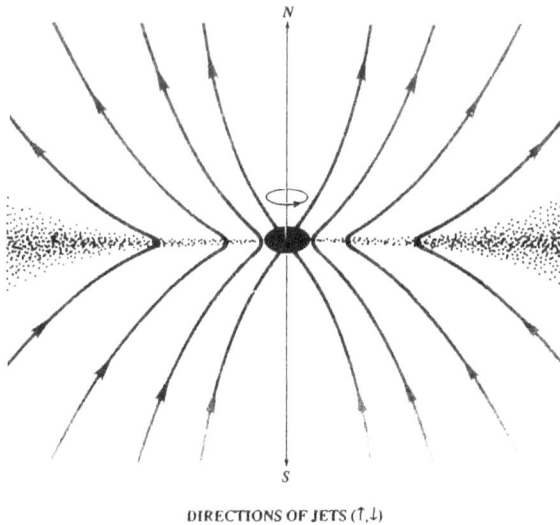

DIRECTIONS OF JETS (↑,↓)

Fig. 10.7. (From *Galaxies and Quasars* by William J. Kaufmann III. Copyright ©1979 by W. H. Freeman and Co. Used by permission.)

force of the hole on opposite sides of the star. Let the mass of the black hole be designated as M, and suppose that the diameter of the star is $2d$. The difference in gravitational force per unit mass on the two sides of the star is then $(GM/R^2)(2d/R) = 2GMd/R^3$ where R is the distance of the star from the black hole. This may be thought of as the force trying to pull the star apart. The gravitational force on the other hand is trying to hold it together; this is, to within a numerical factor, Gm/d^2, where m is the mass of the star. The star will be completely torn apart at the surface of the event horizon of the black hole if $c^6/4G^3M^2 > m/d^3$. For this to happen, the mass of the black hole must be less than $(c^6d^3/4G^3m)^{1/2}$. This can also be written in the alternative form $(M/m) < 2(d/r_o)^3$, r_o being the black hole radius of the star. This limits M to about 1.5×10^8 solar masses, if we take m to be the mass of the sun.

Thus a black hole whose mass is less than that of 100 million suns will be able, by tidal force, to completely tear apart stars in its vicinity; the mass of these stars, being reduced to gas, will then feed the accretion disc from which the quasar draws its energy. When the black hole becomes bigger than this, gas will still be pulled off neighbouring stars by tidal force, which however continues to weaken as the black hole continues to grow.

This conclusion strengthens the hypothesis that quasars are, at their core, black holes, because one of the interesting features of the study of quasars is that they are always found at great distances, more than a billion light years away. They are a phenomenon of the distant past of the cosmos. Why are there no recent ones? Presumably none of the quasars that have ever existed have ceased to exist; they have just lost their energy sources and so become invisible. A black hole without its accretion disc of hot plasma will not be visible; it will have become too massive to continually provide its own energy supply, or will have exhausted all the gaseous matter in its surroundings.

How far back do quasars go? The following table lists the more distant ones.

Table 10.1.

Quasar	Redshift	Speed	Distance (light years)
Cygnus A	5.7%	17,000 km/s	1 billion
3C273 (Schmidt)	16%	45,000 km/s	3 billion
3C48 (Greenstein & Matthews)	37%	1/3 c	5 billion
3C147	55%	0.41 c	7 billion
3C9 (Schmidt)	2.01	0.8 c	14 billion
OH471	3.4	>0.9 c	16 billion

The last figure is close to current estimates of the age of the universe! Note, however, that we are now estimating distance purely by means of the cosmic redshift. If the deceleration of the expansion of the universe is really manifesting itself at these distances, this method may lead us to overestimate those distances.

A study by Schmidt and Bello, an account of which was given

in *Scientific American* ("The Evolution of Quasars", May 1971), clearly established that quasars become more numerous, at a rapid rate, as we go back in time beyond a billion years. However, beyond a redshift of 3, they rapidly disappear. They seem to be a phenomenon of a certain early stage of cosmic evolution, and since that time they have been gradually losing their luminosity, the last of them disappearing from sight some billion years ago. (We shall see later, however, that black holes at the centres of some nearer galaxies may still be observed.) Because it is a particularly nice example of scientific deduction, we shall describe this study in a bit more detail.

10.4. The Distribution of Quasars

Schmidt and Bello based their work on a study of one thousandth of the celestial sphere. In this region, they looked for quasars of the 18th astronomical magnitude (a change of astronomical magnitude of one unit corresponds to a change of brightness of a factor of about 2.5. A change of magnitude of 5 represents a change of brightness of 100). In their section of the sky, they found 20 quasars; assuming an isotropic distribution, this would correspond to 20,000 in the entire sky.

18th magnitude quasars have the same apparent brightness, but they are all at different distances, and so represent objects of quite different intrinsic brightness. Their distances can be determined by their redshifts. The region of the sky being examined can be divided into successive shells of distance, chosen in such a way that if a quasar of magnitude 18 in the fourth shell were moved in to the third shell, it would appear to be of 17th magnitude (i.e. two and a half times brighter). These shells may be characterized, not by the radii of their inner and outer limits, but by the range of redshifts which they represent. The distribution of various quasars among the shells may now be determined, as shown in Fig. 10.8a, which shows the distribution of the 20 observed quasars in the shells; we shall number the shells outward, the innermost occupied one being

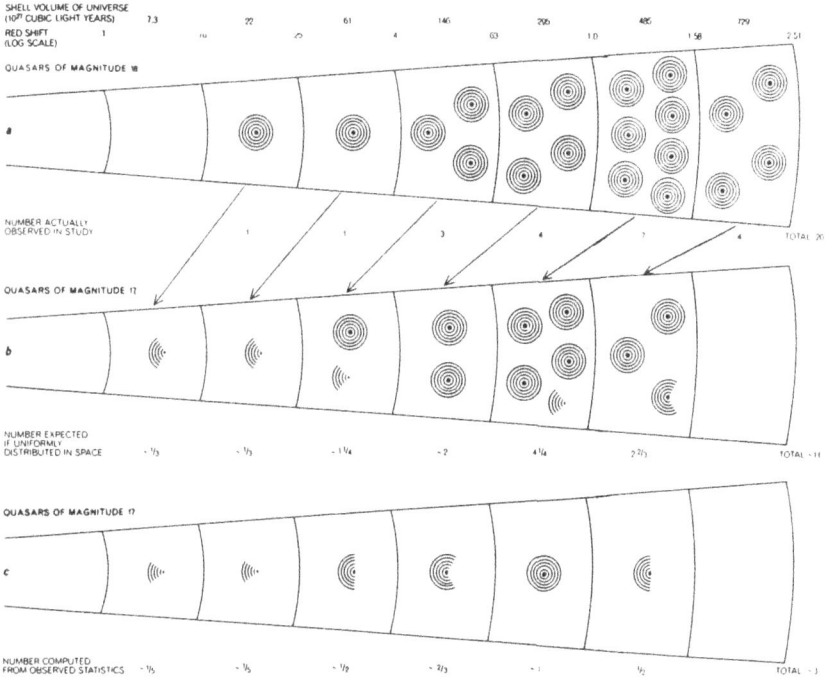

Fig. 10.8. (This figure and Table 10.2 from *The Evolution of Quasars* by Maarten Schmidt and Francis Bello, *Scientific American*, May 1971. Copyright ©1971 by Scientific American, Inc.)

designated I, the outermost VII.

First, we will deduce how many 17th magnitude quasars there should be in the various shells, *on the assumption that the quasars are uniformly distributed*. On the basis of this assumption, the number of 17th magnitude quasars in shell "n" should be the number of 18th magnitude ones in shell "$n + 1$", reduced by the relative volumes of the shells; thus, $1000 \times 7.3/22 = 332$. Figure 10.8b shows the numbers deduced in this way. Adding up the number of the 17th magnitude quasars in all shells, we obtain 10,842. However, only about a third as many are found. When these are distributed among

the shells, the distribution is given in Fig. 10.8c. We now look at
the ratio of these numbers to see what would be expected if the
distribution were uniform. The results are shown in Table 10.2. If
we assume that the density of quasars increases with redshift as
$(1 + Z)^6$, the densities given in the "ratio" column in the table are
quite well reproduced, as shown in the last column.

To illustrate this, consider shell III. The number of 17th magnitude quasars expected in this shell on the assumption of a uniform
distribution is 3000 times the ratio of the volume of the third shell
to that of the fourth. This is $61/146 \times 3000$ or 1253. But the number obtained from the empirical formula is less by $(1.325)^6/(1.515)^6$
$= 0.45$; multiplying this by 1253 gives 560, which agrees with the
number observed. (0.325 is the average value of Z in shell III; 0.515
is that in shell IV.)

We can make other interesting deductions. For example, let us
ask how many 19th magnitude quasars we should expect in shell V.
There are 3000 of 18th magnitude in shell IV. Since the volume of
shell V is $295/146 = 2.02$ times that of shell IV, the hypothesis of
uniform distribution would give 6060. But the density of quasars in
V is expected to be greater than that in IV by $(1.815)^6/(1.515)^6 =$
2.96, so we deduce that the number expected is $2.96 \times 6060 = 17.940$!

The empirical $(1 + Z)^6$ law represents a very rapid increase of
quasar density with distance, and hence a very rapid decrease with
time since the early stages of cosmic evolution. This suggests that
the number of "extinct" (i.e. presently black and invisible) quasars
vastly exceeds the number of visible ones. All the same, if we go back
far enough, no more are found. Their formation was predominantly
a phenomenon of the early universe, yet can only have post-dated or
at most been simultaneous with the formation of the galaxies.

10.5. Are There No Closer Galactic Black Holes?

That black holes may exist at the centres of galaxies much closer
to us than the distant quasars can be illustrated by citing one example, which has been thoroughly studied and seems to be very well

Table 10.2.

Shell	Vol.	Z Range	N_{18}	N_{17} Calc	N_{17} Obs	Ratio	Density corr.
I	7.3	0.10–0.16		332	219	0.66	0.68
II	22	0.16–0.25	1000	361	202	0.56	0.57
III	61	0.25–0.40	1000	1253	560	0.45	0.45
IV	146	0.40–0.63	3000	1980	660	0.33	0.34
V	295	0.63–1.0	4000	4249	1026	0.24	0.25
VI	486	1.0 –1.58	7000	2667	503	0.19	0.18
VII	729	1.58–2.51	4000				

The second column gives the volume of the shell in 10^{27} light years.
Z is the redshift.
N_{18} is the estimated number of 18th magnitude quasars in the whole sky; N_{17} is the same for 17th magnitude.
"Calc" is the number of quasars of the indicated magnitude calculated on the assumption of a uniform distribution.
"Obs" indicates the number actually observed.
"Ratio" is the ratio of the two preceding columns.
The last column is the ratio of the density to that for a uniform distribution.

understood. The galaxy is called M87 and is a part of the Virgo cluster, at a distance of some 60 million light years. It is an unusual galaxy. As long ago as 1918 it was noted to have a brilliant jet of matter shooting out from it. It is a supergiant elliptical galaxy with a very bright nucleus, and in some respects it is similar to the quasar 3C273 (Fig. 10.4b).

The results of two studies, published in 1978, gave information, through investigation of the distribution of brightness across the galaxy, of the corresponding stellar distribution. At the same time, spectroscopic studies based on the observation of Doppler shifts revealed the distribution of stellar velocities as a function of distance from the galactic centre. A theoretical model of galactic motion, based on the same dynamical principles as are used for modelling the solar system, was worked out by I. R. King in 1966. When the results of the observations on M87 were analyzed in relation to this model, a serious disagreement was found. This is illustrated

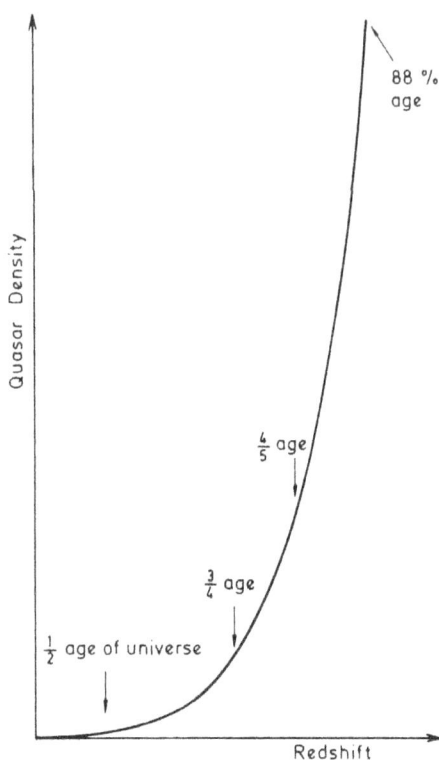

Fig. 10.9.

in Fig. 10.10. Beyond a certain distance from the galactic centre, the facts fitted the model quite well, but in the central region, the brightness was found to rise well above the theoretical expectation. The only interpretation consistent with the observations was that there was a supermassive black hole at the galactic centre, with a mass of approximately 5 billion suns! This hypothesis explained the anomalously high velocities of the stars near the centre of the galaxy, and was also consistent with the existence of a magnetized accretion disk and the jets which would be expected to be emitted in the direction perpendicular to this disk. The existence of a large, dark mass at the core of the galaxy explained a mass/luminosity ratio ten

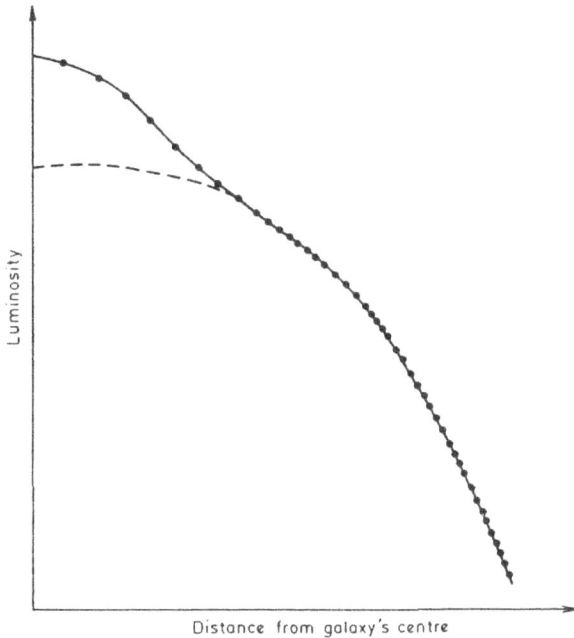

Fig. 10.10. The luminosity profile of M87. (From *Galaxies and Quasars* by William J. Kaufmann III. Copyright ©1971 by W. H. Freeman and Co. Used by permission.)

times greater in the core than in the outer regions.

Although the case of M87 is particularly dramatic, evidence suggests that black holes exist at the centre of many galaxies, including our own. Since we lie within the central disc of our galaxy, and are nearly 30,000 light years from the galactic centre, we cannot hope to see visible light from that centre. However, radio waves can penetrate the gas and dust which fills the greater part of the space between. Thus, radio astronomy takes a "look" at the inner parts of the galaxy.

In our vicinity, stars are relatively scarce; there is about one star for every 400 cubic light-years. Moving toward the centre, we would pass through two spiral arms, 6 thousand and 14 thousand light years further in. These arms have a considerably higher density of stars.

At about 10 thousand light years from the centre, we would be approaching the core region, with billions of closely packed stars. In this inner region, increasingly dense clouds of gas and dust would float among the stars. At 100 light years from the centre, the galaxy's nucleus would radiate with the intensity of sunlight on earth, but between us and the core would be gas and dust illuminated by over 50 million stars lying within a radius of 30 light years. At 3 light years, the 400 cubic light-year zone, which at the sun's actual distance would on the average only contain one star, would now contain 2 million stars buried in swirling clouds of dust.

Finally, at one-tenth of a light year from the centre the heart of the galaxy would appear as a monstrous disk-like vortex of matter whirling in toward absorption in a great black hole of a mass of several million suns. The black hole might have a radius of about 10 million km, or about one half of a light-minute.

This black hole is much smaller than the one at the centre of M87, but M87 is a quite different sort of galaxy, a much denser giant elliptical one. Such galaxies are thought to have arisen from collisions of spiral galaxies, creating much debris which would be accreted to make much larger central black holes.

All this, however, is descriptive astronomy, which we do not wish to pursue too far. What is of most interest to us is that black holes, those strange, seemingly unlikely anomalies of the theory of relativity, once considered the wild fantasies of theorists, now appear to be established realities of the astrophysical world.

Lest we not take for granted the picture of the cosmos which has emerged from all these studies, let us note that the path we have followed is not a random one. Starting with the simple experiments of Oersted and Faraday, nourished by the theoretical insights of theorists from Maxwell to Einstein and beyond, we can trace a continuous thread of development by which Faraday's magnetic experiments, so primitive by modern standards, made inevitable our present awareness of the most grandiose manifestations of the modern cosmos. If one traces the path that we have followed and sees that each step has emerged from the preceding in a natural way, physics would no longer be seen as a collection of diverse "subjects", as taught in schools, but as a single intellectual adventure, full of excitement at every step.

10.6. Appendix: Mass Determination in Double Stars

Assume that, in a double star system, the stellar orbits are circular. If the masses of the two stars are m and m' and the distance between them d, the radii of their orbits are $(m'/M)d$ and $(m/M)d$ respectively, M being the sum of the masses $m + m'$. The speed of the heavier member $(m$, say) is then $2\pi(m'/M)d/T$, T being the orbital period, and its momentum $2\pi(mm'/M)d/T$. Since the change in momentum in half a period is twice this, the rate of change of momentum is $8\pi(mm'/M)d/T^2$. But this must balance the gravitational force, i.e.

$$8\pi mm'd/MT^2 = Gmm'/d^2$$

from which we deduce Kepler's law: $8\pi d^3 = GMT^2$. The argument now goes as follows: the speed $2m'd/MT$ may be determined from the Doppler shift, as may the period. This enables us to determine $2m'd/M$. But d may be obtained from the Kepler law, since $d^3 = GMT^2/8\pi$; we thus find the speed to be $m'(\pi^2 G/2M^2 T)^{1/3}$. Knowing the speed and the period, we can now determine $m'/M^{2/3}$. Finally, if we can estimate the mass of the heavier visible star from a mass-luminosity relation, the mass of the lighter invisible one can be deduced.

For readers wishing to avoid an algebraic argument, its content may be summarized as follows: the relations governing the orbit of the heavier visible stars depend on the masses of the two stars, the periodicity of the orbit, and the distance between them. Two things may be observed directly: the periodicity (from oscillations in redshift as the star orbits) and the velocity of the star (from the magnitude of the redshift). The mass of the supergiant star can be roughly inferred from its luminosity and spectral type.

These three data are not sufficient to determine the four quantities designated, but one more theoretical principle provides the fourth piece of information needed: Kepler's law, which relates the interstellar distance to the period. Thus, in effect, all relevant parameters can be determined.

The argument has been somewhat simplified by assuming circular orbits; its essence is however correct.

Chapter 11
HEAT AND THERMODYNAMICS

There is a tendency to think that the only fundamental problems left in physics are to be found in the subnuclear regime, which is concerned with the ultimate constituents of matter, the "fundamental particles", or in cosmology, which deals with the universe as a whole. We focus our efforts on the discovery of "fundamental laws of nature" from which, presumably, the detailed aspects of the physical world will follow.

Nevertheless, Feynman, in his Messenger lecture on "The Distinction of Past and Future", said: "I must say immediately that one does not, by knowing all the fundamental laws as we know them today, immediately obtain an understanding of anything much ... . Nature, as a matter of fact, seems to be so designed that the most important things in the real world appear to be a kind of complicated accidental result of a lot of laws."

Up until recent times there was a tendency to think of the physical world as Laplace's machine which was such that if one knew, at a given time, the exact positions and motions of all particles in the universe, the laws of nature would permit us with complete precision to reconstruct the entire past and predict the entire future. Modern physics has shaken our faith in such simple certainties; quantum mechanics has shown us that matter is more subtle and elusive than we thought, and relativity has taught us that "past" and "future" have no fixed significance; all the same, old habits remain ingrained. We are only now beginning to understand a principle suggested by

Einstein, that if laws are simple, nature itself is very rich and complex, so that the path from law to reality may be long, complicated, and full of surprises. This is true, not only of "new" and esoteric physics, but even of the Newtonian system in its complex manifestations. The more progress we make toward discovering that "unity of nature" which was the driving concept behind the work of Faraday and all the great scientists who followed, the less the physical world appears to resemble Laplace's "machine". This means that we must beware of carrying reductionism, the notion that knowledge of the tiniest identifiable components of the world contains total knowledge of the whole, beyond its justifiable limits. Since the study of heat and thermodynamics illustrates this problem it must be included in any serious discussion of "modern" physics. It provides some of the clearest indications that "the whole is greater than the sum of its parts".

The proposition is not unique to physics, but appears in other disciplines (e.g. "gestalt" psychology). Those who hope to learn from physics should be aware that it does not teach such simple lessons as is often thought.

11.1. Macroscopic and Microscopic Laws

Despite the preoccupation of Newtonian physics with "particles" as the ultimate dynamical components of nature, classical physics also concerned itself with bulk matter, the elasticity of solids, the flow of liquids, the various properties of gases. Rigid bodies were treated as though they had no internal degrees of freedom, whereas elasticity was treated as a continuous phenomenon, in a "field-like" way. The flow of liquids was also treated as field-like, in terms of continuously varying velocity, pressure, or density. Although the basic picture was one of aggregates of minute particles, the independent motion of these particles was not in general the focus of concern; matter was treated as though it were continuous.

The study of gases was another story, since their constituent molecules seemed to be moving independently. Despite this, gases were treated as having gross properties (pressure, temperature, density), so that the molecular motions were hidden behind these collectively defined properties.

What is important, however, was that simple laws were formulated relating these collective properties, laws not concerned with the independent molecular motion.

Robert Boyle, for example, who was roughly contemporary with Newton, observed that if a gas were kept at a constant temperature, its pressure would vary inversely with its volume (or, its density would vary in proportion to its pressure).

We can illustrate this by imagining a gas in a container, closed at the top by an airtight, very light but frictionless piston.

Suppose now that a one-kilogram weight is applied to the piston. The gas will contract to a certain volume. Now suppose that the weight is increased to two kilograms. It is found that the volume is halved (more precisely, if the weight of the piston is x kg, the added weights should be $(1 - x)$ and $(2 - x)$ kg respectively) (Fig. 11.1). This is a manifestation of Boyle's law.

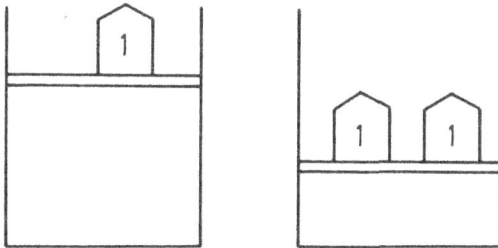

Fig. 11.1.

Bernoulli (1738) explained this phenomenon with the hypothesis that gas consisted of "minute corpuscles" of matter in rapid random motion.

The equilibrium position of a 1 kg loaded piston represented a balance of forces. Gravity acting on the piston and the weight exerts a force of 1 g. On the other hand, consider a gas molecule with a vertically upward speed of v. If the molecule has mass m, it hits the piston with a momentum mv. If it undergoes an elastic collision with the piston, it bounces back with momentum $-mv$. Thus, by conservation of momentum, it conveys an upward momentum $2mv$ to the piston. If N such molecules strike the piston per second, the rate at which momentum is given to the piston is $2Nmv$. The volume of the gas will stabilize at a value for which this quantity is equal to the gravitational force on the loaded piston.

To see how the volume enters, consider the rate N of molecular collisions with the piston. Suppose that the density of the gas is n molecules per cubic centimetre (cc). The number of collisions will certainly be proportional to n. But it will also be proportional to the average vertical molecular speed. Thus, $N = znv$, where z is some constant. Therefore, the vertical upward momentum given to the piston per second is $2znmv^2$. This must balance Mg, M being the mass of the loaded piston and g the acceleration of gravity. Thus, if the total molecular energy (temperature) is not changed, n, the molecular density, must change in proportion to M, that is, to the pressure of the piston on the gas. If M doubles, the density of the gas must be doubled for equilibrium to be restored, which in turn requires that the molecules be squeezed into half the previous volume.

It is important to note that this is a statistical argument. While the force of gravity is constant, the molecular collisions are discrete and random. The result is that the piston will not be held in steady balance, but will tend to oscillate up and down (the effect will, however, be too small to be observable). This is a phenomenon of the same sort as the Brownian motion studied by Einstein in 1905, where the motion of a tiny pollen grain, much lighter than a piston, could be observed in a microscope.

11.2. On the Nature of Heat

The simplest answer to the question "what is heat?" is that it is what a thermometer measures. This is the response of the ultimate empiricist; it is akin to the psychologist's definition of intelligence as being that which is measured by an intelligence test. The behaviourist psychologist who believes that he has learned from physics that the criterion of real science is the elimination of all "subjective", unmeasurable concepts would presumably stop there. In physics, however, such an attitude usually represents an obstacle to understanding; it certainly would not lead to the resolution of the problem we have posed. We now understand heat as energy in the atoms or molecules of which matter is composed.

The characterization of heat as the motion of the microscopic elements of matter was not an original insight of nineteenth century physics. Even Plato recognized a connection between heat and motion: "heat and fire, which generate and sustain other things, are themselves begotten by impact and friction, but this is motion". In 1602, Francis Bacon declared that "heat is motion and nothing else". Von Leibnitz too recognized the connection between motion and heat. He also appeared to have an understanding of the conservation of energy which incorporated heat energy. He was challenged on this point with the example of two inelastic bodies colliding head-on, both coming to rest. Where, he was asked, does your energy go? This was his reply: "It is true that the wholes lose it in reference to their total movement, but it is received by the particles, they being agitated inwardly by the force of the collision. Thus the loss ensues only in appearance. The forces[10] are not destroyed, but dissipated among the minute parts."

The prevailing view of the nature of heat in the later 18th and early 19th century was formulated by the Scottish chemist Joseph Black. According to Black, heat had the character of a fluid, which he called "caloric", and to which he attributed special properties. This theory illustrates two points which recur frequently in the history of physics. One is that, if one sets about to find evidence in support of a theory, even if that theory is wrong, evidence can usually be found. The caloric theory was a rather seductive one, and

---

[10] i.e. energy.

although it was regarded with scorn in the time of Maxwell, it convinced many of the best minds of its time – people like Voltaire, Euler and the Marquis de Chatelet. They all wrote in support of it.

A second and closely related point is that an advance in science is normally made only when one finds evidence in contradiction to an incorrect theory, rather than evidence to support a new one.

The essence of the caloric theory was expressed by John Dalton: "The most probable opinion concerning the nature of caloric is that of its being an elastic fluid of great subtlety, the particles of which repel one another, but are attracted by all other bodies."

This "explains" a number of things.

Consider, for example, a hot body in contact with a cooler one. Since the particles of caloric are closer together in the hot body than in the cooler one, their mutual repulsion will be greater, and they will be pushed into the colder body until they are equally dense in both!

Consider next the boiling of water. In this case the heating of the water forces more particles of caloric into it. At first their attraction to the atoms of matter is stronger than their mutual repulsion. But as more and more caloric particles attach themselves to the molecules of water, these molecules are forced apart until they no longer cohere.

One can also explain why solids contract on cooling. As the number of caloric particles decreases, the repulsive force between the particles of matter due to them does too, and their natural mutual attraction tends to bring them closer together.

Even the production of heat by friction could be "explained" by this theory. The idea was that when a surface was rubbed it was damaged, permitting the particles of caloric to leak out more freely, thus cooling the material. Lord Rumford later pointed out the flaw in this idea. If this were true, he noted, the more persistently one pounded or pummeled an object, the colder it ought to become.

Biographical sketches of all the contributors to the evolution of the theory of heat are given in Angrist and Heppler's "Order and Chaos" (Basic Books).

It is interesting to note that this theory treated heat, which we now know to be a quality of objects, as a *thing* itself. This reflects a tendency of Newtonian physics to put *objects*, particles or rigid bodies, at the centre of the conceptual scheme of physics. Stimulated by advances in the mechanization of production, scientists tended more and more to think of nature as the ultimate machine. What was more natural than that concepts be materialized? So, space was materialized as the aether and heat as caloric, and Newton had materialized light. Such a tendency was frequently considered to be the essence of a "scientific" view of the world. It is still manifested today in the influence of physics on other domains of science: the association of memory with a particular site in the brain, of morality with our genes, of living behaviour with the reactions of molecular chemistry. The attitude was perhaps best illustrated in J. C. Maxwell's reasoning in opposing the concept of biological evolution: living organisms, he said, were made up of particles (molecules). Since these particles did not evolve, he argued, neither therefore could organisms.

One can also argue plausibly that this tendency has affected Western views of social, economic and political phenomena, as manifested in their strong emphasis on individualism.

In recent years books have appeared purporting to see parallels between current physics, now almost as far removed as one can conceive from the mechanistic dogma, and Eastern philosophy (e.g. Capra's "The Tao of Physics", Denis Postle's "Fabric of the Universe"). There do seem to be grounds for seeing similar issues in these two different domains, though one should beware of pushing parallels too far, or of the possibility of using single words out of context to characterize what are not identical concepts. The issues came to me rather vividly, however, on reading Frances Fitzgerald's fascinating book on Vietnam entitled "The Fire in the Lake". The book calls attention to fundamental differences between American and Vietnamese world views, as regard both science and conceptions of individuals and their social relationships:

"For Americans, the close relationship the Vietnamese draw between morality, politics and science is perhaps more difficult to understand than at any other time in history. Today, living in a social milieu completely divided over matters of value and belief, Westerners have come to look upon science and logic alone as containing universal truths. Over the past century Western philosophers have worked to purge their disciplines of ethical and metaphysical concerns; Americans particularly have tended to deify the natural sciences and set them apart from their social goals. Under pressure of this demand for "objective truth" the scholars of human affairs have scrambled to give to their own disciplines the authority and neutrality of science. But because the social scientists can rarely attain the same criteria for truth as physicists or chemists, they have sometimes misused the discipline and merely taken the trappings of science as a camouflage for their own beliefs and values...", and further on:

"In the Vietnamese language there is no word that exactly corresponds to the Western personal pronoun 'I, je, ich'.

When a man speaks of himself he calls himself 'your brother', 'your nephew', 'your teacher' depending on his relationship to the person he addresses.... The traditional Vietnamese did not see himself as a totally independent being, for he did not distinguish himself as acutely as does a Westerner from his society.... He did not see himself as a 'character' formed of immutable traits, eternally loyal to certain principles, but rather as a system of relationships, a function of the society around him. In a sense, the design of the Confucian world resembled that of a Japanese garden where every rock, opaque and indifferent in itself, takes on its significance from its relationship to the surrounding objects."

Western society in the 18th and 19th centuries was strongly attracted (and still is to some extent) to a very materialistic view of the world. Modern computer science distinguishes between "hardware" and "software"; the history of scientific concepts teaches us that the two have sometimes been confused, that a process, a system of organization or a quality has been misinterpreted as a material. If one wishes to approach science critically, one must recogize this sort of acquired prejudice, and must remember that we must sometimes make a conscious effort to force ourselves to "see the world from another point of view".

In the case of the caloric theory of heat, its replacement by a theory based on molecular motion was a rather long and tedious process: The American adventurer Benjamin Thompson (later Count Rumford) became, in the 1790's, a military adviser of the Elector (Prince) of Bavaria. Part of his duties was to supervise the boring of a cannon. He observed that the process appeared to produce inexhaustible quantities of heat. This led him to argue as follows:

"In reasoning on this subject, we must not forget to consider that most remarkable circumstance, that the source of the heat generated by friction, in these experiments, appeared evidently to be inexhaustible.

It is hardly necessary to add that anything which any insulated body can continue to furnish without limitation cannot possibly be a material substance, and it appears to me extremely difficult, if not quite impossible, to form any distinct idea of anything, capable of being excited and communicated, in the manner the heat was excited and communicated in these experiments, except it be motion."

This idea did not gain immediate acceptance. Herschel, writing in 1831 (nearly 40 years later) states that "the nature of heat is as yet unknown". But it is perhaps significant that he discusses it under the heading "imponderable states of matter"!

An important contribution to the understanding of heat was made by a medical doctor, Julius Robert Mayer, in 1840. He was a ship's doctor, who had much experience in bleeding sailors suffering from fever in the tropics. He observed that the blood of patients bled in the tropical regions was much redder than that of patients bled in northern climates. He knew that this redness signified a high oxygen content. He concluded that the body did not need to expend as much energy maintaining its temperature in the tropics as in Europe; thus, there was a relation between work and heat. In a daring generalization, he enunciated the following law:

"An energy ("force" in the original) once in existence cannot be annihilated; it can only change its form; and the question therefore arises what other forms is energy capable of assuming?"

"If potential energy and kinetic energy are equivalent to heat, heat must also naturally be equivalent to kinetic energy and potential energy."

Mayer made another important observation; it concerned the concept of *specific heat*. The specific heat is the quantity of heat energy required to raise the temperature of one gram of a substance by 1°C.

The specific heat of a gas can be measured under two different conditions. In one case, the *volume* of the gas can be kept constant; on the other, one can keep it at constant pressure. Will the specific heat be the same in both cases?

No, said Mayer. For if the gas is kept at constant pressure, it will expand as the temperature is raised. In expanding, it will do work (e.g. in moving a piston). Thus, some of the energy put in will do mechanical work, and the rest will heat the gas. Thus, it will take *more* heat energy to heat the gas one degree when the *pressure* is kept constant than when the volume is kept constant.

This seems obvious enough, but note that the argument is based on the understanding that heat is *energy*, the kinetic energy of the particles of gas.

Mayer could not get his work published in the "respectable" scientific Journals of the time (in this case, the *Annalen der Physik*). His ideas were treated with scorn and ridicule, so he published his work privately at his own expense.

At about the same time James Joule, the son of an English brewer, was studying the relation between electrical, mechanical and heat energy. His most significant work concerned, however, the *quantitative* connection between heat and kinetic energy. A schematic representation of the apparatus he used for these experiments is shown in Fig. 11.2. Using a falling weight to rotate vanes in an insulated basin of water, he was able to calculate the rise in temperature of the water due to the *mechanical* energy transferred to the vanes by the falling weight. Of course, the *gravitational* initial potential of the weight is converted into energy of motion in various forms: there is the kinetic energy of the weight and of the vane system, as well as the kinetic energy of the swirling water, in addition to the heat energy communicated to the water molecules by the agitation. These latter, however, are only transitory. The weight comes to a stop, the vanes stop turning, and the swirling water ultimately becomes still. Of the initial energy, nothing remains but heat. If the surrounding insulation had a very low specific heat, the energy would manifest itself as the heat energy of the water, i.e. in a rise in its temperature.

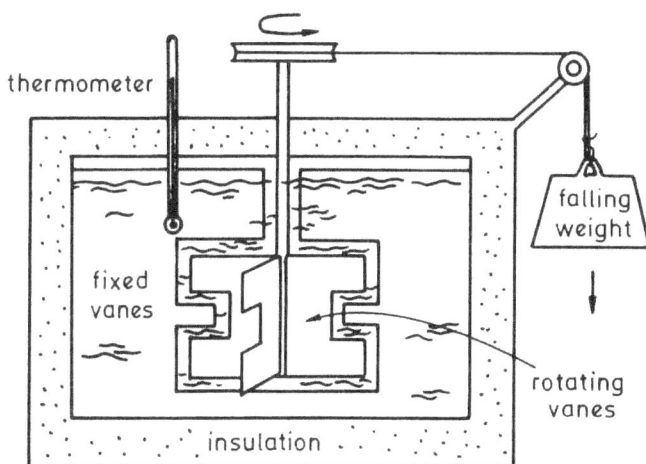

Fig. 11.2.

The result found by Joule shows that great quantities of "hidden" energy can be stored in the molecular or atomic motion of matter. Dropping a weight of 430 kg (nearly half a ton) a distance of one meter would heat one litre of water by 1°C. To make it more dramatic, if we take an iron pile-driver two meters long and half a meter in diameter and drop it five meters, its terminal kinetic energy will only be 1% of its internal heat energy at 20°C ("room temperature").

The existence of such large amounts of latent heat energy in matter is certainly of interest because usable energy is essential both for life and for social organization. What do we mean by usable energy? We mean energy which can be converted into forms which perform specific necessary tasks: run motors, create electricity whose energy can be transported over large distances and then used for light, heat or machinery etc.; in short, energy which can be used for useful work on essential tasks. To what extent is the reservoir of heat energy usable for these purposes? The study of this problem constitutes the science of thermodynamics.

As a result of all these experiments and observations, and many others, the notion that heat was in fact simply atomic or molecular motion slowly gained acceptance. The first general statement of the law of conservation of energy was made by Helmholtz in 1847, when he was 26 years old!

"From a similar investigation of all the other known physical and chemical processes, we arrive at the conclusion that nature as a whole possesses a store of energy which cannot in any way be increased or diminished; and that, therefore, the quantity of energy in nature is just as eternal and unalterable as the quantity of matter."

I have named the general law "the principle of the conservation of energy". This law, which is sometimes also called the first law of thermodynamics, has subsequently become a keystone of all physics; no physical principle carries more authority. That this was not so in Helmholtz's time is evident from the fact that the most prestigious physics journal of the time, the *Annalen der Physik*, rejected for publication the article in which Helmholtz enunciated this principle. The law has, however, been modified in the light of Einstein's principle that matter itself was energy (the famous $E = mc^2$). Contrary to Helmholtz' formulation, we now know that the quantity of energy is not "as unalterable as the quantity of matter"; rather, it is the two together which are conserved and unalterable.

11.3. What is Temperature?

When we speak of temperature, we think of an instrument which measures it, for example, a mercury thermometer. But why do we believe that the length of a column of mercury is a reasonable measure of how hot things are, that is, how much heat energy they contain? If we *add* heat energy to the mercury column, it expands. By careful measurement, we can even verify that the amount of the expansion is proportional to the amount of heat added, within limits. The expansion, therefore, can give us a linear scale of heat content. Note that the significance of the temperature scale depends on an empirically verifiable law. The question of whether the mercury thermometer is the *best* measure of heat energy is open. It is conceivable that other physical phenomena are more accurately correlated with heat content (the voltage in a thermocouple, for example), and therefore more accurate thermometers, but this is simply a technical question.

11.4. The Law of Gay-Lussac and Charles

A second phenomenon embodied in a statistical law for gases is associated with the names Gay-Lussac and Charles. This law states that if the pressure of a gas in a closed vessel is kept constant, its volume is proportional to temperature. In the same situation used to illustrate Boyle's law, heating the gas will result in its expansion, at a rate which is linear in the temperature. This law too can be explained with a molecular model. Heating the gas increases the kinetic energy, and the momentum, of the particles. Thus, each collision of a gas molecule with the piston conveys more momentum to it. This will drive the piston upward. But as it rises, the density of the gas decreases, so that the collisions become less frequent. Expansion takes place until the increase in the effect of each collision is compensated by the reduced frequency of collisions. We note that the momentum communicated to the piston by the gas is proportional to the density of the gas and the average kinetic energy of the molecules. Thus, in equilibrium, the density must be decreased and the volume increased, in inverse proportion to the increase in average molecular kinetic energy. If this latter quantity is proportional to the temperature, the law follows.

A word of caution needs to be added concerning these two laws. In the case of Boyle's law, if the pressure is increased rapidly, heating of the gas may occur. An important qualification of the law is that the temperature must remain constant. Thus, the pressure increase must take place slowly enough that the gas in the vessel maintains thermal equilibrium with its surroundings. Similarly, Gay-Lussac's law remains valid only if the pressure remains constant; that is, if heat is added slowly enough that the expansion may keep pace with the absorption of heat. Technically stated, the two processes are adiabatic.

One also has to note that some of the energy communicated to the gas may appear in a form other than that of translational kinetic energy, it may, for example, cause the molecule to rotate more rapidly. In turn, the momentum of the gas molecules will increase proportionately less than the rise in temperature. The validity of the law thus depends on the assumption that this effect is negligible.

We shall disregard these complications, since our main concern is to illustrate the *character* of statistical laws and not their details.

11.5. Disorder and the Second Law of Thermodynamics

Thus far we have established that heat is the kinetic energy of the motion of the elementary constituents (molecules, atoms) of matter. This energy is *disordered* in that it is the energy of *incoherent* motion; because the motions are as likely to be in one direction as another, no work can be done using them. In the case of a solid, the atoms vibrate *back and forth*. In a liquid, there may be bulk flow, in the sense that the centre of mass of macroscopic elements of the fluid move, i.e. the atoms or molecules of the elements are translated together. Heat energy, on the other hand, is kinetic energy of these particles in the frame of reference which is instantaneously moving with their centre of mass. That is, hydrodynamic flow carries a net momentum, but heat motion does not.

In a gas, the situation is similar; the gas may flow, transporting momentum, and superimposed on this there will be disordered or chaotic motion. The steady flow is a wind or breeze.

The second important fact that we have noted is that the principle of conservation of energy is valid when *heat energy* is taken into account.

Once we know that energy is conserved in the physical world, that is, cannot be created or destroyed, we are confronted with the problem of the transformation of energy from one form to another, e.g. potential to kinetic, kinetic to electric, electric to heat, heat to chemical, etc. If we consider any of the forms of energy other than heat, we can readily recognize that they cannot do work without the generation of heat as a by-product. This happens when a body falls through the air, or if through a vacuum, when its fall is stopped. Heat is produced when water power turns the generator to produce electricity, which in turn produces more heat in the wires carrying it. Nuclear or fuel-fired turbines create mechanical energy which in turn can be converted into electricity, but only with the production of "heat pollution" which, especially in the case of nuclear energy, continues to be generated in radioactive waste product decay.

Since heat appears to be more or less inadvertently produced in all energy transformations in which it is used to do useful work for specific purposes, to what extent can heat energy itself be exploited to do such work?

This question attracted the attention of Sadi Carnot (1796-1832), eldest son of Napoleon's ministry of war. Since he was interested in both economics of industrial organization and in physics, it was perhaps natural that he should be interested in steam engines. In 1823, at the age of 27, he wrote a memoir called "Reflections on the Motive Power of Fire", which attracted relatively little attention until after his death.

It is interesting to note that his book contained the first statement of what we now call the "Second Law of Thermodynamics", well before the "first law" — the law of conservation of energy — was formulated by Helmholtz.

The form in which he stated the law was roughly as follows:

The motive power of heat is independent of the agents employed to realize it; its quantity is fixed solely by the temperature of the bodies between which it is effected — by the transfer of caloric.

Although his work on the motive power of heat was formulated in terms of caloric, unpublished notes left at his death show that he later came to understand that heat was motion. The fact that his work was expressed in terms of the caloric concept created obstacles to its acceptance, first by Joule and later by Lord Kelvin. It remained for Clausius, in 1850, to point out that Carnot's work could equally well be interpreted in terms of the kinetic theory of heat. It was Clausius who gave the idea precise mathematical form, and who introduced the concept of entropy in terms of which the second law can be most succinctly expressed.

Recently, we have come to realize that the underlying concepts with which the second law deals are capable of very broad generalization. Nevertheless, it is useful to look at them in the context of the steam engine, the context in which they were originally developed.

The phenomenon on which the steam engine is based is simply that gases expand when heated and contract when cooled. Why is this so?

Imagine a vessel filled with a gas, divided in the centre by a freely movable piston AB, each side being filled with a gas at the same density and temperature (i.e. in "equilibrium") (Fig. 11.3). Suppose we heat the gas on the left. As a result its molecules will move faster than those on the right, and will have more momentum. Molecules from both sides are constantly bombarding the piston, but the molecules on the left will be striking it with more momentum

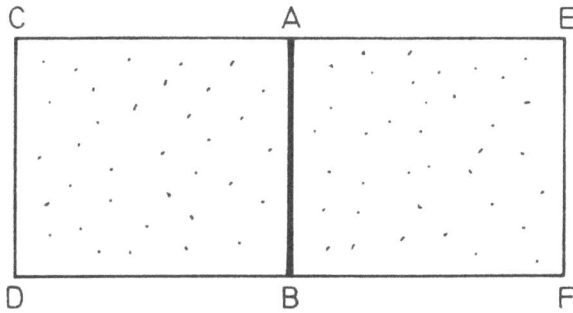

Fig. 11.3.

than those on the right. The piston will then move to the right.

Will this displacement continue indefinitely? Clearly not, because as the piston moves the gas molecules on the left will have to go further between successive collisions on the piston, i.e. their collisions will become less frequent. More momentum will be communicated on each collision, but equilibrium will ultimately be restored thus: as the frequency of collisions on the left decreases, so will the pressure on that side. But that on the right will increase, due to compression, until they come into balance.

A difference in temperature has made it possible for the system to do work.

If the gas on the left is initially heated rapidly to a higher temperature and the source of heat then turned off, the gas on the left will cool as the piston moves. This is because the force of the collisions on the left will do *work* on the piston, so that the colliding molecules will lose energy and their temperature will drop. On the right it will tend to rise for the same reason.

If the wall EF of the right hand chamber is removed, pressure will not build up on the right; the piston will of course again move to the right by the mechanisms we have discussed, and will again cease when the momentum transfer to the piston from collisions on the two sides is equal.

What we have described here is a device for having heat energy, in the form of a temperature difference, do a limited amount of mechanical work. How can we make a *continually* working machine?

Let us first try something very crude: consider a gas in a chamber with a piston on one side, and in equilibrium (same density and temperature) with its surroundings (Fig. 11.4). I have not represented the motions of the molecules on the right, but one must imagine them to be the same as those on the left.

Now let the chamber be heated; the gas molecules are heated (faster motions indicated in (b) by larger arrows). The piston will then move to the right. Mechanical work is done in turning the flywheel.

What happens at the end of the stroke? The flywheel will now do work on the gas. If no energy is lost in friction, or from the sides of the vessel, the flywheel will recompress the gas to its original state, and come to rest. No new work will be done. If there is friction or heat escapes, it will come to rest even sooner. How then could we keep the wheel turning? We could do it if we *cooled* the gas when it was at the end of its stroke, i.e. if we took some "waste" heat away from the system. Then, the piston would be forced in by virtue of *gaining* more momentum from the "warmer" air molecules on the right than that it lost to the "cooler" ones on the left. In fact, what one needs is a cycle of heating and cooling – heating on the outstroke, cooling on the instroke.

Figure 11.5 shows schematically how a steam engine works. Steam is created from water in a boiler, and is fed into an expansion chamber sealed by a piston. The steam expanding against the piston pushes it to the right, causing it to turn a flywheel. As the chamber expands, the steam pressure decreases, which permits an exhaust valve to open, and causes the intake valve to close. On the exhaust stroke, the piston pushes the gas into a cooler, where it is recondensed into water. This water is then pumped back into the boiler, where it is heated for the next stroke.

The engine therefore delivers power continuously. The waste energy is drained away in the cooling part of the cycle, as well as

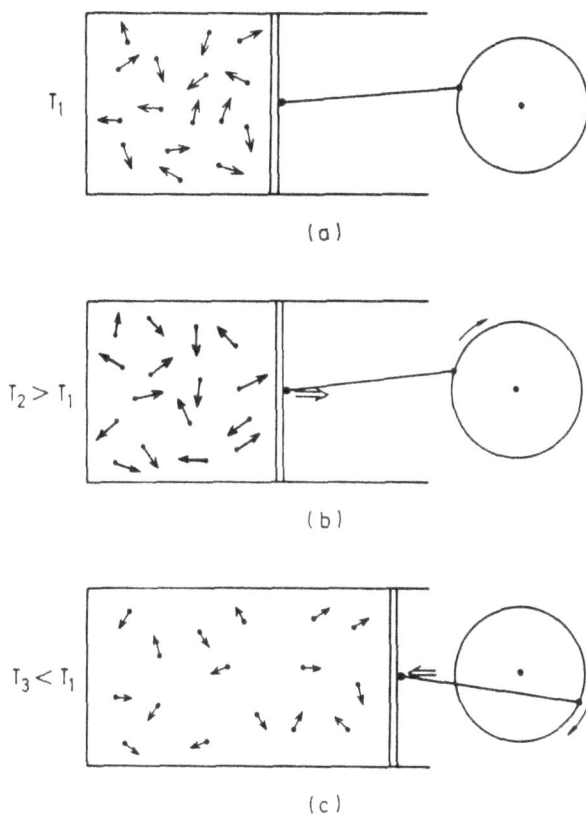

Fig. 11.4.

being lost in friction, heat radiated from the boiler, etc. In short, the energy of the burning fuel is only *partially* converted into the mechanical energy of the flywheel.

This shows that work can be done by heat only by virtue of *differences* in temperature, and only to the extent of the difference. When everything is at the same temperature (the state of "equilibrium"), no heat energy may be converted into mechanical energy.

Disequilibrium, represented in the steam engine by a temperature difference, can do work; at the same time, heat is transferred

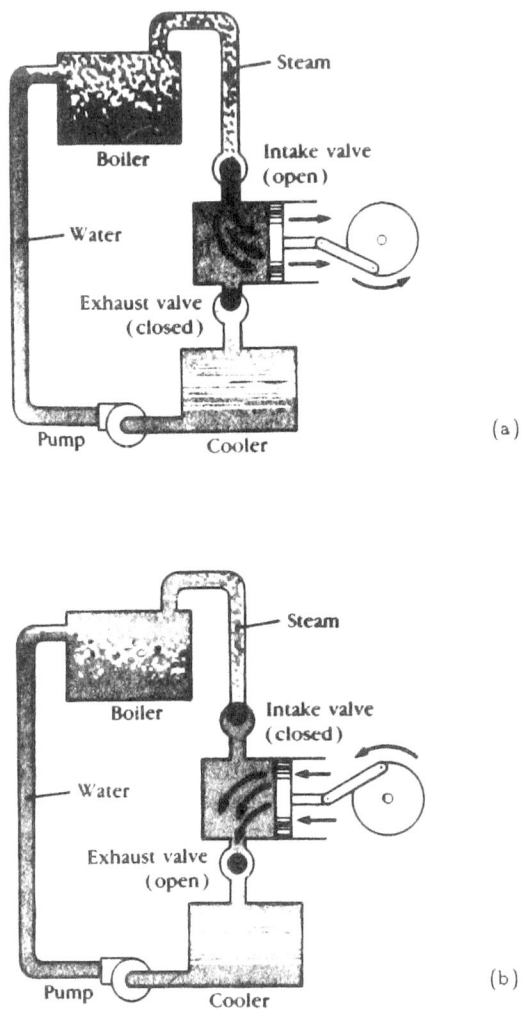

Fig. 11.5. (a) Expansion stroke. High pressure steam expands against a piston, driving a wheel around. (b) Exhaust stroke. Steam, now at low pressure, is pushed out so that new steam can enter the cylinder as in (a). Expanded steam is cooled and pumped back to the boiler to be heated and compressed. (Figure 15-3 from *The Ideas of Physics*, 3rd ed., by Douglas C. Giancoli, copyright ©1986 by Harcourt Brace Jovanovich, Inc., reprinted by permission of the publisher.)

from hotter to colder areas and thus diminishes the degree of dis-equilibrium. The temperatures tend to become uniform. Thus the steam engine evolves toward equilibrium, and continues to work only as long as disequilibrium is re-established by the injection of energy from outside. Similarly, winds blow, electricity flows in power lines, objects roll down hills, in fact all physical changes, represent move toward equilibrium, that state in which no more work is done, no more change takes place. All is uniform and inert. This has been characterized as "the heat death" of the universe. The irony of the situation is that the more desperately we try to create disequilibrium, to do work, to create, the more rapidly do we transform "useful" energy into heat energy, and thus accelerate the inevitable drift toward the equilibrium that is death.

What a depressing picture! Leon Cooper notes this concept "has been associated with the disillusionment of a society no longer reaching toward perfection but running downhill toward mediocrity; a culture with television and without genius, an architecture with suburbia but without Notre Dame".

The case has been stated more positively by Max Born, who affirms that "There is no longer any doubt; all matter is unstable. If this were not true, the stars would not shine, there would be no heat or light from the sun, no life on earth. Stability and life are incompatible". Nature provides us with a rich heritage of diversity and of accumulated energy. We should respect and treasure it!

The principle that the world tends toward equilibrium and that its supply of usable energy is ineluctably depleting itself is known to physicists as the "Second Law of Thermodynamics". The true moral to be derived from it is not that enunciated by Cooper. Neither the evolution of life on earth nor the formation of the galaxies was frustrated by this principle, nor is human "progress". It is not that order and complex structure cannot develop in our world; but rather that a price must be paid for it. Creativity is always possible, but at least its material manifestations inevitably create a larger measure of chaos in its wake. The more intensely we exploit nature for our own ends, the more we accelerate the process. This is part of our experience as a society. It is a phenomenon which we can no more alter than we can change the law of gravity.

Like Born, however, we may turn the argument on its head to draw a positive conclusion. Just as, in a physical system, work can only be accomplished, ordered structures can only be created, by exploiting disequilibrium, so also do societies depend for their dynamism on their inhomogeneities, on the interplay between and the disequilibrium among their elements.

The principle is manifested much more widely than simply in physical processes.

Let us explore this principle further.

First, we can look more closely at the reasons why heat energy plays a special role in physics. A way of illuminating the problem is to think of a waterfall. The water at the top of Niagara Falls, or any power dam, has a higher potential energy than at the bottom; when it falls, that energy is converted into kinetic energy. The *falling* consists of the descent of water molecules from the top of the dam to the bottom. Thus there is a *coherent* motion of all of them together. But at the same time there are much *greater* quantities of energy stored in the *disordered, incoherent* motion of these molecules with respect to each other. At the bottom of the waterfall, most of the ordered kinetic energy of flow is converted into an increase in the disordered molecular motion which is heat. These motions are superimposed on the collective motion which we observe. Relative to an observer falling with the water, these molecules go in all directions, with speeds depending on the water temperature. Precisely the same is true of water flowing through a pipe. The energy of coherent motion can be transmitted to another macroscopic object. The energy of *incoherent* motion cannot, simply because the forces exerted by the molecules in their collisions with the object cancel out. They can only convey their motion to the molecules of the object, in which incoherent motion is again engendered.

What then happens when two bodies at different temperatures

interact? Energy in macroscopic form can be communicated from one to the other, and useful work done. Consider a simple case in which two gases of the same density at different temperatures are separated by an insulating barrier which is suddenly removed. Suppose the hotter gas is on the left. Then the molecules crossing the invisible barrier from left to right are moving faster than those crossing from right to left. There is consequently a net macroscopic flow of particles from left to right. More will cross per second from the hotter gas because the molecules on that side are moving faster; each molecule will also carry more momentum. To be more quantitative, the average number arriving at the division from one side per unit time will be proportional to the average speed v of the molecules on that side. They will carry an average momentum mv, so that the *momentum* crossing from that side will be proportional to mv^2, i.e. *to the average kinetic energy per molecule and hence the temperature on that side*. A net flow of momentum will then be set up from the hot to the cold side which will be proportional to the difference in the temperatures on the two sides. This net transport of momentum permits work to be done, just as in a waterfall or flow in a pipe.

Let us summarize the above discussion, and indicate some of its consequences.

1. It is possible to convert heat into other kinds of energy only by exploiting temperature differences, thus, heat energy may not be totally converted into other sorts of energy.

2. In exploiting temperature differences of different parts of a system, that temperature difference is diminished; thus, the amount of "useful" energy diminishes.

3. Systems tend towards uniformity, a situation in which all parts of the system are at the same temperature. We call this "equilibrium".

We must be careful in defining "the system" above. Some heat energy is always "wasted" in the attempt to convert heat into useful work. It succeeds only in raising the temperature of the surround-

ings. The possibility of useful work being done depends on this ex-
pulsion of heat to the surroundings. It cannot take place unless the
surroundings form a cool reservoir. If the temperature of this reservoir rises above that of the machine from which one needs to expel
the heat, the machine can no longer work, unless *another* machine
(refrigerator?) works to cool the reservoir.

A brief formulation of the above principles is that all physical
systems tend to evolve from states of greater order to states of greater
disorder.

We must, however, define precisely the term "order". The concept of entropy, introduced by Clausius, serves to quantify the idea
of *order*.

Something that is "ordered", or "orderly", is not *disordered* or
random. Randomness is the converse of order.

What do we mean, for example, by saying that the positions,
or the motions, of the molecules of a gas are *random*?

We mean that the positions and motions of the molecules are
as likely to have any value as any other. The state of any molecule
is independent of the states of any other; there are no correlations
between them. (We must remember, however, that the equal probability of all configurations is subject to obvious macroscopic constraints; all gas molecules stay in the container, and their total energy is fixed.)

The situation is similar to flipping a coin. In any given toss, the
probability that the coin lands head up is the same as the probability
that it lands tail up. On the face of it this is a curious assumption.
Presumably, Newton's laws of motion apply to the coin-flipping process. If we always flipped the coin exactly the same way, it should
always come up the same way. Why, then, do we believe that the
result of the toss is unpredictable, *totally* unpredictable, in fact? It
must be that the result is *very* strongly dependent on the conditions
of the toss, so that a *very* slight difference in these conditions (the
"initial conditions") can *completely* change the result.

For the molecules of a gas this is even truer. A *very* slight change in conditions at some instant can make a *very* important difference in the subsequent state. The result is that, for a system of this complexity, detailed prediction of the evolution of a system on the microscopic scale is not possible.

11.6. A Question of Philosophy

We started with the realization that complex systems cannot be described in microscopic detail. It might be thought that this problem could be overcome by using more powerful computers, but the difficulty is more fundamental, since we can never specify the initial conditions of a system with infinite accuracy. One might think that this is not very important, because a tiny variation in initial conditions would only make a correspondingly tiny Σdifference in what followed. But this becomes less and less true the more complex the system is. It is precisely this fact that allows us to assume that the flip of a coin is a random process, or that the pattern of motion of gas molecules has a random character.

A minute change in the initial conditions of a system would at the very beginning make very little difference. Think, however, of the molecules of a gas. A molecule of air at room temperature and normal pressure makes about ten billion collisions per second. At each collision the uncertainty in its own state will be added to the uncertainty associated with the molecule with which it collides; that uncertainty will be passed on to the next collision of each of these molecules, and so forth. In other words, at each collision there is a loss of information with regard to the state of each molecule involved, with the result that in a very short period of time all initial information about the system will be lost. The motions will be completely random. If we consider the random deviation of the path of any particle from the one it would follow if the initial conditions were known with infinite accuracy, at each subsequent collision that deviation would be increased by the amount of the accumulated deviation up to that time (the deviation of the particle with which it collides). Thus, the deviation follows an exponential growth law.

Randomness can be achieved in a fraction of a second, due to the accumulated effect of the ten billion collisions.

Because of the accuracy of the statistical physics of macroscopic phenomena, we find ourselves in a strange situation: we understand the macroscopic world if we discard *all* microscopic information. It is precisely the *impossibility* of explaining the macroscopic world in terms of microscopic determinism that makes it appear deterministic. We have come a long way from Laplace's mechanistic model of the physical world. Quantum mechanics will take us even further in revealing the element of chance at the core of things, in the behaviour of matter at the atomic and subatomic levels.

11.7. Some Remarks About Probability

We have great difficulty in grasping very large numbers, and even more in comprehending the statistics of such numbers.

Consider, for example, public opinion polls. The population of Canada is about 25 million. Pollsters use samples of a thousand or so to predict the behaviour, or to sample the thinking of these millions. A typical poll of the voting intentions of a "random" sample of about 1000 voters will be claimed to be accurate to within 4%, nineteen times out of twenty (which, incidentally, does not prevent newspapers from publishing headlines, or articles, attributing significance to a two percent change in the relative standing of parties from one poll to the next).

Oscar Wilde once said that people are often quite willing to believe the impossible, but much less likely to believe the improbable. The difficulty lies in understanding the different levels of improbability. Confidence in public opinion polls seems quite high, though there must frequently be an error of five percent or so. On the other hand, the number of people who buy lottery tickets is high, though the probability of winning a large sum of money is so low as to be negligible. (There is a popular lottery nowadays called the 6/49, where the trick is to choose six numbers from 1 to 49. The probability of success is about one in *fourteen million!*)

That the problem is more subtle than it first appears is illustrated by examining political polls more closely. In a parliamentary system, the question of accuracy takes on a quite different aspect if one considers that individual candidates are elected by only a fraction of the electorate. Consider, for example, my home province of Quebec. Assuming a sampling of a thousand voters and about 120 constituencies, only about eight voters per constituency are polled. If the population is roughly equally divided, there is, in a given constituency, a probability of about 28% of an error of 50% or more in the prediction of the result.

Clearly, one must use statistics with great care!

A simple and informative "laboratory" for experimenting on probability is to be found in the repeated flipping of a coin. Without knowing precisely why, we assume that the two results, "heads" and "tails", are equally probable.

Suppose that we flip a coin ten times. The probabilities of the different possible distributions of head and tails are proportional to the number of ways in which these possibilities can be realized:

10–0 or 0–10	1 way in 1024 = 0.2%
9–1 or 1–9	10 ways in 1024 = 1.0%
8–2 or 2–8	45 ways in 1024 = 4.4%
7–3 or 3–7	120 ways in 1024 = 11.7%
6–4 or 4–6	210 ways in 1024 = 20.5%
5–5	252 ways in 1024 = 24.6%

There is a probability of about 11% that the distribution differs from the even one by 3 parts in 5, a 60% deviation. If one considers 100 tosses, there is a roughly similar probability of a deviation of at least 9 parts in 50, an 18% deviation. Although the *absolute* deviation is greater for the larger sample, the fractional one is less.

If the coin is tossed a thousand times, the fractional deviation from equal numbers of the two results is reduced to 6%.

There is in fact a simple formula for any number of trials. Let us assume a value of 1 to each head, −1 to each tail. If one makes

N throws, and adds up the score, one obtains the deviation S from equality, which corresponds to a score of zero. Let us call the results of the trials $a_1, a_2, a_3, \ldots a_N$ where each a is either $+1$ or -1. Then the departure from the mean over N trials is $S = (a_1 + a_2 + a_3 + \ldots + a_N)$. The square of this is $S^2 = (a_1^2 + a_2^2 + a_3^2 + \ldots + a_N^2) + 2$ (the sum of all products of two different a's). But the products are just as likely to be positive as negative, so the average value of the second term is zero! The first term has the value N, since each term is $+1$. Therefore the average deviation of S is the square root of N, or $N^{\frac{1}{2}}$. We call this quantity the *root mean square deviation* (Table 11.1).

Table 11.1.

N	Rms deviation (d)	d as %
10	3.2	32
100	10	10
1000	32	3.2
10000	100	1.0
1000000	1000	0.1
10^6	10^3	0.1
3×10^{19}	5.5×10^9	1.8×10^{-10}
10^{23}	3.2×10^{11}	3.2×10^{-12}

3×10^{19} is the approximate number of molecules in 1 cubic centimetre of gas; 10^{23} that in a similar volume of solid material.

Departures from averages are almost infinitesimally small in macroscopic systems. The bulk density of particles is constant from sample to sample to an extraordinarily high accuracy.

11.8. Particle Density, an Example of a Statistical Quantity

Particle density is the number of particles per unit volume. In mechanics, this is normally treated as a *field* quantity, varying continuously from point to point. However, at the scale of the average distance between particles, this concept makes no sense. We need a sufficient *number* of particles to give a meaning to the concept of density.

The *mass* density is proportional to particle density for a pure material, in which all molecules are identical.

A volume of gas of 50 cubic centimetres at normal temperature and pressure contains about 1.35×10^{21} molecules. This is the standard molecular density of all gases under standard conditions. For a variety of macroscopically identical samples of gas under standard conditions, the fluctuation in the number of molecules is about 3.7×10^{10}, which represents a *fractional* fluctuation of about 3 parts in 10^{11}. It is insignificant.

We cannot define a density in *too* small a volume. Let us consider a volume big enough to contain, on the average, 10 or 100 particles; these volumes will be those of cubes with sides of 0.7×10^{-6} and 1.6×10^{-6} cm respectively. The dispersion in the densities of different volumes of this size in a *macroscopically uniform* gas is about 30% in the first case and 10% in the second. In either case the density function is very poorly defined.

For a more precise definition, we must choose a larger region. An accuracy of $1/10$ of a percent requires a volume containing about a million molecules. This is the volume of a cube with sides about 3×10^{-5} cm.

Note now the dilemma into which we are forced. We cannot define density accurately for regions whose dimensions are less than about 10^{-4} cm. That is to say, we cannot describe non-random variations of density over lesser distances. Yet we formulate our theory in field-like terms, according to which we wish to define density as a continuously varying quantity. Below this intermediate range of distance, macroscopic variables become imprecisely defined. That is true not only for density, but also for the other parameters of the theory of fluids, like pressure and temperature. Yet the number of molecules involved in a volume of this size is about 30 million, far too many to permit analysis at the microscopic level. There is a sort of "uncertainty principle" constraining us here: there is a transition region in which a microscopic description is not feasible because of complexity and a macroscopic one is uncertain because of statistical fluctuation.

One may argue that this creates no real difficulty in practice. It is interesting, however, to note the existence of this sort of complementarity principle which is inherent in statistical mechanics, which puts *theoretical* limits on the applicability of the theories which we are forced to make about problems in this area. Obviously not only quantum mechanics limits our capacity to be omniscient.

11.9. Order and Disorder; Entropy and Information

Heat energy is the energy of the random motion of molecules. We call such a motion "disordered". But disorder need not only be a characteristic of motion. The spatial distribution of gas molecules is also random, or disordered, compared to the distribution of atoms in a crystal, which follows a law of regularity. How, in a quantitative theory, can we describe orderliness, or disorderliness? *Disorder* can be described mathematically in terms of randomness, or equal *à priori* probability of all conceivable possibilities.

We have seen that a mechanistic description of a gas at the molecular level is not possible. Even if we knew the complete state at a given time, the molecular motions would rapidly become more and more independent of that initial state.

This statement does not in itself justify the assumption of randomness, though it is suggestive. Randomness was defined in terms of equal *à priori* probability of all possible states, and probability in turn implies repeated samplings of macroscopically identical systems. Suppose we had many identical containers of the same gas, all at the same temperature and pressure.

No correlation would be expected between the states of individual molecules in various vessels. Thus, the study of the system at a microscopic level cannot be essential for the determination of the macroscopic properties of the gas. Is the situation properly described by taking the individual molecular configurations to be random? Any attempt to *prove* this is as difficult as proving that the probability of heads and tails in the toss of a coin is random. We therefore adopt the procedure, by now familiar, which Feynman describes for the evolution of physical laws. We *guess* at a law, and then test the consequences of our guess. To the extent that the consequences conform to physical reality, we accept the validity of our guess.

Testing the new statistical law is a much more critical process than we might at first think. According to the random law, the results of macroscopic measurement will *not* always be the same. This is the same as to say that, in 100 tosses of a coin, we will not always get 50 heads and 50 tails. There will be fluctuations about the average, and *the probability of a given fluctuation is also predictable*. Thus, the verification of the law puts stringent requirements on what we should observe.

Note, however, that to sustain the assumption of randomness it is necessary to make many observations. If the assumption enables us to predict the *probability distribution* for the results of measurement, it is clear that the verification of the assumption becomes more sure the more accurately the distribution is measured. (For this reason the estimates made in official and unofficial studies of the probability of certain sorts of accidents in nuclear reactors are almost meaningless. The data are simply insufficient for any justification.)

So much for randomness or total disorder. One way of approaching an understanding of order is by looking for the opposite extreme to randomness, or disorder. For example,

— A disordered situation is one which may be realized in a large number of ways, since "everything is possible". An ordered situation is one which may be realized in only one way, or in a very small number of ways. Order is characterized by the exclusion of most possibilities.

— A disordered system is one in which there is no correlation between its elements (e.g. the molecules of a gas). An ordered one is one in which that correlation is high. Consider, for example, the difference between a gas and a crystal. In the former, there is no correlation between the variables (position, velocity) of one molecule and another. In a crystal the positions of the molecules are highly correlated, the distances between the atoms are fixed and the positions of the various atoms are strongly correlated.

— Since, in an ordered system, many conceivable situations are *excluded*, the probabilities of the remaining ones are enhanced. If all but one configuration is excluded, its probability is 1.

All of the above constitute a basis for the quantification of the idea of order. Let us take a very simple example. Suppose we wish to locate four points inside a square (Fig. 11.6).

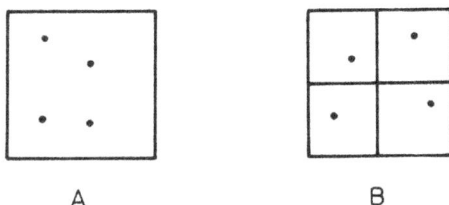

A   B

Fig. 11.6.

First, we might put them in by chance (A). This can be done in an infinite number of ways.

Or we may divide the square into four smaller squares, as in B, and put one point in each. Relative to A, this excludes many possibilities. There are 24 ways of doing this: there are 4 possibilities for the first point, 3 for the second, 2 for the third, and only 1 for the fourth, i.e.

$$4 \times 3 \times 2 \times 1 = 24 .$$

But the *total* number of ways you can distribute the 4 points among the 4 squares is $4^4 = 256$. Thus, B is $24/256 = 0.094$ time as probable as A.

Note, however, that this "order" is rather arbitrarily defined. It is similar to having 52 cards in "order" ranging from king down to ace with the suits in the order spades, hearts, diamonds, clubs. The *order* is there because of our convention about the relation between the different suits and values of the cards.

Finally, we can prescribe the situation in which there is one point in each small square, and each is at the *centre* of its square (Fig. 11.7). This situation is completely ordered, infinitely more so than B, because in B each point could be *anywhere* in its own small square.

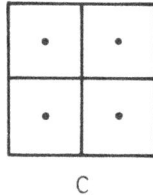

C

Fig. 11.7.

Entropy can be defined in terms of the "number of ways" in which a given configuration can be realized. The *probability* of such a configuration is, using the randomness hypothesis, *proportional* to the numbers of ways in which it can be realized; this is how the connection between *probability*, entropy and disorder can be under-stood.

If, now, we have *two* systems, the probability that they be in the state they are is the *product* of the probabilities of the states of the individual systems. This may be expressed as:

$$P_{12} = P_1 P_2 .$$

The *logarithm* of the product of two quantities is the *sum* of the logarithms of the separate quantities. Therefore

$$\log P_{12} = \log P_1 + \log P_2 .$$

The logarithm, in short, is an *additive* quantity. If we define the *entropy* of a system as $k \log P = S$ where k is some arbitrarily chosen constant, the entropy of a pair of systems is the sum of their separate entropies, a desirable characteristic.

This is a very general definition of entropy. In the context of the theory of heat and thermodynamics, we can give a more restricted one, as follows: suppose that a quantity of energy ΔQ is added to a system. It cannot all be converted into useful work; it produces some heat, i.e. some disorder, so that the entropy of the system increases by an amount ΔS. Let us define the temperature T of the system so that

$$\frac{1}{T} = \frac{\Delta S}{\Delta Q} \ .$$

Does this correspond to our intuitive notion of temperature? Let us show by two illustrations that it does.

1. Consider two systems (1 and 2) which spontaneously exchange energy, an amount of energy ΔQ flowing from system 2 to system 1. Since for each system

$$\Delta S = \frac{\Delta Q}{T} \ ,$$

$$\Delta S_1 = \frac{\Delta Q}{T_1} \ , \quad \Delta S_2 = -\frac{\Delta Q}{T_2} \ .$$

The *total* change in entropy becomes

$$\Delta S = \Delta S_1 + \Delta S_2 = \Delta Q \left(\frac{1}{T_1} - \frac{1}{T_2} \right) \ .$$

Thus, if $T_2 > T_1, \frac{1}{T_1} > \frac{1}{T_2}$ and the flow of energy from system 2 to system 1 increases the entropy, in conformity with the notion that entropy is a measure of disorder. In fact, the second law now states that, when two systems at different temperatures are brought into contact, heat flows from the system of higher temperature to that of lower temperature.

2. Consider an engine which runs by using a hot source, and expels its excess heat into cooler surroundings. Suppose the heat

put in is ΔQ_2, and that expelled to the surroundings (the "waste" heat) is ΔQ_1. Then, by conservation of energy, if work W is done by the engine,

$$\Delta Q_2 = \Delta Q_1 + W .$$

If the engine were 100% efficient, ΔQ_1 would be zero and we would have

$$W_{\max} = \Delta Q_2 .$$

Thus $\frac{W}{W_{\max}}$, the efficiency of the engine, is

$$\frac{\Delta Q_2 - \Delta Q_1}{\Delta Q_2} = 1 - \frac{\Delta Q_1}{\Delta Q_2} .$$

Now let the entropy change in the energy of the surroundings due to the addition of an amount of heat energy ΔQ_2 into the machine at temperature T_2 be

$$\Delta S_2 = -\frac{\Delta Q_2}{T_2} .$$

The entropy due to taking *out* an amount of heat energy ΔQ_1 at temperature T_1 is

$$\Delta S_1 = \frac{\Delta Q_1}{T_1} .$$

Since the total entropy must not decrease,

$$\frac{\Delta Q_1}{T_1} \geq \frac{\Delta Q_2}{T_2}$$

so that

$$\frac{\Delta Q_1}{\Delta Q_2} \geq \frac{T_1}{T_2}$$

and the efficiency of the machine is

$$< \left(1 - \frac{T_1}{T_2}\right)$$

which clearly depends on the temperature difference of the input heat and the heat drawn off.

This, too, is in conformity with what we said earlier about heat engines.

The definition of the temperature by the equation

$$\frac{1}{T} = \frac{\Delta S}{\Delta Q}$$

is the most precise and rigorous one we have. To demonstrate that this is effectively equivalent to the temperature measured by thermometers is more difficult; in fact, this correspondence is the criterion for a good thermometer.

11.10. The Idea of Information

Like many other words used in a technical sense by scientists, the word information can be given a precise quantitative meaning.

Let us use a simple example to illustrate how we can quantify information. Suppose we take a square, and wish to locate a particular point in it to some desired accuracy, we can do it as follows (Fig. 11.8):

1. We divide the square in half vertically, and specify whether the point lies in the top or the bottom half. This is one "bit" of information. As in the case of flipping coins, it is a matter of choosing between two possibilities.

2. Next, we divide it in half horizontally, and specify whether the point lies on the left or the right of the dividing line.

With these two bits of information, we have placed the point in a new square with one quarter of the area of the original one.

We now go through the same process again, dividing this new smaller square in two both horizontally and vertically. Two more bits of information serve to place it in one of these new squares, now 1/16 the area of the original. This process can continue until we have established the position of the point within the required accuracy.

Suppose this required N bits of information. The probability that a point, placed at random in the original square, is in the region

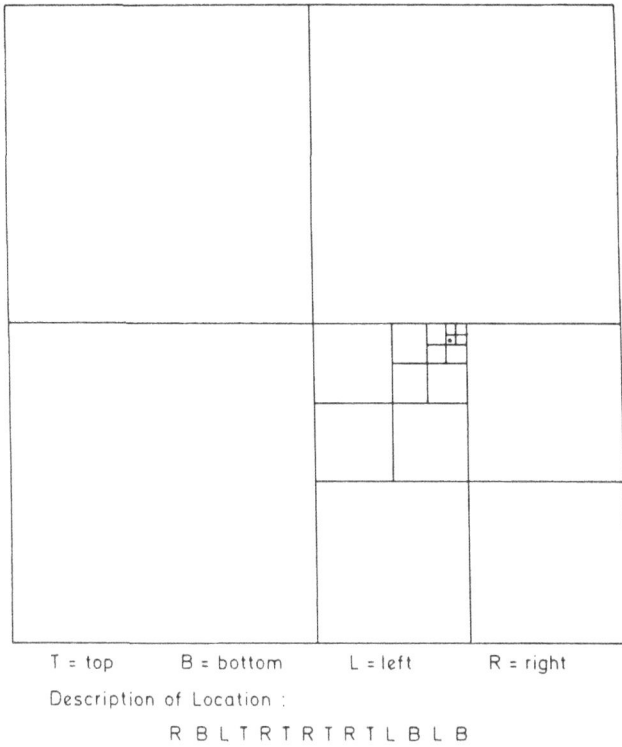

T = top   B = bottom   L = left   R = right

Description of Location :

R B L T R T R T R T L B L B

Fig. 11.8.

determined by the N bits is $\frac{1}{2^N}$, since each extra bit decreases the volume available to the point by a factor of 2.

If we wish to specify the position of a second particle, in the same way using N' bits of information, the probability that the two points together lie in the regions specified is

$$\frac{1}{2^N} \times \frac{1}{2^{N'}} = \frac{1}{2^{N+N'}} .$$

The situation is described by $N + N'$ bits of information.

At the start the location of the points was purely random, and $N + N' = 0$. This is the state of no information. The greater the amount of information given, the less is the position of the particles random. Greatest randomness is equivalent to maximum entropy and minimum information. Adding information decreases the randomness and thus the entropy. Thus, entropy and information are complementary; the greater the one, the smaller the other.

Furthermore, the amount of information associated with the two particles is the sum of the amounts of information associated with them separately; information, just like entropy, is additive.

It is interesting to note the large amount of information in an ordered pattern. If the positions of a large number of points were to be specified, we would need the same amount of information for each.

But suppose we prescribed a pattern, such as the one that is represented by a square grid of points (a 2-dimensional square lattice; Fig. 11.9). We can specify this lattice by the statement that, given any point, any other may be reached by a series of horizontal and/or vertical displacements one centimetre away.

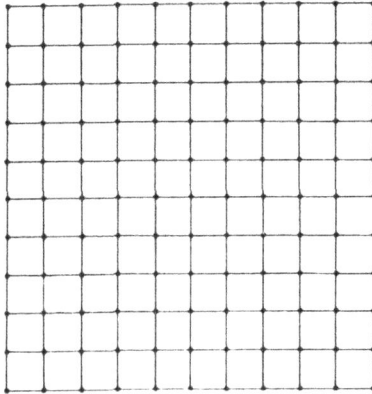

Fig. 11.9.

The information about the 100 points is conveyed by specifying the correlations between the positions of the different points. One way of specifying order is by giving these correlations. The various points so specified are in fact not random at all; we have a very low statistical probability or low entropy.

The number of bits of information is, in fact, mathematically proportional to the logarithm of the probability of the configuration determined by that information.

Let us take as an example the heating of water, from ice to steam.

Figure 11.10 shows what happens if a quantity of ice well below the freezing temperature (−20 K, for example) is heated at a slow but constant rate.

(a) At first, the temperature will rise steadily, until it reaches the melting point (0°C).

(b) As it melts, the temperature remains constant; the energy added is used to break the bonds which hold the water molecules in their places in the crystal. It creates disorder but not heat energy.

(c) Point B represents the point at which the ice is completely melted. After B the temperature of the water rises regularly as more heat is added.

(d) At point C the water has reached the boiling temperature. From this point till D, the added energy frees the individual water molecules from the attraction of their neighbours; steam and water remain in thermal equilibrium.

(e) At point D all the water is boiled. Thereafter, any additional energy raises the temperature of the steam.

The Second Law of Thermodynamics states that, at each stage of the system as it evolves through continuous heating, the entropy is increasing. What we shall now do is to illustrate this increase with a "phase diagram", which will reveal how the order of the system will decrease while the randomness will increase.

In the phase diagram, the state of the system can be described by giving the position of each molecule, represented by its three spatial coordinates, and its velocity, given also by three quantities representing its velocities in three different directions, the coordinate directions. This is a comprehensive description, for once the

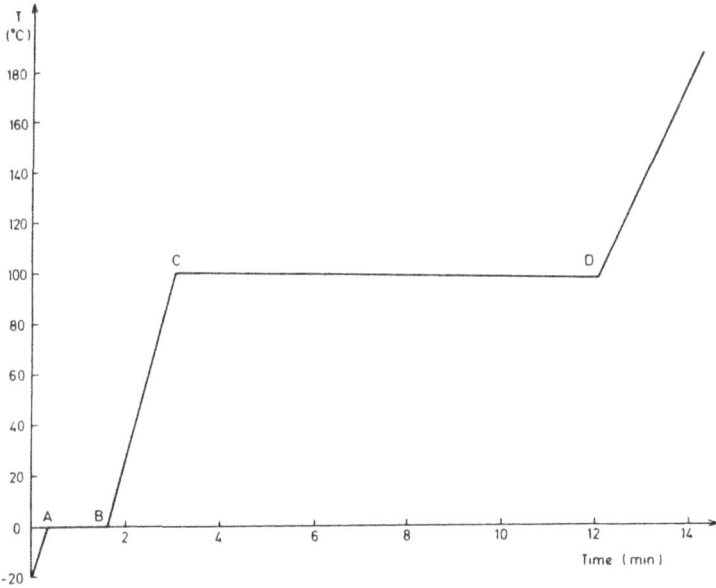

Fig. 11.10. 100g of ice at $-20°C$ initially, heated by 1000 watt hotplate. Melting starts at A, is complete at B; boiling starts at C, is complete at D.

position and velocity of a particle are known, the laws of motion will determine unambiguously its motion under any given forces. Any further information is redundant.

The six quantities given may be imagined as coordinates in a six-dimensional "space". That we cannot envisage six dimensions is unimportant.

Our system has enormous numbers of water molecules. Following the six coordinates for all of them would be enormously complicated. Instead we shall follow them in a *statistical* way; that is, we can consider the statistical distribution of any one of them. Even that requires six dimensions, since there are three directions in space. But nothing distinguishes one such direction; the situation with respect to the other two will be similar. In this way, we can represent things on a two-dimensional graph. In one direction, we plot the position of a molecule (measured from some arbitrary base point); in the perpendicular one, its velocity in that direction.

Consider ice at the absolute zero of temperature. It is in the form of a crystal; each molecule is in a given position (we ignore the internal structure of the molecule for the sake of simplicity). It will also be at rest. We can therefore characterize the typical molecule by

(displacement 0, velocity 0) .

On the phase diagram (Fig. 11.11), this is represented by a point at the origin of the coordinates.

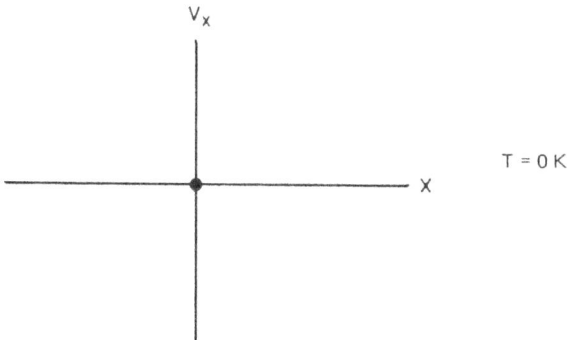

Fig. 11.11.

There is no randomness whatsoever in either variable.

Let us add energy to the system. The molecules will undergo small vibrations about their initial positions. They will have small velocities. Their phase points will be distributed through a small region in both x and v_x.

We shall indicate this by the hatching on the diagram (Fig. 11.12).

As more heat is added, the region of phase space accessible to the particles will grow in both directions (Fig. 11.13).

When the melting point is reached, two things happen:

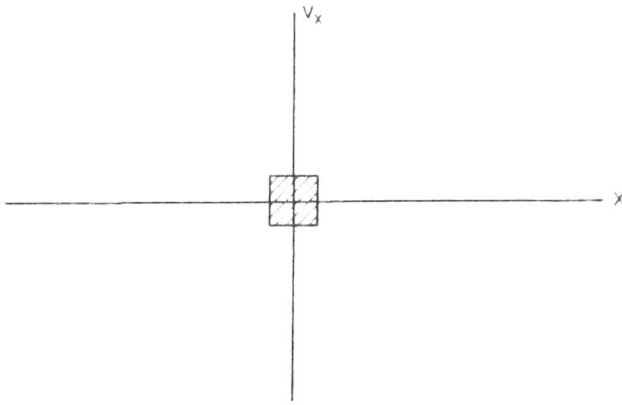

Fig. 11.12.

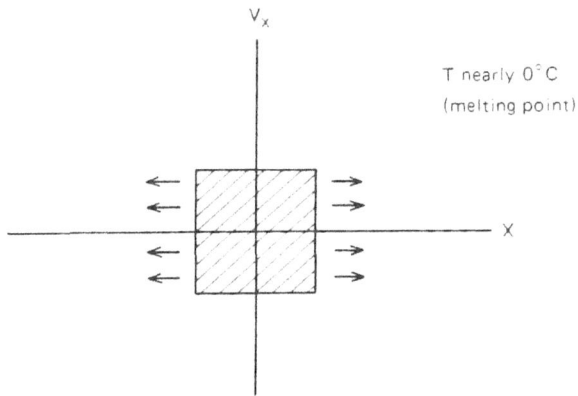

Fig. 11.13.

— First, the molecules no longer oscillate about fixed positions, but are free to move throughout the liquid. That is, they may move over a much longer range of x.

— The temperature does not rise until the melting is complete, so that mean velocities do not increase. The accessible region of phase space increases as shown by the arrows, i.e. in the direction of x.

By the time melting is complete, x has increased to cover the whole volume of the liquid (Fig. 11.14). The velocity limits remain essentially unchanged.

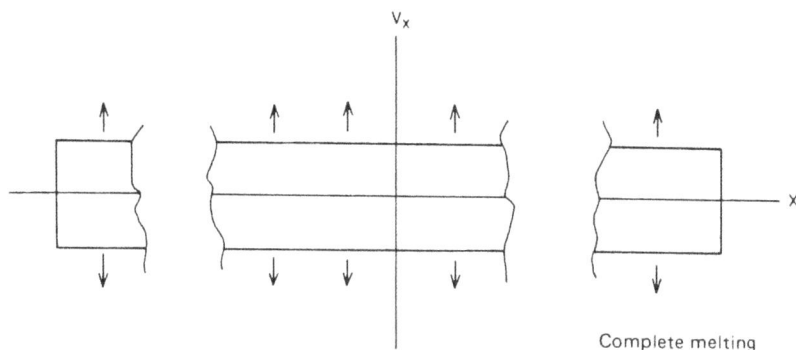

Fig. 11.14.

As the temperature now increases, the range of velocities of the molecules increases again, as shown by the arrows in Fig. 11.14, and this continues until the boiling point is reached.

Figure 11.15 shows the situation just before boiling; the molecular velocities are now those corresponding to a temperature of 100° C. During the boiling process, the temperature again remains constant at 100° C, but more and more of the molecules escape, so that in effect the range of x available to them increases almost without limit.

Figure 11.16 shows the situation at the completion of boiling; further heat now only raises the temperature and hence the molecular velocities, so that the accessible region of phase space now grows steadily in the vertical direction.

What is demonstrated throughout the whole evolution of the system is that, as heat is added, the volume of phase space accessible to the water molecules constantly increases. This increase in volume is directly related to an increase of entropy. Information

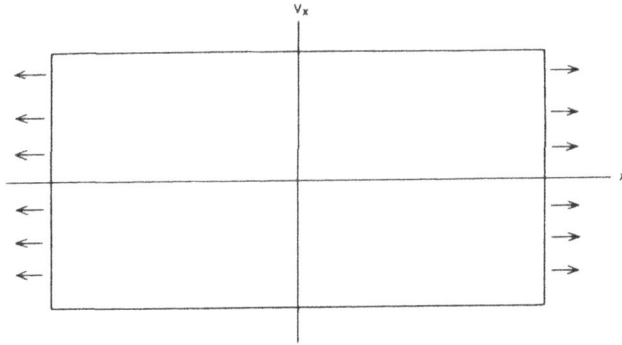

Fig. 11.15.

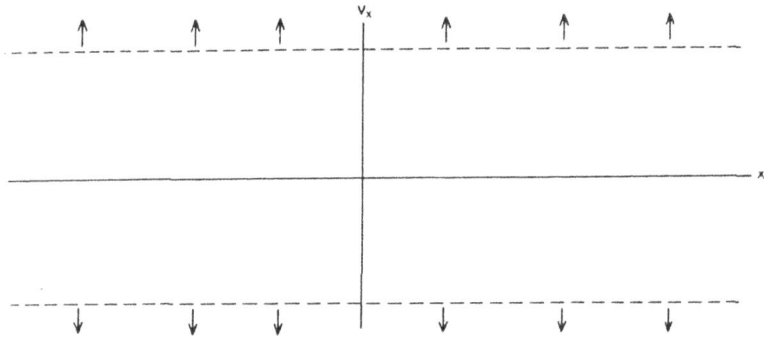

Fig. 11.16.

is constantly lost with increased heating, both with regard to the positions and velocities of the molecules, until a state of complete disorder is reached.

11.11. The Second Law and Time's Arrow

The observation, in conformity with the second law, that all systems appear to evolve from states of greater order (lower entropy) to states of greater disorder (higher entropy) is so profound a part of our experience of the physical world that it appears to determine our sense of the direction of flow of time. That this is so may be illustrated by imagining that a moving picture is taken of any natural phenomenon, and then playing the film backward. If, as appears to be the case, the laws of physics are invariant with respect to time reversal, any sequence of events in the physical world taking place according to the laws of physics should have the property that, if they took place in the reverse order in time, it would be in conformity with those same laws. Imagine some examples:

(a) A book is slid along the top of a table, coming to rest. In reverse, a book, originally at rest, starts to move with increasing speed back toward the hand that originally launched it.

(b) Smoke is pouring from a smokestack. In reverse, it flows back *into* the smokestack.

(c) Water is poured from a glass into a basin. In reverse, the water rises up from the basin and collects in the glass.

(d) A ball is thrown and shatters a window. In reverse, the fragments of glass gather themselves up and reform the solid pane, while the ball flies back into the hand of the thrower.

The reverse scenarios are so unnatural and contrary to experience that it is hard to believe that nothing in them violates the basic laws of physics. This is nevertheless the case. Why, then, do we know that we will never see them in nature? How, in short, can time-reversible laws of nature generate non-time-reversible phenomena?

We can provide a simple, superficial answer. What happens in nature is a consequence not only of the laws of physics, but also of the initial conditions of the system on which they operate. The laws say that if things start in such and such a way, they will evolve in a manner which is rigorously deducible from the laws. Thus, if some of the things which can in principle happen do not happen, it must be because the initial conditions which would have generated them do not exist in the real world.

While this is a plausible answer, it is not very enlightening. It shifts the question to another front; why are such initial conditions so unlikely to be realized? The key to the answer is that, in the natural evolution of things, some energy is dissipated as heat, in uncorrelated motions of enormous numbers of molecules. It would be difficult to imagine that, at some time and place in the universe, identical molecules could be in the same configuration, and with these motions precisely reversed. But only in such a case would the reverse phenomenon be seen.

But what exactly makes this so improbable? The answer lies in both the incredible number of molecules involved, and the consequent difficulty of organizing them with precisely the right correlations, and in the extreme sensitivity of the subsequent behaviour of systems to minute changes in initial conditions (recall Section 11.6).

To illustrate the former point, consider a large number of identical vessels full of a certain gas, all at the same temperature, density and pressure. Suppose that the volume of these vessels is 1000 cc; if the gas is air at normal atmospheric pressure and room temperature, each vessel will contain about 2.7×10^{22} molecules. The detailed molecular state of these molecules will be quite different in all vessels, and will be constantly changing. Each molecule will make about 50 million collisions per second, so that in a time of one ten-millionth of a second the molecules will be randomized.

How long a time will it take for a given state to arise in such a flask of gas? Poincaré showed that, in a totally isolated system, anything can happen, and in principle will happen an infinite number of times. But how long will we have to wait for a given configuration? Poincaré found that the time would be something like $10^{5 \times 10^{22}}$ seconds. But the age of the universe is only about 6×10^{14} seconds. Poincaré's time is quite unimaginable, being inconceivably larger than the age of the universe.

In short, if we wait for a given configuration, it will simply never appear.

Nor is it enough to come close. As pointed out in the article by David Layzer, Borel showed that the change in gravitational potential caused by displacing 1 gram of matter by 1 cm at the distance of the star Sirius would, in one microsecond (one millionth of a second!), substantially alter the microscopic state of a macroscopic volume of gas.

What this latter point illustrates is that, at the molecular level, no system is isolated. It is of course evident that the displacement of 1 gram a distance of 1 cm on Sirius could not affect any macroscopically measurable physical quantity; if it did, physics would be impossible. Rather, this illustration demonstrates the total irrelevance of the detailed molecular state of a system. For it is evident that knowledge is not possible (or information does not exist) concerning the micro-state of any macroscopic subsystem of the universe. At the microscopic level, information must exist only at the level of the whole universe if at all. And since it seems that no part of the universe can carry as much information as the whole, such micro-information is simply not available.

The Second Law of Thermodynamics, then, has to do with the constant loss of *macroscopic* information in macroscopic systems which are effectively isolated.

In some sense, this proposition is self-justifying. The physical world is full of instances of the growth of order out of disorder. It is manifested in the phenomenon of life, but also in the familiar physical phenomenon of crystallization. Whenever we are confronted with such a phenomenon, which appears at first sight to violate the Second Law, we extricate ourselves by invoking some *extended* system which has the property that, however much entropy may be lost in some identifiable subsystem, there is a concomitant greater increase in its larger environment. That is, we extend the system until the Second Law can be verified.

It is of course not a trivial statement to say that it is always *possible* to find an extended system for which the Second Law is satisfied; especially if the subsystem and its larger environment are required to be coupled to each other by understandable interaction processes.

The sustenance of living organisms, for example, involves the creation of greater disorder in the environment. The more complex the form of life – the greater the degree of order that it represents – the greater is the cost. Animals feed on plant or other animal life, leaving disordered excrement in its place. For heat and shelter we cut trees and burn petroleum products, or use dam rivers for electricity. The more sophisticated and complex our social organization, the more damage is done to the environment. The problem of pollution is a direct reflection of the action of the Second Law.

Since animal life is ultimately dependent on plant life, we can focus our attention next on the world's flora. The prerequisite for plant life is *light*. If "artificial" light is used, i.e. if light is produced by the expenditure of useful work, we are back on familiar ground where a classical manifestation of the Second Law is at stake. However, almost all of the plants of the world use sunlight, which stimulates the process of photosynthesis. Thus, sunlight is the essential component in the production of plants. (It is important to note that the fossil fuels on which we largely depend for our energy needs are a product of the sunlight of eons past.) A by-product of plants is the breaking down of carbon dioxide to make oxygen, which is the major energy source for animals. Animals burn oxygen to create carbon dioxide, which is in turn necessary for plant growth.

If we are to understand the Second Law in a world of living organisms, we must expand our system to include the sun. The sun itself is a manifestation of the Second Law; it is constantly burning its nuclear fuel and will ultimately "run down". Nevertheless, the solar radiation impinging on the earth is an essential source of negative entropy, that is, it can be used to make more ordered chemical combinations out of less ordered ones.

This is a very comforting conclusion, full of optimism for the human species. While the entropy of the sun increases, we can draw on its energy, but also on its negentropy, to sustain the burgeoning life on earth. This process has a cost in entropy production, but that cost is part of the entropy associated with the burning of the sun. Since we have no control over the processes taking place in the sun, we might as well use them to the best of our ability to bring order out of chaos on the earth. This is the secret of life on earth.

The situation is beautifully stated by Donald Clayton in his book "The Dark Night Sky":*

"The earth radiates just as much energy as it receives, but it receives it from a high-temperature source (sunlight) and re-radiates it as low temperature energy (infrared) into the cold emptiness of space. The entropy increase (disorder) associated with this overall process is large and allows a smaller amount of entropy decrease (order) to emerge as life. In that sense, life is like the air conditioner, although we may not be fond of the comparison. In both cases, the restricted event that at first seems contrary to the Second Law is really part of a much larger machine that does not violate it. There's poetry in that. Life is not isolated from the universe.... One might have thought that all the earth does is keep us warm, but the situation is more dramatic. If the earth were placed inside a large incubator whose walls were at the average temperature of the earth's atmosphere, the earth would be just as warm as it is now, but life would not survive. We would be radiating energy at the same temperature we receive it, and there would be no overall entropy increase that would allow a local entropy decrease (life) to occur. Slowly plant life would wilt and die and the earth would become a warm randomly disordered grave."

Some readers may have seen a film entitled "The Silent Running" a few years back, in which this point is vividly made.

What we have been describing is an "organizing principle" in the physical world. There are others. On a more mundane level, consider the process of crystallization from a solution (for example, copper sulphate, the photographer's "hypo"). If the substance is totally dissolved in water at a sufficiently high temperature, and then slowly cooled, it will spontaneously crystallize out. The solvent is clearly less ordered in the liquid solution than in the crystal, that is, order appears spontaneously.

---

* Quoted with the permission of Donald D. Clayton.

Now if the experiment is carefully done, the crystallization will be found to be accompanied by the release of some heat, in such a fashion that there is a total entropy increase. But it is worthwhile to consider a bit further the process by which entropy *decreases* in one part of a system, and increases even more in another. The Second Law of Thermodynamics tells us that it *can* happen, but not *why* it happens.

The problem is one of competition between entropy and energy. Crystallization lowers the energy of material which is crystallizing, by permitting a stronger binding of the atoms. But it also lowers the entropy, so that it tends to be inhibited by Second Law considerations. At a fixed temperature, adding energy ΔQ to a system increases the entropy, which is a measure of disorder by $\Delta S = \frac{\Delta Q}{T}$. Thus, the added energy is in the form of an amount $T\Delta S$ of *heat energy*. If a system has a total energy U, and TS is its heat energy, we may call the rest "free energy", i.e. energy in forms other than heat. Systems tend toward states in which this quantity is as small as possible. Crystallization should then take place if *this* quantity is minimized. Since U decreases on crystallization, as does S, the question is whether the whole quantity $U - TS$ decreases or not. Let the energy decrease be ΔU, the entropy decrease ΔS; the change in $U - TS$ is then

$$-\Delta U + T\Delta S \ .$$

If the temperature were too high this quantity would be positive, i.e. the free energy would increase. In this case crystallization will not take place.

Of course, since the entropy of the system plus its environment must increase, the entropy of the environment must increase; i.e. heat energy in sufficient amount must be taken up by the environment.

From this example, we see that chemical binding is another factor in creating order in the physical world; another "organizing principle".

11.12. The Puzzle of the Cosmological Arrow of Time

As pointed out by David Layzer in his article "The Arrow of Time" (*Scientific American*, December 1975), the most dramatic violation of the Second Law of Thermodynamics appears to take place at the cosmological level. According to currently accepted cosmology, the universe started with the "Big Bang", an infinitely condensed state from which all matter has ever since been expanding. In fact, the initial state of the universe is like a "white hole", the time inverse of the *collapse* of all matter into a cosmic singularity. At sufficiently early times, all the components of the universe must have been in equilibrium, i.e. in a state of maximum entropy. Subsequently, however, ordered structures began to arise, which gave rise ultimately to stars and planetary systems, to galaxies and to systems of galaxies. All this presumably took place under the action of gravity, which, on the cosmic level, has acted as the organizing agent. In any case, we seem to be dealing here with a "cosmological arrow of time", characterized by the appearance of order out of disorder, that is, pointing in the opposite direction to the thermodynamic arrow.

Once again, it is essential to emphasize initial conditions. The initial conditions of the universe as a whole seem to have favoured initially the increase of order in the universe. Such initial conditions, as we have already indicated, are very special. In this case that statement is evident. The existence of our universe depends on them.

Why is it that the very early universe was in fact in equilibrium? To understand this, consider two time scales. One is the time scale of equilibration, defined as the time taken for a random fluctuation of cosmic matter to dissipate. This was very short because of the almost infinite density of "matter" (more properly, energy), and its very high temperature, so that all elements of the system were in interaction with, and collided with, each other. Thus equilibrium could be attained in a period of time in which there was negligible expansion of the system. In short, the characteristic equilibration time was short compared to the characteristic expansion time.

However, as the universe expanded, matter became both less

dense and cooler, and collisions became less frequent. Calculations done by Layzer and his students suggest the following sequence of events:

After about 15 minutes of expansion, the universe crystallizes (or freezes) into an alloy of liquid hydrogen and helium. However, due to continued expansion this system "fractures" or shatters into fragments of approximately planetary size. The particles of these fragments then lose their crystalline structure and become gaseous.

From this point on the role of gravitation becomes crucial. Within a gas, there are inevitably density fluctuations, which act as centres of gravitational attraction around which matter condenses. But within the system of fragments, too, there will be fluctuations, which in turn will create clusters of fragments, and so forth. While the theory of galaxy formation, or of the formation of galactic clusters, is not well understood in detail, there seems little doubt that it is the result of gravitational agglomeration due to fluctuations on various scales.

We thus have a model, albeit somewhat sketchy, of how the initial chaos generated order in the cosmos, and how that order might have developed into the stars, galaxies and clusters of galaxies that we observe today.

It is evident then that this cosmic violation of the Second Law has its origins in initial conditions. But while the probability of the spontaneous appearance in any part of our universe of "initial conditions" which lead to the spontaneous generation of order out of disorder is inconceivably small, we cannot apply such an argument to the universe as a whole simply because we cannot attach any meaning to the idea of "probability" in connection with the origin of our universe. To put it differently, even if our origins were "improbable" it would be irrelevant, since the question can only be posed once that "improbable" has happened.

Chapter 12

THE ROAD TO QUANTUM THEORY

The dawn of the twentieth century set the stage for a new age
of physics. The mood of the time was one of optimism, of stirrings
and innovation in many areas of society, of belief in the power and
promise of science. In retrospect, however, it was a period characterized by the lack of a clear conception of the direction in which physics
was headed. New discoveries destined to play an important role in
the future physics were being made, and had obvious importance,
but could not be fitted into a unified framework.

The concerns of the time centred around two issues. One involved problems created by the elusive character of the aether, and
had sprung from the great success of Maxwell's electromagnetism;
these problems could be defined in terms of reconciling the new electromagnetic theory with classical mechanics. The other concerned
the nature of matter, and in particular, the bridging of the gap between the observable and familiar physical world and the atomic
images which, despite their successes in certain areas, such as the
theory of gases, remained at best controversial and at worst vague
and ill-defined.

Atomic theories went back to the times of ancient Greece and
Rome. Matter was believed to be made up of small, indivisible,
indestructible and invisible elements ("atoms"). Anaxagoras (500–
428 B.C.), Democritus (460–370 B.C.), and Epicurus (341–270 B.C.)
are some proponents. The full flowering of the theory came with the
"De rerum natura" of Lucretius (99–55 B.C.). The idea remained

sterile, however, because of its purely philosophical nature, until the 17th century, when it was revived in the context of studies of the gaseous state of matter. Notable in this regard is Robert Boyle (1626–1691), whose gas law we have already discussed. The fact, however, that Boyle believed in corpuscles but not in the possibility of a vacuum, showed that he did not have a proper understanding of the difference between a gas and a solid. Not until the 18th century did Bernouilli (1700–1782), a mathematician, propound something like the modern theory of the nature of a gas, and explain pressure as being due to the bombardment of the walls of the containing vessel of a gas by "molecules".

Meantime, research in chemistry was providing evidence (albeit circumstantial) of the atomic nature of matter. Dalton (1766–1844) proposed the "law of multiple proportions", based on the observation that when two elementary substances formed more than one compound (e.g. H_2O and H_2O_2), the combining weights were in the proportion of simple integers. He also found that, at a given temperature and pressure, two volumes of hydrogen combined with one volume of oxygen to form two volumes of water (steam). Avogadro (1776–1856) interpreted observations of this sort in terms of the interaction of molecules, e.g. that two molecules of hydrogen combined with one molecule of oxygen to form two molecules of water. He also enunciated the following simple and striking law: that equal volumes of all gases at a given temperature and pressure contain the same number of molecules, a law from which Dalton's observation follows.

To illustrate these principles, let us consider the three gaseous compounds: H_2O, CO_2, CH_4, and look at the proportions by weight and by volume involved. By weight, one part of hydrogen combines with eight parts of oxygen. By volume (of gas) two parts of hydrogen combine with one part of oxygen. It follows that the "unit weight" of oxygen is 16 times that of hydrogen. This unit weight can be taken to be, on some appropriate scale, the weight of the molecule.

In the case of carbon dioxide (CO_2), six parts of carbon by weight combine with 16 parts of oxygen. The volume of oxygen used in the reaction is equal to the volume of CO_2 produced. The unit weight of CO_2 is 22/16 times the unit weight of oxygen. Thus, the weight of carbon in CO_2 is 6/16 of the weight of oxygen. So the unit weight of carbon, the weight of its reacting unit, is six times that of hydrogen. Once again, that of oxygen is seen to be 16 times that of hydrogen.

Consider finally methane (CH_4). By weight, 12 parts of carbon react with four parts of hydrogen. But this time, the volume of methane made is only half the volume of hydrogen. It follows that the mass of a methane molecule is eight times that of a hydrogen molecule. Since 3/4 of the mass of the methane molecule is carbon, the unit mass of carbon (which is in fact the mass of an atom of carbon) is six times the mass of the unit weight of hydrogen (the hydrogen molecule).

We see, then, that the data on masses and volumes of the reacting materials hydrogen, oxygen and carbon are nicely verified on the assumptions of the molecular model of gases, each element having a definite reacting weight.[11]

Chemistry aside, physics itself had provided a substantial body of evidence supporting the molecular hypothesis. For instance, Robert Brown, the Scottish botanist, found by careful observations through a microscope that tiny grains of pollen suspended in water manifested a seemingly random jiggling motion; it was, however, not until the time of Einstein that this was satisfactorily explained in terms of their collisions with molecules of the liquid. Also, Rayleigh, Maxwell, Clausius and Boltzmann had all contributed to the formulation of a successful theory of the properties of gases on the basis of the molecular hypothesis.

One of the most important elements in laying groundwork for modern physics was the discovery of the electron, which had its origin in the study of high-voltage discharges in gases. The history of this

---

[11] We have been rather careless here in the use of the words "mass" and "weight". One should in fact have used mass rather than weight throughout; unfortunately, the use of the term "molecular weight" has been fairly general in chemistry.

field goes back to Michael Faraday, who investigated the conduction of electricity through gases. Discharges through a gas gave rise to a glow. Faraday found that reducing the density of the gas enhanced rather than inhibited the glow phenomenon. The effect appeared to be a continuous and not an intermittent one, and persisted at the lowest pressures that Faraday could realize. His attempts to create an effective vacuum were, however, unimpressive by modern standards.

Twenty-five years after Faraday's research, the German mathematician Plucker had the idea of bringing a magnet close to the discharge tube, to see what its effect on the glow would be. He observed some deflection of the discharge. Ten years later Plucker's pupil Johann Hittorf repeated the same experiment with a much better vacuum, made possible by the intervening development of the mercury pump. Hittorf was also able to show that, by placing an object in front of the cathode, he could cast a shadow of it, thus showing that the discharge originated in the cathode.

A vigorous argument followed over the true character of the "cathode rays". German scientists, led by Hertz, contended that the evidence pointed to the rays being waves rather then particles. The prevailing view in England was that they consisted of electrically charged particles; the most prominent exponent of this view was J. J. Thomson.

The argument was ultimately settled in favor of the particle view by the French physicist Jean Perrin, who was able to "capture" the negative charge and measure it.

The "discovery" of the electron is usually attributed to Thomson in 1897. Figure 12.1 shows the sort of apparatus he used. The rays were emitted by the cathode by virtue of a voltage applied between C and A. Passing them through slits in A and B created a well-defined beam.

Let us see what can be learned by deflecting the beam (a) with an electric field applied between the plates D and E and (b) with a magnetic field applied across the tube from back to front. If we

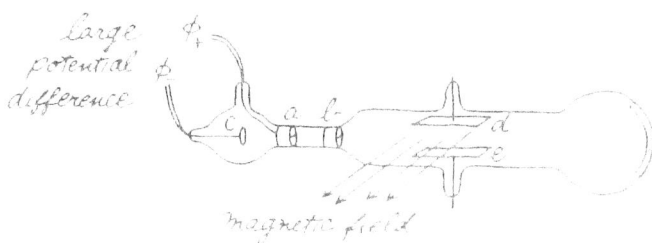

Fig. 12.1. Diagram of Thomson's apparatus. (From *An Introduction to the Meaning and Structure of Physics* by L. Cooper. Redrawn for Cooper's book from W. F. Magie, *A Source Book in Physics*, Harvard University Press, Cambridge. Copyright 1935, 1963 by the President and Fellows of Harvard College. Reprinted by permission of the author and the publishers.)

designate the charge on an electrified particle by e, its mass by m, the electric field by E, and the particle acceleration by a, the motion in an electric field is only governed by the equation

$$ma = eE .$$

The acceleration is vertical and determines the deflection, which takes place in the vertical plane. Thus, a as well as E will be known, so that e/m can be deduced. The force due to a magnetic field B depends on the particle velocity, and is equal to $(v/c)eB$ so that the acceleration it induces is given by

$$ma = eBv/c .$$

c is of course the velocity of light. In this case we can determine $(e/m)(v/c)$. We are now able to deduce v/c, and thus the velocity of the particles.

Having determined the ratio e/m, we now ask, how we can determine the charge e and the mass m separately? This was the goal of a famous experiment by the American physicist R. A. Millikan.

The idea of Millikan's experiment was as follows. A voltage was applied between two parallel conducting plates. A small hole was made in the upper plate, and fine oil droplets were sprayed above it. Some of the drops then fell through the hole; of these some were charged by shining ultraviolet light on the gas between the plates, thus liberating their electrons. This enabled the oil drops to pick up charge. They would then be subjected to an upward force due to the charge on the plates; the amount of that force would depend on the charge of the drop. But the drop would also be subject to a gravitational force downward, proportional to its mass. By adjusting the voltage between the plates, one could bring the forces into balance, and the drop would remain suspended. By observing what voltage brought about this balance, the ratio charge/mass of the drop could be determined. Then, by turning off the voltage and letting the drop fall freely, the mass of the drop could be determined separately, and its charge could finally be deduced.

Millikan observed that this charge was usually a multiple of a basic unit, which he assumed to be the charge of a single electron, the smallest quantum, so to speak, of charge.

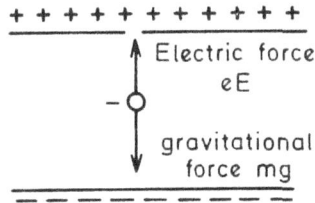

Fig. 12.2.

Something more precise needs to be said about the determination of the mass, since we have already learned that the rate at which objects fall in a vacuum is independent of their mass. In Millikan's experiment, however, the drop was falling not in vacuum, but in air, so that air resistance had to be taken into account. In fact, Millikan's drops fell with constant speed, indicating that the forces acting in the absence of a voltage on the plates were in balance. The gravitational force is mg, where m is the mass of the drop and g the acceleration of gravity. On the other hand, the frictional force due to the air is proportional to the speed of the fall of the drop, i.e. is kv, where k is a constant depending on the density of the air and the cross-section of the drop. From the equality of these forces

$$mg = kv$$

the mass may be deduced.

The mass of the drop is of course of no interest; what is important is that, knowing it, we may deduce its charge. Again, of course, the charge is that of the drop. The results, however, indicate the existence of an integral charge, which is then interpreted as the charge of the electron which it has picked up.

The story of this experiment is a fascinating one, and carries interesting implications concerning the nature of discovery in physics. At the same period in which Millikan was doing this experiment, the Austrian physicist Ehrenhaft was also testing the hypothesis of the existence of the electron. This does not seem to be quite the same thing as "measuring the charge of the electron"; the difference becomes clear if we look at the results of Ehrenhaft's experiments, since he found not merely singly and doubly charged particles, but ones whose charge appeared to lie between the two. Such a result, if true, would deal a severe blow to the whole concept of an electron as a particle with fixed properties.

The result of this discrepancy was a running controversy between Millikan and Ehrenhaft concerning the very existence of the electron. The fact was that Millikan, despite rather categorical statements to the contrary, had not by any means always found his results in agreement with his conclusions. What is clear is that he had a predisposition to believe that the electron, with its fixed charge, existed, and history, happily, proved him right. What he did, however, was to discard anomalous results, attributing them to spurious effects. His notebooks contain interesting marginal notes:

"Publish this surely, beautiful!"

"Error high, will not use"

"Might omit because discrepancy"

"This is almost exactly right and the best one I ever had!"

"Very low something wrong"

"Will not work out"

"Beauty, one of the very best"

"Excellent!"

For those who are taught that science rests on a foundation of complete objectivity, Millikan's approach may appear shocking. Millikan, however, had no doubt that the evidence for the electron hypothesis was overwhelming. If an experimental run did not give the expected result, something was wrong with the run, so it was legitimate to omit it; it was not an observation of the quantity with which he was concerned. Ehrenhaft, on the other hand, true to the principles of "honest" research, included all of his results, giving them all equal weight. In this way, he arrived at an erroneous conclusion, that the electron did not exist.

What this story proves is that propositions like the familiar one of Popper, that observations tending to refute a theory or an hypothesis have more scientific value than those providing support, must be properly qualified. In a broad sense, Popper's proposition is correct, but only as an approximation; refutation in science is necessary but not sufficient to arrive at the truth. This illustrates the hazard of philosophical generalization in science. Reality has more dimensions than such generalizations can portray; the words intuition, insight, imagination bear witness to a more "subjective" reality through which the generalizations must be interpreted. We must not beg the question, for example, of what constitutes a "refutation". A non-science student in one of my classes remarked that the most important thing he had learned about science was that one must always pay attention to detail. It is a message that is often lacking in "philosophical" interpretations of science.

Let us return to Millikan's result. Does it have a wider significance than the determination of a number? The answer is categorically positive. The determination of the charge of the electron, coupled with the already determined value of e/m, permits us also to calculate the mass of this newly established particle. The astonishing thing about this mass is its small size. It is thousands of times smaller than the mass of atoms! What, then, is the nature of the atom, and what role does the electron play in it? Before trying to answer this question, we have to look at some other discoveries.

First, there was the discovery by Roentgen of X-rays. The discovery was an accidental one, made in the course of experiments with a Hittorf tube. Operating a tube covered with black cardboard, in a completely darkened room, he noticed that a nearby screen, covered with barium platino-cyanide, was made to fluoresce. Fascinated and overawed by what was clearly a new form of radiation, he tried out various experiments. He found that his new rays passed through different objects, that photographic plates were extremely sensitive to them, and that the rays could not be deflected in a magnetic field. The discovery caused great commotion all over the world. The ability to see one's own bones, or solid objects embedded in the body seemed miraculous and incomprehensible. It was also shocking to Victorian sensibilities. The medical value of the discovery was however immediately appreciated.

Roentgen was quite puzzled about the nature of his "X-rays". He speculated that they might represent longitudinal vibrations in the aether, somewhat like elastic waves in a solid. It took over 15 years, and much work by others, before X-rays were understood to be nothing but Maxwellian electromagnetic waves of very short wavelength and high frequency.

Of even greater importance was the discovery of radioactivity. It held the key to the structure of matter, and its consequences were to mark one of the turning points in human history.

12.1. The Key to Pandora's Box; the Discovery of Radioactivity

The story begins, again, with an accident. Henri Becquerel, (1852–1908), whose grandfather and father had been distinguished scientists, had succeeded the latter in the chair at the Musée

d'Histoire Naturelle, and had become a professor at the Ecole Poly-
technique. He was an expert on fluorescence and phosphorescence,
and was intensely interested in Roentgen's X-rays. He was curious
as to whether intensely fluorescent bodies might not emit X-rays.
To explore the possibility, he exposed uranium, known to be fluorescent, to intense sunlight, and looked for penetrating radiation,
which would be detected by a photographic plate. During a spell
of bad weather which lasted a couple of days, he put his samples
in a drawer, along with undeveloped photographic plates. When he
later developed them, expecting to get only a very feeble darken-
ing, he found instead silhouettes of great intensity. Even in a dark
drawer, with his samples surrounded by black paper, the radiation
penetrated strongly without having been exposed to sunlight.

    Although his discovery turned out in the long run to be much
more important than Roentgen's, it attracted relatively little attention. Becquerel continued his investigations with uranium. Marie
Curie, in Paris, was examining various heavy minerals to see whether
there might be other radioactive elements. Her first discovery was
that thorium was also radioactive. Then she found that certain
minerals displayed an intensity of activity three or four times greater
than could be explained by their uranium and thorium content. This
led her to search for the element responsible. At his point her husband, Pierre Curie, joined her in the daunting task of isolating the
most strongly active component of these minerals. The first fruit
of this search was the isolation of radioactive polonium. But the
ultimate step was heroic. It involved starting with a ton of pitch-
blende residues and extracting finally small amounts of an intensely
radioactive element, radium, a million times more active than uranium. The work, carried out by primitive methods and in complete
ignorance of the dangers involved in the task, ultimately took its
toll on the health of Marie Curie; nonetheless, she survived to the
age of 67, though her later years were plagued by persistent and
little-understood illness.

In 1903 the Nobel prize was awarded jointly to the Curies and Henri Becquerel. Marie Curie was 36 and her husband 44.

It is interesting that the speech given by Pierre Curie at the Nobel Prize ceremony foreshadows many of our contemporary concerns. In it he says: "It is conceivable that radium in criminal hands may become very dangerous, and here one may ask whether it is advantageous for man to uncover natural secrets, whether he is ready to profit from it or whether his knowledge will not be detrimental to him. The example of Nobel's discoveries is characteristic; explosives of great power have allowed men to do some admirable work. They are also a terrible means of destruction in the hands of the great criminals who lead nations to war. I am one of those who believe, with Nobel, that humanity will derive more good than evil from new discoveries." Such was the scientific optimism of that time; today we have lost much of our innocence.

The next chapter in the story of radioactivity centers around Ernest Rutherford. The 24-year old Rutherford arrived in Cambridge in 1895, to work with J. J. Thomson. Their first project was to measure ionization produced by the then newly discovered X-rays, but they soon turned their attention to the radiation from uranium. Rutherford observed two sorts of radiation — alpha rays, not well understood but suspected by him and the Curies of being electrically charged particles emitted at high velocity, and beta rays, which were soon identified as electrons. A third kind of radiation, discovered in France and designated gamma, was found not to be affected by magnetic fields and was therefore not charged; the radiation was more similar to X-rays, though even more penetrating. Although we now know it to consist simply of very high frequency electromagnetic waves, the original terminology persists.

In 1898 Rutherford left Cambridge to take up a post at McGill University. One of his first projects there was to try to establish the nature of alpha rays. Though at first they appeared not to be deflected by magnetic fields, it was found on closer investigation that they were and Rutherford was soon convinced that they were helium ions. This was an astonishing hypothesis, in that one of its consequences would be a change in the atomic number, and thus of chemical properties of the emitting substance. In short, it appeared to be the realization of the alchemist's dream, the transmutation of one element into another!

Let us put the novelty of the process in another light. Chemical processes involve the interaction of atoms; the energies involved are of the order of a few electron volts. Chemical processes are also very sensitive to the environment (e.g. temperature) because thermal energies are comparable with the energies involved in reactions. The processes of radioactivity were correctly perceived by Rutherford to take place within the atom, and involve energies which may be hundreds of thousands times greater; they are therefore independent of macroscopic surroundings. They do not vary with temperature.

Shortly after Rutherford's arrival at McGill, a position opened in the chemistry department, which was filled by Frederick Soddy. Soddy became Rutherford's scientific collaborator. The team of Rutherford and Soddy carried out the pioneering work on the process of radioactivity, and in particular formulated the law of radioactive decay. The law states that the intensity of radioactive emission from a sample at any given time is proportional to the amount of radioactive material present. This can be expressed in a different way: each atom of radioactive material is just as likely to decay in any given infinitesimal interval of time as in any other.

Expressed in this form, the law has dramatic and unexpected consequences. Perhaps the most striking consequence is that it is quite impossible to predict when any given atom is going to decay. This appears to represent a breach in the supposedly inexorable laws of causality, though this aspect was not as evident then as it is now. It represents one of the earliest manifestations of the famous "uncertainty principle" which subsequently generated much controversy. The law is perhaps most vividly illustrated by comparing radioactive decay with human mortality. Actuarial tables spell out how the mean expectation of life diminishes as we grow older. When we are adolescents we can look forward to more than fifty years of life but at seventy, that expectation usually shrinks to about a dozen years. But the death of a radioactive atom is just as likely at one age as another. One that happens to survive to a ripe old age has the same expectation of life to come as one that is newly born.

Here, for the first time, we have evidence of a possibility that so disturbed Einstein. It appears that God does play dice with the universe!

Such a law is called an exponential decay law. The death rate at any time is proportional to the number of survivors at that time. This is the converse of the exponential growth law which we discussed in the first chapter, in which the rate of growth of a population at any given time is proportional to the population at that time.

A convenient measure of the decay rate of an atomic species is its half-life, the time needed for half the atoms to decay. A typical graph of the probability of survival to a given time is shown, as a function of time, in Fig. 12.3.

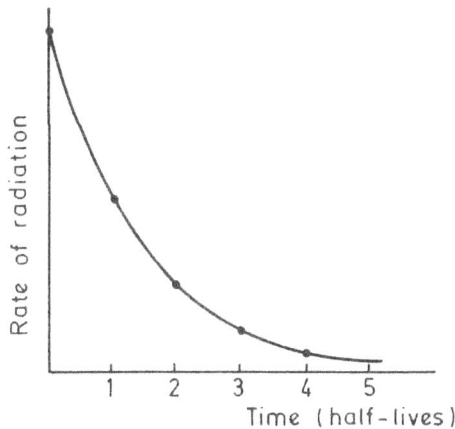

Fig. 12.3. Half of the nuclei decay in each interval shown.

12.2. The Structure of Atoms

In the foregoing discussion, we have always spoken of the radioactivity of atoms. We have, however, sidestepped the question of the nature of those atoms. How does the electron fit into the atom? The prevailing picture was one proposed by J. J. Thomson, who envisaged the atom as consisting of a homogeneous, positively charged "electric fluid", in which the electrons were embedded, something like the currants in a currant bun. Because of the electric forces, these electrons were imagined to arrange themselves in some energetically favorable pattern. If the atoms were excited, they would vibrate and emit electromagnetic radiation.

We must remember that the electron is very light compared with the atom as a whole; the positive "fluid" would then have to contribute most of the atom's mass.

Rutherford might have doubted this model, but, like any good scientist, he was interested in putting the idea to test. He had an idea about it, which became a prototype for one of the standard methods of particle physics. The test consisted of firing energetic particles into the atom, and observing how they were deflected. Interestingly, he did not do this experiment himself, but proposed it to two of his students, Hans Geiger and Ernest Marsden. It is reasonable to deduce from this fact that he did not expect the experiment to turn up anything radically new. If so, he was mistaken. No less was at stake than the discovery of the atomic nucleus.

A schematic diagram of the experimental arrangement is shown in Fig. 12.4. Alpha particles in a collimated beam impinge on a gold foil and are scattered in all directions. The numbers detected at different scattering angles are recorded. Using the Thomson model, only a small scattering would be expected (a) because the positive charge is uniformly distributed, so that the forces exerted on the alpha-particle largely cancel each other and (b) because the electrons are so light that they could not appreciably deflect the incident particle.

Something quite different was found. Most incident particles

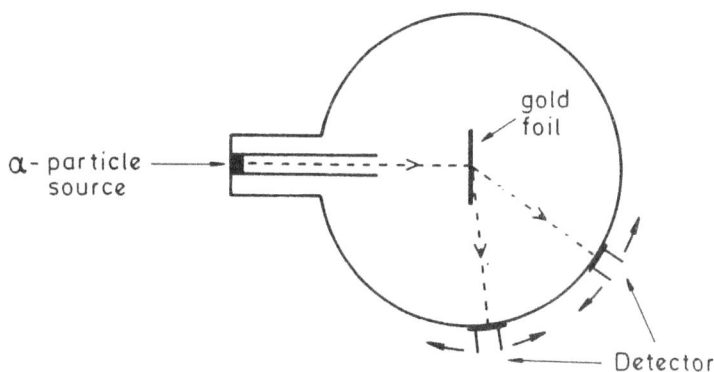

Fig. 12.4.

were only deflected through small angles, but a few were scattered at large angles; in fact, some almost recoiled backwards! What does this signify? First, something heavy, heavier than the alpha particle itself, must be responsible for this backward scattering. Secondly, this heavy object must be very small, since only a small proportion of the incident particles encountered it almost head-on. By analyzing the scattering intensity as a function of angle of scattering, one could deduce the size of the scattering object.

The result was astonishing. It showed that positive charge was concentrated in a region whose radius was of the order of one ten-thousandth that of the atom (which means a volume of one million millionth that of the atom). Into this minute space was concentrated about 99.95% of the mass of the atom. The "planetary" model of the atom emerged, and nuclear physics was born.

12.3. The Problem of Black-body Radiation

In our discussion of thermodynamics and statistical physics, we focussed our attention on phenomena involving material particles. We know, however, that energy is constantly exchanged in nature between matter and radiation. If we heat a lump of metal, at a certain temperature it will begin to glow. The colour of the light it emits will at first be a dull red. (Actually, before we can see radiation we can feel "heat" (energy) coming from it; this is actually invisible infrared radiation). As we heat it up, it goes through a cycle of colours — bright red, orange, yellow and finally white. It finally becomes "white hot". The lump of metal absorbs energy from its hot surroundings; it also radiates energy in the form of electromagnetic waves. Thermodynamics suggests that net energy flows from the hotter to the colder part of the system, depending on the temperatures difference. Does radiation then have a temperature? If so, how can it be determined? How, for that matter, can it be visualized?

If radiation is in equilibrium with its surroundings, it should be at the same temperature as those surroundings. One way of obtaining equilibrium is to imagine a box whose inner surface is painted black, so that it can absorb radiation at detectable frequencies. It is then heated to the desired temperature, where it is maintained long enough that the radiation in the interior can come into equilibrium with the sides. The radiation inside will then have the same temperature as the box. If we now make a small hole in a side of the box, the escaping radiation will be that of the temperature in question. This is called "black-body radiation" (Fig. 12.5).

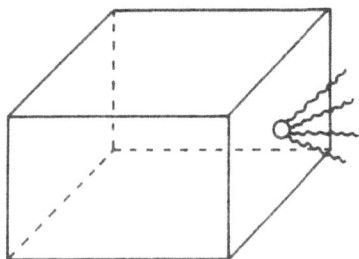

Fig. 12.5. Box filled with radiation in equilibrium.

The problem of the spectral constitution of black-body radiation interested both theoretical and experimental physicists near the turn of the century: Rubens, Pringsheim and Lummer on the one hand, Lord Rayleigh, James Jeans, W. Wien and especially Max Planck on the other.

Figure 12.6 shows curves giving the intensity of radiation as a function of wavelength at temperatures of 1600, 2400 and 3600 K. Aside from the increasing overall intensity of the radiation with increasing temperature, the peak of the curve, which corresponds to the frequency at which the radiation is most intense, becomes higher the higher the temperature (increasing frequency corresponds to decreasing wavelength).

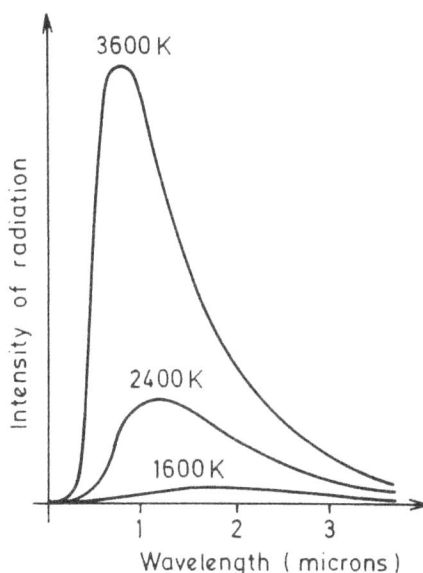

Fig. 12.6.

How would one go about making a theory of the frequency spectrum of black-body radiation? Rayleigh and Jeans addressed themselves to this problem. The problem is discovering how many "degrees of freedom" the radiation in the box had in any given fre-

quency range. The key to the question was to ask how many waves
in a given frequency interval, but traveling in different directions,
could exist inside the box. Taking the box to be rectangular with
sides a, b, and c, and using the fact that opposite faces were identical, we can state that an integral number of waves must fit into the
box in each direction. If we define wavenumber as the number of
wavelengths per unit length, which is the inverse of the wavelength,
then the wavenumber in each direction is an integer divided by the
dimension of the box in that direction.

What we can do, then, is represent each wave by a point on
a diagram, which we show in two rather than three dimensions for
simplicity. The points form a regular rectangular lattice, the spac-
ing of the points being $1/a, 1/b$, and $1/c$ respectively in the three
directions. These points are uniformly distributed over the diagram,
with one point in every volume $1/abc$ in three dimensions. The wavelength of each wave is given by its distance from the base point or
the "origin".

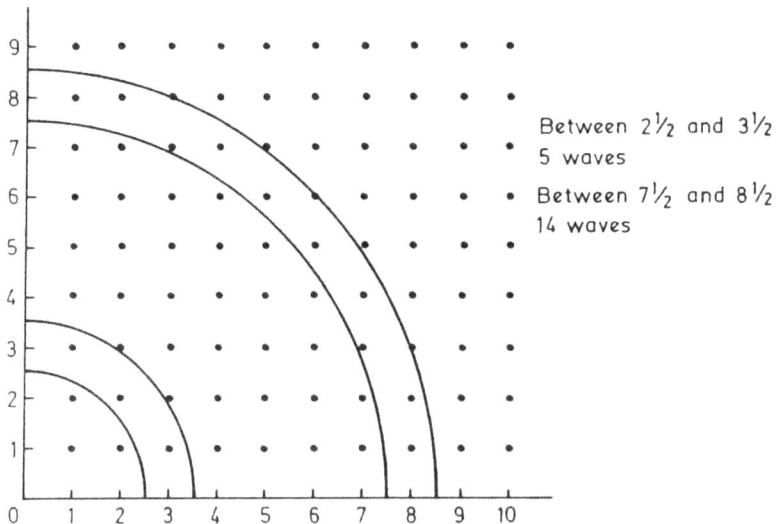

Between $2\frac{1}{2}$ and $3\frac{1}{2}$
5 waves

Between $7\frac{1}{2}$ and $8\frac{1}{2}$
14 waves

Fig. 12.7.

Figure 12.7 shows that there are more waves in the enclosure at high wave number than at low wave number. Let us trace out two regions corresponding to equal intervals of wave number; one between radii 2.5 and 3.5 of the graph, and the other between radii 7.5 and 8.5. The areas of the two regions are in the ratio

$$(3.5^2 - 2.5^2)/(8.5^2 - 7.5^2) = 6/16 = 0.375 \ .$$

The number of points lying in these regions should be roughly in the same proportion, since the points are uniformly distributed. Inspection indicates that the ratio is $5/14 = 0.357$. In three dimensions the corresponding ratio is smaller, since we are concerned with relative volumes, which are proportional to the cube of the radius, rather than areas, which depend on the square. The ratio is then about 0.124.

Since, at a given temperature, each degree of freedom should get its "fair share" (kT) of energy, the prediction of the theory of Rayleigh and Jeans is that the intensity of radiation will rise indefinitely at large wave numbers, or frequency; that is, at short wavelengths. This is reasonable enough as a consequence of the model, since it appears that one can in fact fit in indefinitely many waves of short wavelength. It appears to be precisely at this point that the theory goes astray; experiments show that this high intensity at short wavelength does not occur (Fig. 12.8).

Max Planck, who had become professor at the University of Berlin in 1889, had a deep interest in thermodynamics and a predilection toward problems of a fundamental and general character. The fundamental character of black-body radiation is manifested in the fact that the frequency spectrum of the radiation does not depend on the radiating material; the law depends only on the character of the radiation itself. Wien had proposed an empirical formula, which at least fitted some available data, and Planck wanted to find a theoretical basis for it. His point of departure was that radiation was created by independently oscillating atoms in the surface of the material; the problem was to find the energy and entropy distributions

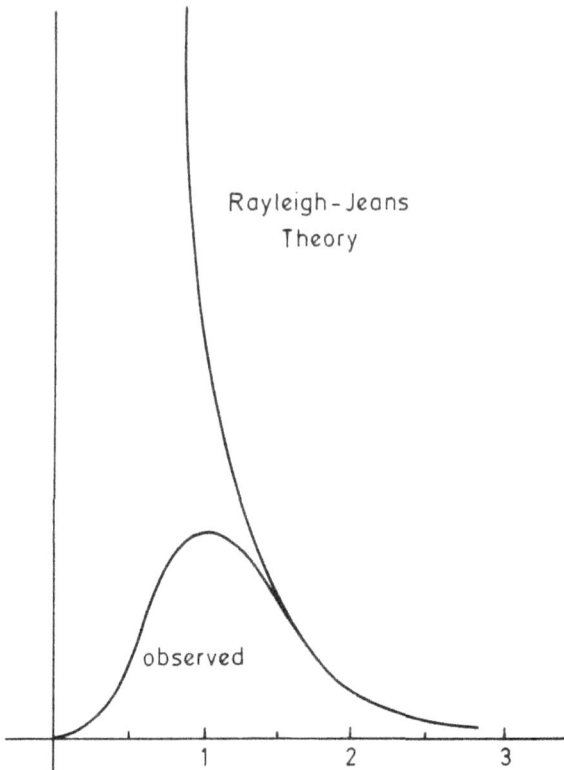

Fig. 12.8.

of these oscillators. The counting of possible configurations which is at the basis of the calculation of entropy is a straightforward problem for discrete systems, but becomes more obscure for continuous ones. Planck therefore, for convenience of calculation, divided the scale of energy into minute but discrete elements whose energies he took to be multiples of a small unit e^*. He found, however, that to agree with the experiment, he had to choose these intervals to be proportional to the frequency of the oscillator, i.e. e^* had to be proportional to the frequency f. The proportionality constant could also be determined by recourse to the experimental data. Following this line of argument, Planck obtained a formula for black-body spectrum which was a refinement of Wien's. Planck expected to obtain a result independent of the size of the energy intervals and thought the discrete intervals were merely an artifact of the calculation. For such intervals to have a fundamental physical significance had revolutionary implications, which were not lost on Planck. Incredulous at the significance of what he called his "act of desperation", he told his son that he had made a discovery worthy of Newton. Nevertheless, he was to spend years trying to circumvent the arbitrary assumption of the "quantum" of energy. At the end he confessed the efforts had led him to the conviction that "the quantum of action has a much more fundamental significance than I originally suspected". In fact, it was more fundamental than he could have imagined!

We should recall the context of Planck's hypothesis; he introduced the quantum into a consideration of the sources of the radiation, and not of the radiation itself. This did not seem entirely consistent with the fact that the radiation spectrum was independent of the radiating material, as we have already noted. It was left to Einstein, in that crucial year of 1905 in which he gave the world the special theory of relativity, $E = mc^2$, and the theory of Brownian motion, to put Planck's quantum in the proper perspective. For he addressed directly the problem of the entropy of radiation, and this led to the idea that the energy of radiation came in quanta proportional to its frequency. He considered the quantum of light energy itself to be absolutely fundamental, despite its apparent contradiction of Maxwell's wave theory of light. Another fundamental problem remained: the reconciliation of the wave and quantum theories.

12.4. Photoelectric Effect

Einstein's belief in the new "photon" theory of light was so strong that he applied it to another puzzling phenomenon — the recently discovered photoelectric effect (Fig. 12.9). It is this work which finally won him the Nobel prize in 1921. The effect, first

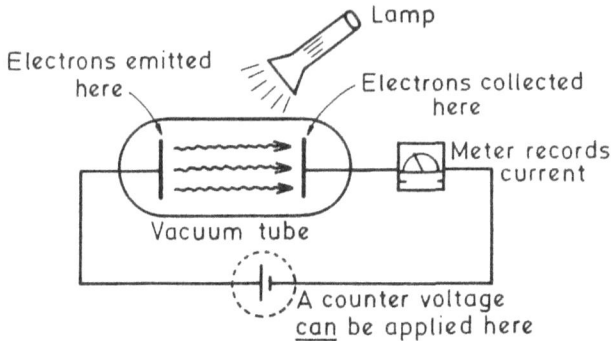

Fig. 12.9.

observed by Hertz, is this: if a metal surface is bombarded by ultra-violet light, electrons are ejected. Subsequent experiments showed several curious features:

(i) When the light intensity was increased, more electrons were emitted, but their energy was not increased.

(ii) The higher the frequency of the light at a fixed intensity, the more energetic were the emitted electrons.

(iii) The electrons were emitted from the instant the light was turned on; there was no delay.

What should the results be, according to electromagnetic theory? Consider first the effect of increasing the intensity of the light. In the Maxwell theory, more intense light means a stronger electric field acting on the electrons in the solid; they should therefore be accelerated to a higher energy and so be emitted at higher velocities (Fig. 12.10a).

What would one expect to be the effect of using light at higher and higher frequencies? At higher frequency, the electric field of the radiation would change sign more and more frequently. At low enough frequencies, the electrons could follow these oscillations, but with increasing frequency the inertia of the electrons would inhibit them from continuing to do so; the electrons would then pick up less energy from the field and would be emitted at lower, not higher,

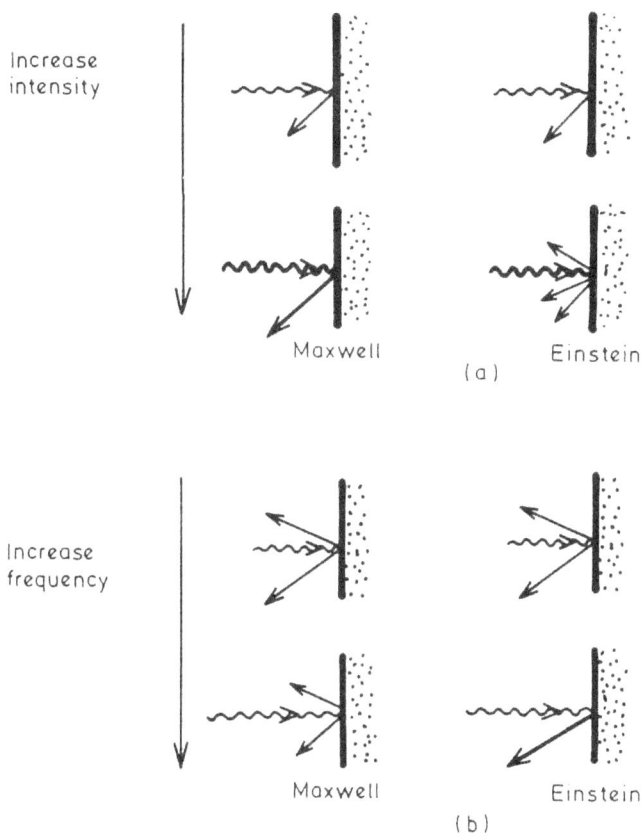

Fig. 12.10.

energy (Fig. 12.10b).

Finally, electrons would not be expected to be emitted instantaneously; it would take some time before they could be accelerated to a high enough energy to escape from the solid surface. Once again, this seems not to be the case.

Suppose now that we look at the problem from the viewpoint of the quantum theory of light.

(i) At a given frequency, a greater intensity of light simply means more photons, but their energy remains the same. Thus, more electrons will be emitted, but their energies will be unchanged.

(ii) If the frequency is increased and the intensity (energy flux per unit area) is kept constant, the photons will, by Planck's hypothesis, have a higher energy, but the numbers will be fewer. Thus, fewer electrons will be emitted at higher speeds.

(iii) Since the light comes in energy bundles, which can only be absorbed whole, the activation of the electrons can be almost instantaneous; there need not be a time delay.

Thus in every respect, Maxwell's theory makes the wrong prediction, and the photon theory the correct one.

Does this mean that Maxwell's theory is wrong? No more so than Newtonian mechanics. It is simply, in its classical form, inadequate for the treatment of phenomena involving single quanta. It is necessary to quantize electromagnetic theory, remembering Bohr's correspondence principle that in the limit of highly excited quantum states, the quantum and "classical" results must approach each other more and more closely. The photoelectric effect shows how radically different the two theories can be in the true quantum regime.

It might, from the above account, seem that the photoelectric experiments should immediately have convinced everyone of the validity of the photon hypothesis. Unfortunately, that is far from the truth. This cannot be attributed, either, simply to the reluctance of physicists to accept radically new ideas. For one thing, the experiments undertaken by Hertz and later Lenard were difficult, and could well not appear definitive enough to justify overthrowing very basic and well-established concepts. The weight of accumulated evidence in favour of Maxwell's theory made it difficult to accept with confidence evidence of possible shortcomings. It is not surprising that only someone with such unfailing instincts as Einstein's could entertain the possibility that there might be a way to reconcile such seeming contradictions. Though Einstein may have made other, even more radical, contributions to changing our world view, the linking of his Nobel prize to his work on the photoelectric effect is one of the striking examples of his daring and insight.

The photon controversy dragged on therefore for many years,

and grudging acceptance of the quantum view of light came only after the quantum idea had accumulated a number of other successes. The most important of these were in the field of spectroscopy, so it is to this that we now turn our attention.

12.5. The Problem of Discrete Spectra

Isaac Newton is best known for his work on mechanics, but his interests extended to all areas of physics. In 1666 he published a classical work on "The Phaenomena of Colours" in which he observes that when light is passed through a prism, it is broken up into a band ("spectrum") of colours, just as in a rainbow. The colours also represent a spectrum of wavelengths or frequencies, the splitting being due to the varying properties of light of different frequencies in a material medium. It was not until 1802 that W. H. Wollaston noted that one could observe dark lines within the visible spectrum. Shortly after, Fraunhofer, by combining a prism, a small slit and a viewing telescope (a "spectrometer") was able to detect a large number of vertical lines of this sort across the spectrum.

In 1859, Kirchhoff and Bunsen (of the famous Bunsen burner) discovered that two spectral lines observable in the spectrum of a candle flame fell at precisely the same frequencies as ones found in sunlight. This led Kirchhoff to surmise that spectroscopy might provide a source of information concerning the composition of the sun. This surmise is based on the assumption that certain spectral lines were characteristic of specific chemical elements. This notion was strengthened when it was observed that bright lines from the flame were enhanced when salt was poured on the flame. From this he concluded that there was sodium in the atmosphere of the sun (Fig. 12.11).

The bright lines in the candle flame in fact coincided with dark lines in the solar spectrum. The existence of dark lines in its spectrum appeared to indicate that the cool outer part of the sun's atmosphere absorbed light at the frequencies in question, light which had been produced deeper in the sun's atmosphere.

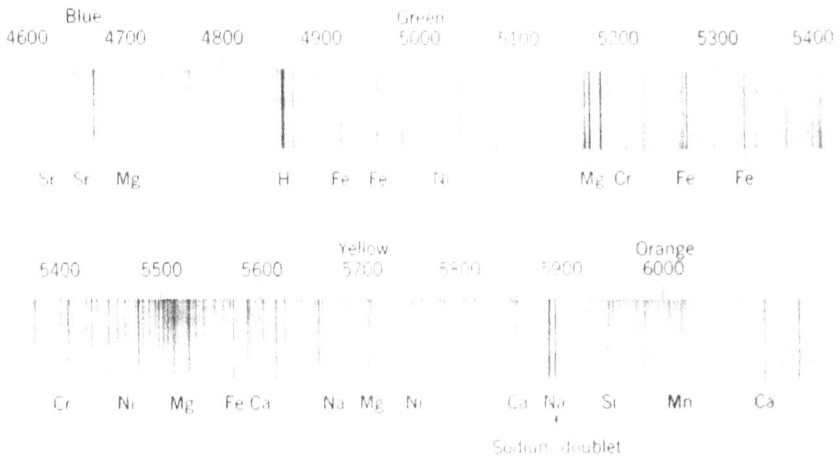

Fig. 12.11. (From *Discovering Astronomy* by W. W. Jeffries and R. R. Robins. Copyright ©1981 by John Wiley and Sons, Inc.)

These observations led Kirchhoff to the formulation of two fundamental laws of spectroscopy: first, that each chemical species had a characteristic spectrum, a sort of fingerprint for that species; and secondly, that each species could absorb light at the same frequencies which it emitted. The importance of these laws is evident; they provide a method by which one can determine the qualitative composition of even complicated compounds or mixtures.

The spectrum of hydrogen appeared particularly simple, and was therefore subjected to intense study. In 1885, Balmer found a simple mathematical relation between the four strongest lines in the spectrum; extrapolating from the formula, he found that it predicted almost the entire spectrum quite well. Balmer's formula for the wavelengths of the lines was

$$\text{wavelength} = \text{constant} \times n^2/(n^2 - 4) \ .$$

In terms of frequency this can be written as

$$\text{frequency} = \text{constant} \times \left(1 - 4/n^2\right) \ ,$$

or equivalently,

$$\text{frequency} = \text{constant} \times \left(1/4 - 1/n^2\right) ;$$

where the second constant is 4 times the first. The quantity n here is an integer equal to 3 or greater.

This particular series of spectral lines were the first observed because they fell within the visible spectrum; the $n = 3$ line corresponds to a wavelength of about 6500 Angstroms (1 Angstrom = 10^{-8} cm). The others have somewhat greater frequency or shorter wavelengths, going to a limit for large n of about 3400Å. The 6500Å line lies in the red end of the visible spectrum; the others lie progressively nearer the blue end.

This series of lines is called the Balmer series. In the infrared region, other lines appear, whose frequencies are given by the formula

$$\text{frequency} = \text{constant} \times \left(1/n^2 - 1/n'^2\right) .$$

The constant is the same as for the Balmer series. n and n' are integers, and $n' > n$.

All terms of the series may be expressed as sums or differences of two others; for example

$$\left(1/4 - 1/n'^2\right) - \left(1/4 - 1/n^2\right) = \left(1/n^2 - 1/n'^2\right)$$

or

$$\left(1/n_1^2 - 1/n_2^2\right) + \left(1/n_2^2 - 1/n_3^2\right) = 1/n_1^2 - 1/n_3^2 .$$

This is a manifestation of a general principle governing all spectral series, in any chemical species whatsoever. It is called Ritz's combination principle. It states that either the sum or the difference of the frequencies of any two spectral lines of a chemical species is the frequency of another line of its spectrum.

Figure 12.12 shows prominent lines in the visible spectra of sodium, mercury, helium and hydrogen. The patterns for the various elements are so different that the spectrum provides a very

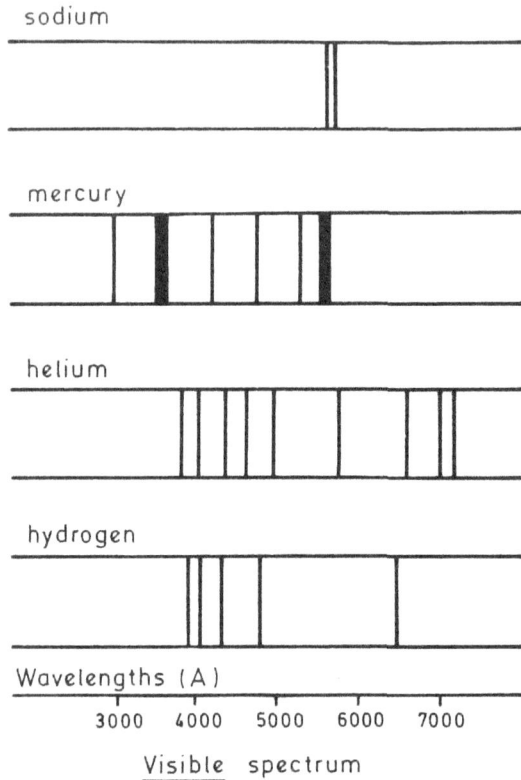

sodium

mercury

helium

hydrogen

Wavelengths (A)

3000   4000   5000   6000   7000

Visible   spectrum

Fig. 12.12.

convenient fingerprint for distinguishing them and so detecting their presence in a mixture.

In retrospect it is rather surprising that the spectral information which displayed such a simple and elegant pattern did not provoke as much curiosity as one might have expected. Surely clues were present there, obscure perhaps but nonetheless highly suggestive, of the nature of atoms. Thomson's "currant bun" model gave no insight into the source of the simplicity, yet even that fact might have generated more scepticism about the model. Until Rutherford pointed the way to a "planetary" model, the work on spectroscopy was not linked to other studies which were leading inexorably to the idea of the quantum, and seems to have played no role in that process. It was only when the quantum idea had taken hold, and when Einstein was already promoting the photon model of light, that Bohr was led to adapt the quantum notion to spectroscopy and to the detailed description of the nature of the atom. Perhaps too many pieces of the puzzle had to come together simultaneously, and only a mind as profound as Bohr's was capable linking all the elements in a coherent pattern. One cannot help wondering, however, especially in the light of what followed, whether a shuffling of the cards of history might not have given us quantum mechanics as a product of spectroscopy alone. One could argue that all the clues were there.

Chapter 13

FUNDAMENTAL PRINCIPLES OF
QUANTUM THEORY

The accumulation of spectroscopic evidence that atomic radiation showed frequency peaks, along with Einstein's introduction of the light quantum (which was first called a "photon" only in 1926), was strong evidence that atoms existed in discrete energy states. The hydrogen atom, for example, had strong peaks in the emission and absorption of light at frequencies proportional to

$$\frac{1}{n_1^2} - \frac{1}{n_2^2} \, ,$$

where n_1 and n_2 were integers. This could be interpreted thus: hydrogen atoms existed in energy states $-R/n^2$, and when a hydrogen atom in a state n_2 made a transition to a state n_1, it would emit a light quantum of energy

$$R \left(\frac{1}{n_1^2} - \frac{1}{n_2^2} \right)$$

and hence of frequency equal to this quantity divided by h, the Planck constant.

But how could this be? The hydrogen atom might be likened to a sort of solar system held together by electric rather than gravitational forces, the negatively charged electron playing the role of a "planet" orbiting around a positively charged proton "sun". But planets in solar systems can have arbitrary energy. Furthermore, it was known from Maxwell's electromagnetic theory that accelerating charges emitted a continuous spectrum of electromagnetic radiation. Worse than that, radiation was predicted to be so strong that the electron lost energy very rapidly, and would in fact spiral in to the nucleus in a minute fraction of a second. Nothing in a world built in this way would be stable!

Even if one could think of a mechanism for transitions between atomic states in which the atom suddenly lost a discrete quantity of energy, other questions remained. What determined *when* the atom would make such transitions, or in what direction the atom would emit its quantum of radiation? The new quantum ideas seemed to pose a myriad of new and seemingly unfathomable puzzles.

At least one of them, however, was not new, as Einstein was quick to observe. The problem of *when* an unstable entity would decay has already appeared in Rutherford's law of radioactive decay. In radioactivity, too, it was impossible to predict *when* a given nucleus would decay.

In 1913, Bohr put forth a radical "explanation" of the spectrum of the hydrogen atom. It could scarcely be called a "theory" since it contained elements in flat contradiction to already proven theory; in particular, Maxwell's theory of electromagnetism.

Ignoring the problem of stability, Bohr proposed that the condition determining the discrete states was that their angular momentum (i.e. the angular momentum of the electron around the proton) should be an integral multiple of $h/2\pi$. He further contended that "experiments indicate very strongly that electrons can rotate in atoms without emission of energy radiation".

Consider circular orbits. They must be determined by a balance of "centrifugal force" (mv^2/r) and electrostatic attraction $(-e^2/r^2)$. Putting the angular momentum $mvr = n\hbar$ we obtain a relation between the radius of the orbit and its speed v. Thus, in the equation

$$\frac{mv^2}{r} = \frac{e^2}{r^2} \, ,$$

we can substitute for the speed:

$$v = \frac{n\hbar}{mr} \qquad (\hbar = h/2\pi)$$

to get an equation for the allowed radii of the orbits

$$\frac{m}{r}\left(\frac{n\hbar}{mr}\right)^2 = \frac{e^2}{r^2} ,$$

from which it follows that

$$r = n^2 \frac{\hbar^2}{me^2} .$$

The quantity $\hbar^2/me^2$ is called the "Bohr radius" and determines the size of the atom. It is a bit more than half an angstrom (1 angstrom $= 10^{-8}$ cm).

The speed of the electron in its orbit is seen to be

$$v = \frac{e^2}{\hbar n} .$$

Then the *energy* of the electron can be calculated:

$$E = \frac{1}{2}mv^2 - \frac{e^2}{r}$$

$$= \frac{1}{2}\frac{me^4}{\hbar^2 n^2} - e^2 \frac{me^2}{\hbar^2 n^2} = -\frac{1}{2}\frac{me^4}{\hbar^2 n^2} !$$

In this way Bohr obtained "quantized" orbits whose energy is proportional to $1/n^2$. He also obtained the proportionality factor, which is known as the Rydberg constant R

$$R = \frac{1}{2}\frac{me^4}{\hbar^2} .$$

At first sight this looks quite impressive. But we must remember the price of this success: a standing conflict with Maxwell's electromagnetic theory. Another problem was that the only atom which could be treated in this way was hydrogen; there was no prescription for determining the quantum states of atoms with more than one electron.

And finally, we know now that the lowest energy state of a hydrogen atom is not one with an angular momentum $\hbar$, but rather one with no angular momentum at all. This comment is related to another electromagnetic consequence of the model, that the circulating electron creates a magnetic flux through its orbit, which is proportional to the angular momentum and thus increases as the energy (through the quantum number n) increases. The atom, in short, becomes a magnet, the various states having different magnetic strengths (moments).

So while Bohr's model gives the correct value of the energy of the lowest energy state of the atom, it attributes to it the wrong angular momentum and thus the wrong magnetic properties. The fact that in reality not only the ground state but a series of excited states have no angular momentum is inconsistent with the orbiting electron model.

Progress in the understanding of atomic spectroscopy was slow in the ten years following the publication of Bohr's paper, a fact probably due in no small part to a general reluctance among physicists to accept the idea of photons. Bohr himself was one of the most reluctant, and in fact published a paper, with H. A. Kramers and J. C. Slater in which, to avoid the photon concept, it was proposed to abandon the strict application of the laws of conservation of energy and momentum!

The photon controversy was ended, however, in 1928 with the publication of a very important paper by A. H. Compton on the scattering of light by electrons, which showed clearly the validity of the photon concept. Compton's experiment consisted of scattering X-rays by the electrons in the molecules of certain gases. Because the energy of the X-rays was very high compared with the binding energies of the electrons, the latter could be treated as effectively free and at rest.

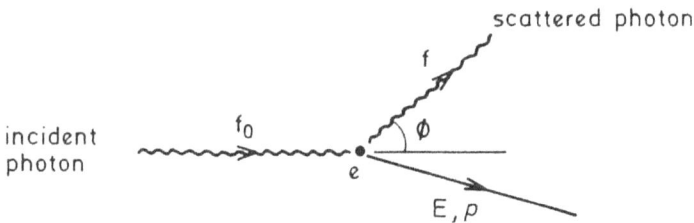

Fig. 13.1.

Considering the scattering to be that of photons of energy hf_0, he analyzed the collision in terms of the laws of conservation of energy and momentum. Since the electron recoils, the photon loses energy. Due to the Planck relation $E = hf$, it must also decrease in frequency. The amount of this decrease depends on the angle of scattering ϕ, in a manner that can be calculated directly.[12] Compton's experiment accurately verified the predictions of that calculation. A "classical" calculation would give quite a different result, since the energy lost to the electron would in this case decrease the *amplitude* (or intensity) of the radiation rather than its frequency. The experiment, then, gives a *definitive* verification of the existence of photons.

Almost immediately after the publication of Compton's results, the young French physicist Louis de Broglie put forth a hypothesis which opened the floodgates to a series of developments which would result, in the short span of six years of frenetic development, in the emergence of quantum theory in its definitive form.

"After long reflexion in solitude and meditation, I suddenly had

---

[12] For those interested in details, we include here a proof of the relation which Compton's experiment verified. We will take f_0 to be the frequency of the incident radiation. The frequency after collision, which we shall call f, will be accompanied by a momentum hf/c which will have two components, one in the direction of the incident photon (hf_1/c) and the other transverse to it (hf_t/c). The target electron will also recoil, with similar momentum components (p_1 and p_t).

the idea, during the year 1923, that the discovery made by Einstein in 1905 should be generalized by extending it to all material particles and notably to electrons" (from the 1924 doctoral thesis of de Broglie).

The logic of the step was evident. If light waves could be quantized ("behave like particles") why should not "particles" (electrons) sometimes have a wavelike character?

Supposing that they did, what would be the properties of the waves?

An electromagnetic wave has a frequency, which is its energy divided by h. It also has a wavelength λ equal to c/f, or $hc/hf = h/p$, p being its momentum. De Broglie proposed the same relations for electrons: that the frequency of electron waves be E/h and their

---

The equation of conservation of energy is then

$$mc^2 + hf_0 = E + hf ,$$

where E is the electron energy after collision $[(mc^2)^2 + (pc)^2]^{1/2}$ Momentum must be conserved in both transverse and longitudinal directions:

$$p_t = hf_t/c \quad \text{and}$$

$$hf_0/c = hf_1/c + p_1 .$$

Using the fact that $p^2 = p_1^2 + p_t^2$ we get

$$p^2 c^2 = h^2[f_t^2 + (f_0 - f_1)^2]$$
$$= h^2(f_0^2 + f^2 - 2f_0f_1) .$$

But also $E^2 = (mc^2)^2 + h^2(f_0 - f)^2 + 2mc^2h(f_0 - f)$. Since $E^2 = (pc)^2 + (mc^2)^2$ we can deduce after some cancellation that

$$mc^2h(f_0 - f) = h^2f_0(f - f_1)$$

is the relation between the frequency of the incoming radiation and that of the scattered radiation.

This relation is often expressed in terms of wavelengths $\lambda = c/f$. If we divide the equation obtained by $mchff_0$ and put $f_1/f = \cos \phi$, where ϕ is the angle of scattering of the photon, we get $\lambda - \lambda_0 = (h/mc)(1 - \cos \phi)$.

wavelength h/p, p and E being respectively the particle's momentum and energy. (It was by this time a well-known relativistic result that, just as $[x, ct]$ were the components of a vector in Einstein's space-time, so were $[p, E/c]$ and $[1/\lambda, f/c]$.)

In terms of space-time vectors, de Broglie's hypothesis is that

$$[p, E/c] = h[1/\lambda, f/c] \ .$$

The relation $E = hf$ comes from the equality of the time components of these space-time vectors, while $p = h/\lambda$ comes from that of the *space* components. But we saw in our discussion of the special theory of relativity that what is a *temporal* effect in one frame of reference can be a spatial effect in another. Thus, the two relations

$$E = hf$$

and

$$p = h/\lambda$$

derive logically from each other.

What precisely is the significance of these "electron waves"? No one, not the least de Broglie, was very sure. One hypothesis was that they were waves that *guided* the "particles" in some unspecified way. Whatever one said about this question, however, one possible test, or verification of the hypothesis suggested itself. After all, one of the most striking characteristics of light waves (or any known waves, for that matter) was the phenomenon of *diffraction*.

Suppose that a beam of electrons impinges on a crystal, the first two layers of which are portrayed in Fig. 13.2. Part of the beam is reflected at A in the first layer, another part at B in the second layer. The latter part travels an extra distance (CB + BD). If d is the distance AB between the layers, CB and BD are of length $d \sin \theta$, where θ is the angle of reflection.

Suppose now that the extra distance $2d \sin \theta$ is one wavelength of the electrons. Then, the two waves reflected from the successive layers will reinforce each other on reflection. This happens when

$$2d \sin \theta = \lambda \ .$$

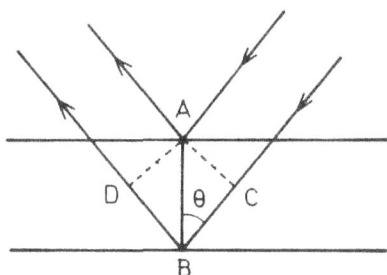

Fig. 13.2.

One in fact gets reinforcement when the length is any integral number of wavelengths:

$$2d \sin \theta = n\lambda \qquad\qquad n = \text{an integer} .$$

If, on the other hand, $2d \sin \theta$ is *half* a wavelength, or any integer number plus half, the reflected waves will cancel each other out. They will be *out of phase*.

Suppose, for example, that d were 1.5 angstroms, -1.5×10^{-8} cm, and θ were 30°. We would then expect a strong reflection when

$$\lambda = 3 \times \frac{1}{2}\text{Å} = 1.5\text{Å} .$$

This would correspond, according to de Broglie, to an electron momentum of 4.4×10^{-19} g cm/s, or an energy of 67 electron volts (an electron volt is the energy gained by an electron accelerated by an electric potential difference of one volt, so the production of 67 electron volt electrons is quite a simple matter).

A change, either of the angle of the beam relative to the surface, or of the electron energy, would result in a weakening of the reflected beam.

Let us note, however, what the argument signifies. In 1914, Sir William Bragg had demonstrated the diffraction of X-rays by crystals. But X-rays are electromagnetic waves. The fact that the same phenomenon could be demonstrated for a beam of electrons would be the most direct evidence conceivable for the wave character of electrons; thus, for de Broglie's daring hypothesis. The experiment was done by Davisson and Germer in 1926. They found diffraction patterns remarkably similar to those found for X-rays; furthermore, the electron wavelengths were found to be exactly as predicted by de Broglie.

At this point, physics entered a new era. The old "particle" picture of the ultimate constituents of matter was dead. Matter had become *waves* – fields – and physics was launched into the quantum era.

Before pursuing the consequences of this new discovery, let us note that it provides a sort of rationale for Bohr's idea that what characterized the "quantum states" of atoms was the quantization of angular momentum.

Try and imagine an electron in a Bohr orbit as a wave. This is only possible if the circumference of the orbit is an integral number of wavelengths, i.e. if R is the orbit radius

$$2\pi R = n\lambda = nh/p \qquad \text{(de Broglie)} .$$

This implies that

$$Rp = nh/2\pi .$$

The left hand side of this relation is the angular momentum of the electron around the nucleus; we have now "derived" Bohr's condition in a most unexpected way.

13.1. Remarks on "Wave-particle Duality"

As Bohr repeatedly insisted, a great deal of confusion has been generated about the quantum theory by the careless use of words. Quantum mechanics has been said to "violate causality"; waves have been said to "collapse" to particles upon observation ("collapse of the wave packet" is a philosopher's favourite); "reality" has been claimed to be created by "observers" or "observation"; "phenomena" are said to be disturbed by observation, and matter is said to display a "wave-particle duality", that is, to act sometimes as wave and sometimes as

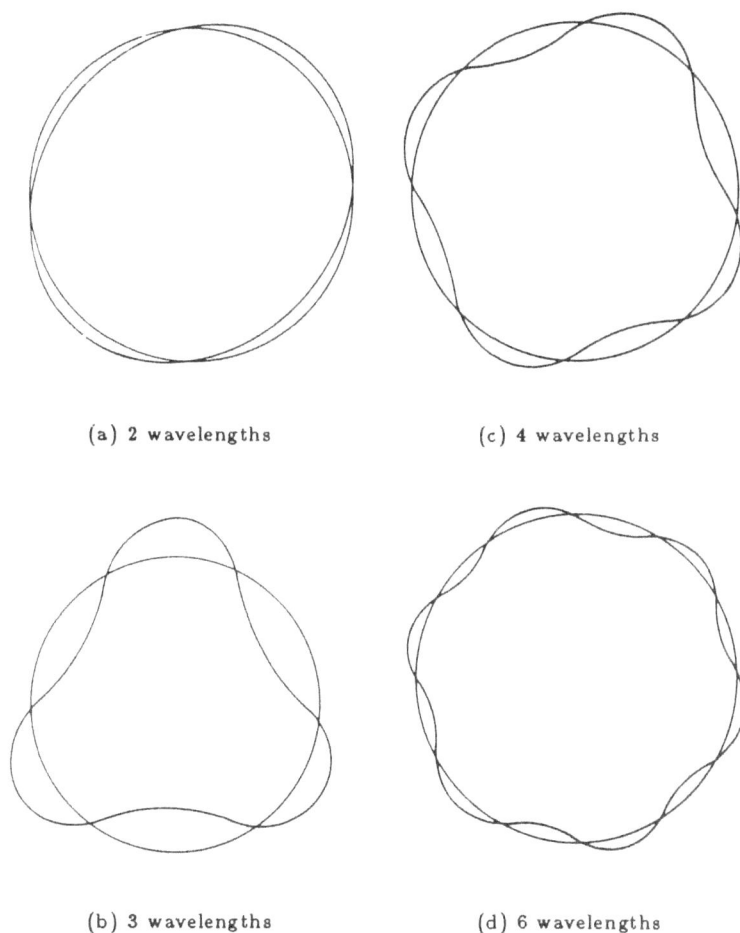

(a) 2 wavelengths

(c) 4 wavelengths

(b) 3 wavelengths

(d) 6 wavelengths

Fig. 13.3. "Angular" de Brogile waves.

particle. In these statements, the words "causality", "observation", "observer", "reality", "phenomena" and "particle" are not used in their conventional sense, so that the images conveyed by them may be misleading and give a false impression of the significance of the quantum theory. Even the word "uncertainty" (as in "uncertainty principle") has misleading connotations. It is unfortunate that some philosophers, as well as some physicists (including Einstein himself), to say nothing of popularizers of science who wish to titillate the reader with the supposed paradoxical character of quantum physics, do not take greater pains to be precise in the use of these terms. We shall subsequently deal with these questions, but for the moment, we will confine our attention to "wave-particle" duality.

A useful starting point in the discussion is to affirm, as Feynman is so fond of doing, that nature behaves *in a quantum fashion*. This fashion is quite strange to us, because it manifests itself on the level of the elementary components of matter. We have not been endowed, being, relatively speaking, "macroscopic" creatures made up of "astronomically large quantities" of these components, with sensory organs to perceive how matter behaves on this "microscopic" scale. We can only surmise, from macroscopic observation, what is going on at the level of atoms, molecules and the like. But our language is based on our experience, so we only have our macroscopic images to describe the microscopic world. In short, we carry deep within us prejudices based on our observation of the world at the human scale. The images, in fact, are all deeply rooted in the world of physics as observed at that scale. This is, in essence, Newtonian physics.

But the new physics is — as in the case of relativity — very novel, outside our direct experience and the language and images rooted in it. The problem is how to describe something fundamentally and profoundly *new* in terms of the language of the reality of the old and familiar? Can we do it without obscuring or distorting the novelty of our new perceptions?

We inevitably use analogies. But it is important to realize that an analogy has limitations. It cannot convey the *totality* of the new reality. We can only use *partial* analogies. What we lose we can only try to recapture by multiplying analogies, elements of which may even appear to contradict each other. But nature is not paradoxical, or self-contradictory, or "dual". These qualities exist only in our minds. Nature is — simply nature. In the old comic strips Popeye says "I am what I am and that's all that I am". We must, to use an

Einsteinian phraseology to express a very non-Einsteinian sentiment, stop telling God what he can do.

So we had an image in our minds of a "particle". It was an infinitely small object. At the atomic level, this concept seems to be wrong. Particles cannot diffract. Waves can diffract, so a particle is "wavelike". But can it sometimes (paradoxically) be particle-like? No, in the sense defined, it cannot be.

What, then, do we mean by particle-like? Is Planck's quantum of electromagnetic energy "particle-like"? Not at all. The electromagnetic quantum has a frequency, which means that it has extension in time, and a wavelength, which implies extension in space. Our most sophisticated quantum view is this: we first imagine a Maxwellian electromagnetic wave with a given frequency. "Quantization" now means that the wave may not have an arbitrary amplitude or an arbitrary energy. As a wave, it may contain only quantized amounts of energy. It is these waves with quantized amounts of energy that constitute what we call "photons". They are not particle-like at all.

However (at first sight at least) it seems particle-like because it can only transfer the *whole* of its energy (not part of it) to another object. It may, for example, be absorbed by an atom. This may be interpreted as a "particle-like" characteristic. From our macroscopic viewpoint, it may even appear to be absorbed "at a point" — where the atom is. But of course the atom is *not* a point. It is sheer prejudice even to say it is *small*. Small compared to us? It is very large compared with, say, an atomic nucleus. The photon cannot behave any more like a point particle than an atom can. So again, the photon does not have the character of a classical particle.

What then remains of "wave-particle duality"? To sustain it, we would have to complicate, in a way that we cannot fully specify, our concept of "particle". In the process, the dualism of wave and particle is undermined.

To follow this line of thought here would lead to a seemingly endless chain of nuances.

Why must nature seem so perversely complicated? If we look
at things the right way, as we have seen in the case of relativity,
nature can appear very simple. It all depends on how we pose our
questions.[13] There was nothing simple about aether. Aether was full
of paradoxes! Perhaps the concept of an electron, or a photon, as
an entity with dual character, sometimes particle, sometimes wave,
is as much fiction generated by our confusion as was the aether.

Quantum theory itself, on the other hand, has an Einsteinian
simplicity, which seems to manifest itself fully only when it is expressed in the clearest and most unambiguous of our languages, that
of mathematics. It seems that this is the language in which nature
is speaking to us. It is the ultimate arbiter of all our confusions.

13.2. Early Approaches to a Quantum Theory: Schrödinger and Heisenberg

Schrödinger and Heisenberg independently of each other attempted to generalize and formalize quantum mechanics.

Schrödinger's approach was based on an analogy with problems
that had played a fundamental role in 19th century classical fields.
These problems concerned "continuous" systems of matter, which
could be described in a field-like manner. They were *vibrating* systems: the vibration of a violin string or of a drumhead, vibrations
of air in pipes or through the air, elastic vibrations of rods, ... Lord
Rayleigh had written a treatise on these problems, entitled "The
Theory of Sound", in which the mathematical techniques of partial
differential equations were used in a highly developed form.

The connection with quantum problems is easy to identify. The
problems of atomic systems were characterized by the existence of
discrete quantum "states" of very definite energies. How to find
these energies? Since Planck had identified *energies* of systems with
their *frequencies*, and de Broglie had extended this relationship to

---

[13] See "Ask a Silly Question" by V. F. Weisskopf and H. Feshbach in *Physics Today*, October 1988.

all sorts of "particles", especially electrons, definite *energy* states of systems were connected with definite *frequency* states.

One phenomenon that had become very familiar for continuous systems was that if they were constrained in some way they could vibrate continuously only at definite and distinct frequencies. A first example was the violin string or the piano string. If the string was held at two points, these frequencies were determined by the following simple consideration: periodic vibrations or waves were set up on the string, whose nodes were at fixed points of the string. Consider for example a string of length L. When a string is very ("infinitely") long it vibrates in a wavelike pattern.

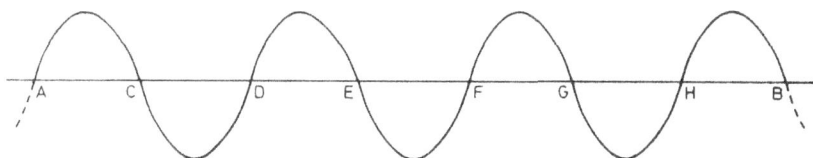

Fig. 13.4.

If the ends are "fixed" the wave-pattern must "fit" between these two ends, which means that there must be, in the length of the string L, either N wavelengths or $N + \frac{1}{2}$ wavelengths of the wave, where N is any integer. Imagine first though a string of infinite length.

Just as for light, the frequency of the wave and its wavelength are related. In fact, the wavelength λ and its frequency f are such that

$$f\lambda = u ,$$

where u is the velocity of the propagation of the wave. This "velocity" is the phase velocity, that is, the velocity with which a crest or a peak of the wave moves. In fact, the whole *pattern* of the wave is transported at this speed.

Let us fix on a point, say F. If the wave is moving to the right we see that, as the wave passes through, the amplitude of the wave becomes negative until the pattern has moved by a quarter wave. At a time $\frac{1}{4}\lambda/u$, the displacement at F has reached a maximum on the negative side. After another similar lapse of time, i.e. at $\lambda/2u$, the displacement will return through zero. At $t = 3\lambda/4u$ the amplitude at f will reach a positive maximum. At $t = \lambda/u$ it will be back to zero, having completed one cycle of vibration. Thus the point F — and in fact all other points — will go through a complete cycle of vibration in a time λ/u, i.e. it will vibrate with a frequency

$$f = u/\lambda \ .$$

Thus $f\lambda = u$, the velocity of the wave propagation. The same result is true for light, except that in this case $u = c$.

What happens now on the string of finite length? Mathematical analysis shows that, as a given point of the wave reaches the end, it generates a *reflected* wave whose amplitude is opposite to that of the incident wave, and this *reflected* wave propagates in the opposite direction, i.e. *away* from the end. At the end, the incident and reflected waves always cancel each other out, so that the end conditions are always satisfied.

In Fig. 13.5, we show the pattern of vibration resulting from this picture. What we get is a *standing* wave, where *each point of the wave* vibrates with the same frequency. The five curves shown give a picture of the pattern of displacement of the points at times which are twentieths of the vibration period. If the period $T = 1/f$, for $T > 1/4f$ until $T = 1/2f$ the string goes back through these five to the moment at which it is straight; then it goes through the same cycle with the directions of all displacements reversed. The pattern ("normal mode") shown is that of wavelength equal to the length of the string. That is not the lowest frequency mode. The lowest frequency mode has twice as great a wavelength and half as great a frequency. Its pattern would be like that of the left half of Fig. 13.5, sketched out horizontally over the total length of the string. The presence of a node (fixed point) always signifies a higher energy.

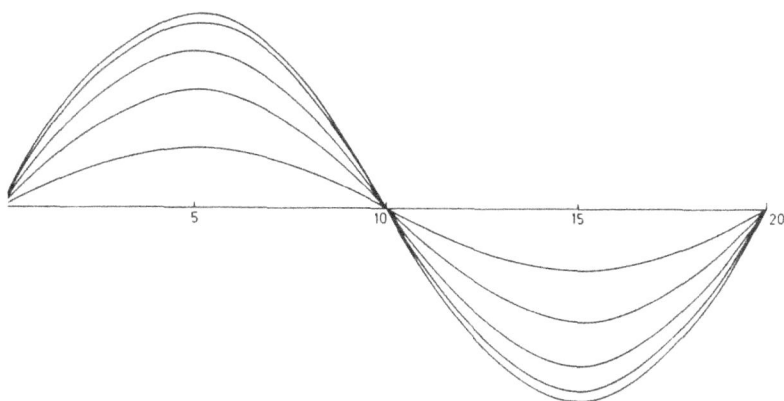

Fig. 13.5. Wave vibration drawings represent the displacement at successive equal intervals of time.

The frequencies are

$$f = \frac{un}{2L} \qquad (n = \text{any integer}) \ .$$

By Planck's hypothesis, the corresponding energy states are integral multiples of a basic energy. A complete quantum treatment gives equally spaced energy levels, but the lowest state turns out to have energy $\frac{1}{2}hf_0$, where $f_0 = u/2L$.

Another example of an oscillating system whose normal mode frequencies had been calculated is the vibrating drumhead. This system is more complicated because it vibrates in *two* dimensions rather than one. It is fixed at its bounding circle. The frequencies no longer have any simple relation to each other. Table 13.1 gives the values of the six lowest ones; they are $\xi u/a$, u being the speed of sound on the drum and a its radius. We give the values of ξ, as well as the number of angular and radial nodes.

In mode I, all points of the drum vibrate in phase; the amplitude is greatest at the centre and diminishes at the edges.

The diagrams for the other modes (Fig. 13.6) are to be interpreted as follows: the white regions and the shaded regions are

Table 13.1.

	ξ	radial nodes	angular nodes
I	2.405	0	0
II	3.832	0	1
III	5.136	0	2
IV	5.520	1	0
V	6.380	0	3
VI	7.015	1	1

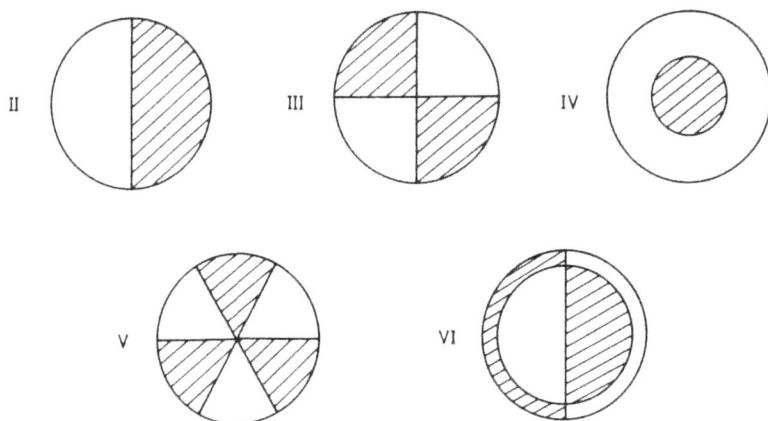

Fig. 13.6.

opposite in phase. When the white regions are displaced upward, the shaded ones are displaced downward, and vice versa.

The fact that all parts of the drum (or of the string in the preceding example) vibrate in unison may be shown in a demonstration in which the drum is excited to vibrate by a source of sound of a frequency which can be changed continuously. Little vibration is observed at frequencies which are not close to their "normal" ones. If one puts the drum in a dark room and illuminates it with a stroboscopic light whose frequency can also be varied, one can simply

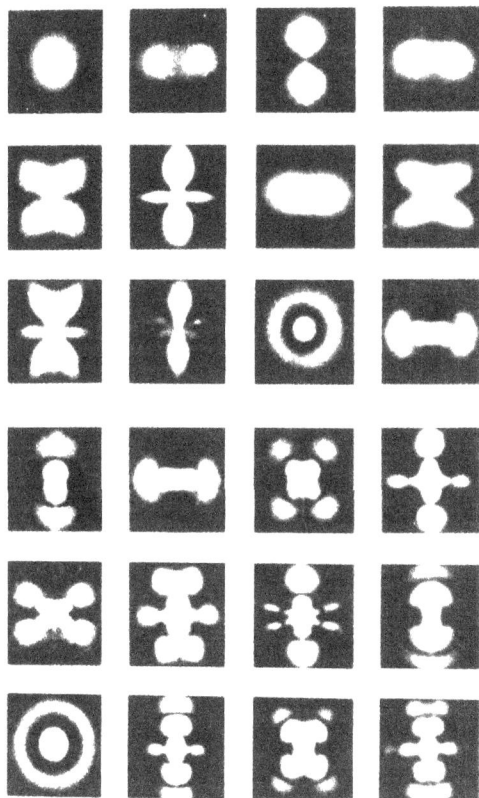

Fig. 13.7. These diagrams show the patterns of the normal modes (energy states) of the electron in the hydrogen atom (they represent two-dimensional sections by planes through the nucleus). The more complicated the patterns, the higher the energy of the corresponding state. There is a direct relation between energy and the number of nodal surfaces in three dimensions, but these are not easily identifiable in two-dimensional sections. (From *The Mystery of Matter*, ed. E. B. Young, Oxford University Press, 1965.)

demonstrate the "normal frequencies". First, the driving frequency is varied until the drum vibrates with substantial amplitude. If the strobe frequency is then set near the normal frequency, the drum appears to vibrate very slowly. With a fine adjustment of the frequency,

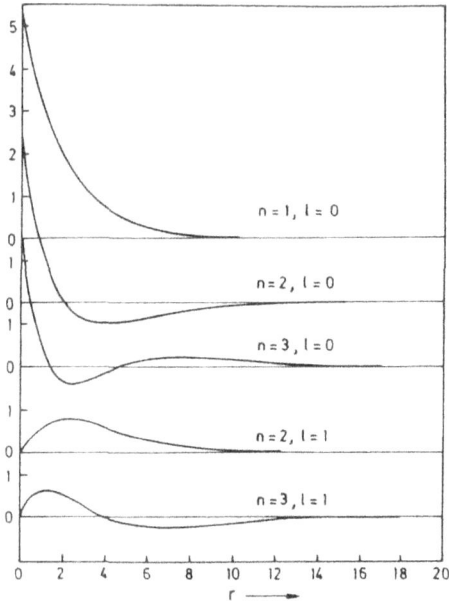

Fig. 13.8. The graph shows, as a function of the radial coordinate, the wave functions of some of the lower energy states of hydrogen atom electrons. The energy is to the first approximation a function only of n; ℓ characterizes the angular momentum in units of $\hbar$.

the drum may be "frozen" in a pattern of displacement which corresponds to that of the vibration. This is a demonstration that *all points of the drum vibrate with the same frequency*. The stroboscopic light shows the state of the drum at intervals of one period, so the drum is always illuminated at the same point in the cycle of vibration. If the frequency were not the same for all points, the *whole* pattern of the vibration could not be brought to apparent rest at a single frequency. To put it differently, there is a phase coherence of the vibration of all parts of the drum.

How is this related to the hydrogen atom? What varies in this case is de Broglie's (or Schrödinger's) "wave function". If the electron is bound to the nucleus (by the electric force of attraction) it is like a three-dimensional version of the vibrating drum. The poten-

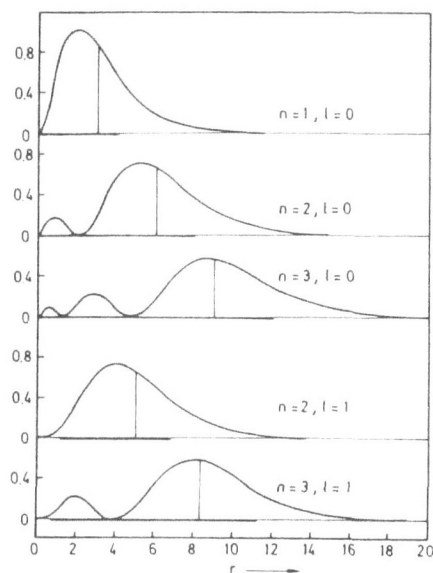

Fig. 13.9. Plots of total electron density as a function of the radial coordinate for the states shown in Fig. 13.8.

tial energy of the electron in the electric field of the nucleus is not analogous to a *uniform* drum but rather to one which is more rigid near the centre; the speed of sound is greater there. Mathematically, the problem is complicated but soluble. We shall show shortly how to get an approximate solution by quite simple arguments.

Before going any further, we must first come to grips with the so-called "Uncertainty Principle".

13.3. The Uncertainty Principle

This principle is based on a principle related to waves which is well known to classical physicists.

If, as de Broglie suggested, the free electron is like a *wave*, it is totally unconcentrated in space. One doesn't ask the question, *where* is a wave; a wave is everywhere! It is totally *unlike* the classical physicist's point particle.

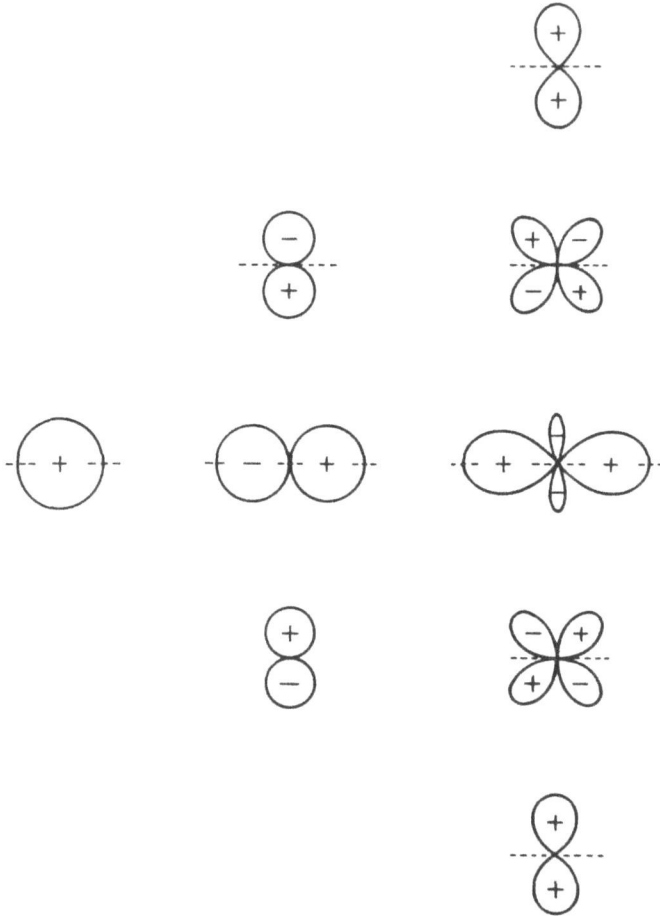

Fig. 13.10. Angular dependence of the wave functions of several states of low
energy in hydrogen. These diagrams represent the cross sections by a plane
through the symmetry axis, which is in the horizontal direction. States depicted
in the three vertical columns are ones with 0, 1, and 2 units of angular momentum
respectively. Regions in which the wave function has different signs are shown,
so that angular nodal lines may be identified. (Dicke and Wittke, *Introduction
to Quantum Mechanics*, ©️ 1960, Addison-Wesley Publishing Co., Inc., Reading,
Massachusetts, p. 147. Redrawn with permission.)

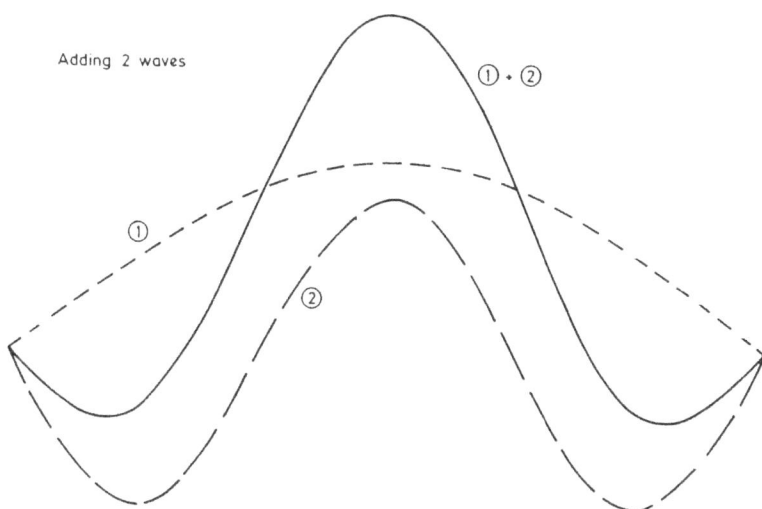

Fig. 13.11.

Imagine a sound wave with a well-defined frequency. Our ear detects it as a well-defined musical note which remains constant. But we also know, if we are stereo fans, that there may be "noise" in the form of almost instantaneous "clicks". These are localized in time and space. How do we describe them in terms of normal modes of vibration, that is, patterns of vibration of a fixed frequency? We shall show that we can do it by superimposing waves of many frequencies.

Curve (1) of Fig. 13.11 shows a half-wave of a given wavelength λ (which corresponds to a frequency of u/λ). Curve (2) shows a "harmonic" of three times the frequency of (1). If we add them together we get the curve (1) + (2). This is considerably more "localized" than the original pure frequencies.

Let us now add more waves of still shorter wavelength (higher frequency). If we use five wavelengths $\lambda, \lambda/3, \lambda/5, \lambda/7$ and $\lambda/9$ in approximately chosen proportions we obtain curve (3) of Fig. 13.12,

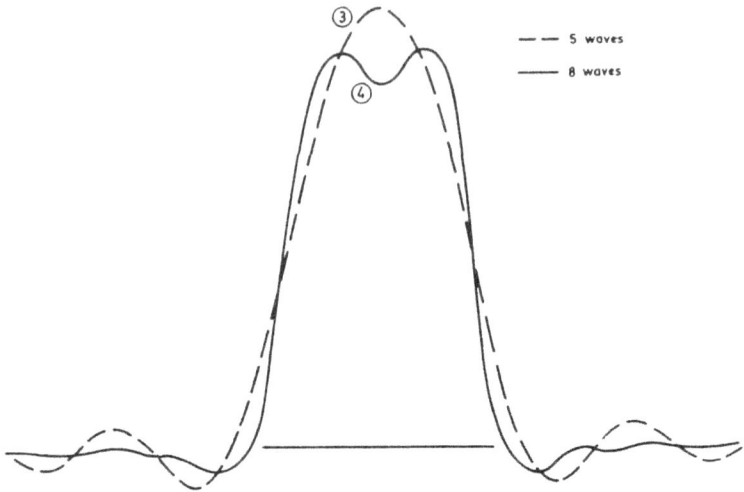

Fig. 13.12.

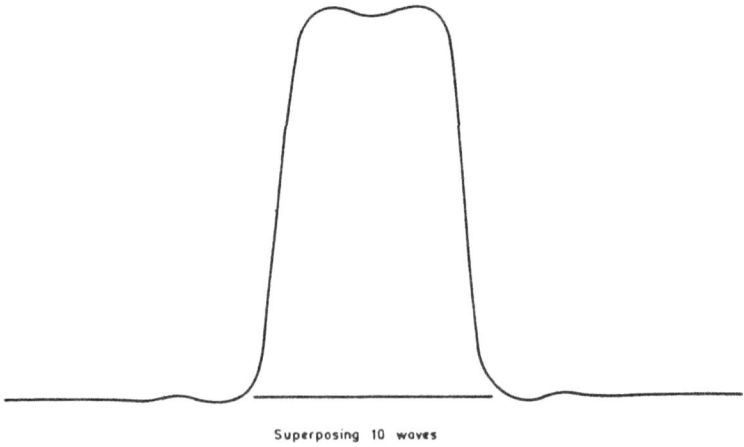

Superposing 10 waves

Fig. 13.13.

in which the disturbance is still better localized. Adding three more with wavelengths $\lambda/11$, $\lambda/13$ and $\lambda/15$, we get curve (4). Adding still two more wavelengths $\lambda/17$ and $\lambda/19$ we get Fig. 13.13.

These waves were chosen to create a sound of constant strength over a finite distance (or time) — a sort of "click". In fact, with ten waves, only a tiny "ripple" remains outside the finite interval of the disturbance. But the moral is that the more we want to *localize* the disturbance, the higher the frequencies (the shorter the wavelengths) we must mix in.

Now imagine doing this with an electron wave. If we want to localize the electron, we must combine waves with high frequencies; according to de Broglie, with high energies. *It costs energy to localize an electron* (or any other quantum particle). This is the source of the "degeneracy pressure" which stabilized the electrons in a white dwarf star, or which did the same for neutrons in the neutron star.

The reason for the phenomenon may be expressed thus: if we want to localize a wave-like phenomenon within certain limits, we must combine together periodic waves which create a *complete* destructive interference outside those limits! The more the localization required, the more demanding is the requirement of destructive interference. It requires the use of ever more waves of ever increasing frequency. In the limit, to have a *completely* localized phenomenon – an instantaneous "click" – we must combine waves of *all* frequencies, without limit, with equal amplitude. The frequency distribution is constant. We call this "white noise". In the quantum problem, it takes an infinite amount of energy.

In relativity we said that a particle could never be accelerated to the speed of light, because to do so would take an infinite amount of energy. Now we can say that no electron, or other particle can be localized, because to do so would also take an infinite amount of energy.

We have, then, a remarkable conclusion, which follows from the observation of the diffraction of electrons (i.e. from the phenomenon of interference of electron waves): the position of an electron cannot be determined. This conclusion follows from the *nature* of electrons: they are of such a character that they do not exist at a single point.

It is not that Nature has conspired to hide the position of electrons from us. It is simply that they always have extent in space. They are squeezable, to an extent, but not indefinitely.

When the debate was still raging, before 1932, about the constituents of nuclei, it was shown, once quantum mechanics was understood, that electrons could not exist inside the nucleus. To squeeze them into nuclear dimensions would require an energy of about 4×10^8 electron volts. Even for the heaviest nuclei, the attraction between the electron and the nucleus was only slightly more than 10^7 electron volts.

Let us use our example of the construction of a localized wave, along with de Broglie's principle, to illustrate the "Uncertainty Principle".

Suppose that we want to superpose (add) waves in such a way as to represent an electron "field" localized within a region of length a (between $-a/2$ and $a/2$). We can start with a wave of *infinite* wavelength, which gives a constant value everywhere. To make the electron intensity zero outside the region, we must use waves which are negative outside; this means waves a quarter of whose wavelength is less than $a/2$, i.e. whose wavelength is less than $2a$ (Fig. 13.14).

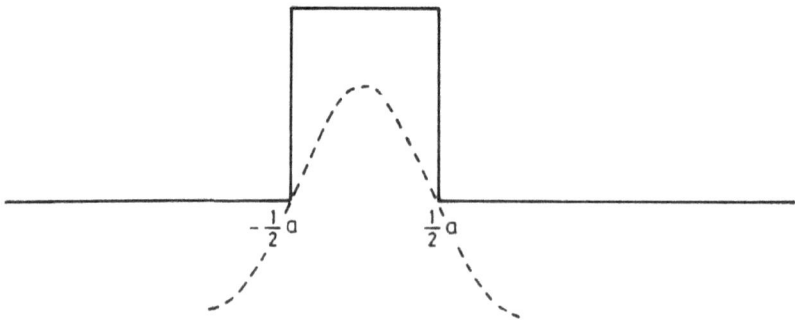

Fig. 13.14.

An infinite wavelength corresponds to a momentum of zero, while a wavelength of $2a$ signifies a momentum of $h/2a$. The "uncertainty" of the momentum can then not be less than $h/2a$.

$$\Delta p > h/2a \ .$$

But the uncertainty in position is $a = \Delta x$. Thus we obtain

$$\Delta p \Delta x > \frac{1}{2} h \ .$$

Anyone familiar with the special theory of relativity might expect a second quite astonishing principle of the same sort involving energy and time:

$$\Delta E \Delta t > \frac{1}{2} h \ .$$

What precisely does this mean? Over short periods of time, the *energy* of a particle is not sharply defined! Of course, this must follow from the Planck relationship $E = hf$, for we cannot precisely define the *frequency* of a wave over a finite time. The argument parallels the preceding one. Suppose we want to localize a disturbance in time (between $-T/2$ and $T/2$). If we know nothing outside this time interval, we must have destructive interference of our waves outside. This requires waves of negative amplitude outside the interval T, i.e. waves whose period is less than $2T$, or frequency greater than $1/2T$. Thus

$$\Delta f > 1/2T = 1/2\Delta t$$

or

$$\Delta f \Delta t > \frac{1}{2} \ .^\dagger$$

---

†These may not be the most general or most precise limits we can place on "uncertainties". It can be shown rigorously that true limits in general are

$$\Delta x \Delta p \geq \tfrac{1}{2}\hbar, \quad \Delta E \Delta t \geq \tfrac{1}{2}\hbar \quad \text{where } \hbar = h/2\pi \ .$$

At first glance this seems to contradict the law of conservation of energy. However, it is not unreasonable. If an electron never interacted with anything else, it would be a wave of infinite extent in space and time. If it comes into interaction with something else, both its momentum and energy will be altered. What the principle does imply is that surprisingly, if the interaction takes place over a limited region of space or a limited interval of time, various outcomes will be possible. We are reminded of Rutherford's law of radioactive decay, in which the moment at which a given nucleus decayed was totally unpredictable. This is consistent with the energy of the decaying nucleus being fixed. Over a fixed time interval, its energy becomes indeterminate – it may or may not decay.

We cannot deny that these kinds of behaviour are not strange. They are contrary to our classical intuition. But they are strange only because they occur in a new context; they become natural when we accept the wavelike character of matter.

13.4. Feynman's "Classical" Paradox

Richard Feynman, in his Beatty lectures at McGill used a phenomenon of classical optics — the law of reflection at a mirror — to show that it is not only quantum mechanics which can confront us with apparent paradox — if we do not pay careful attention to the way we describe phenomena.

Imagine a light source at O', and someone's eye at O (Fig. 13.15). Light is emitted in *all* directions. Feynman's contention is that light can get from the source to the *eye using any path whatsoever*. The law of reflection, that the angle of reflection is equal to the angle of incidence, is satisfied for the path $O'AO$. Light may, however, take other paths, being reflected, for example, at A', or X, or X'.

These paths involve different distances. Because light is wavelike, the effect of different paths is that the light following them will arrive with different phases, simply because they will involve different multiples of the wavelength (we assume monochromatic light).

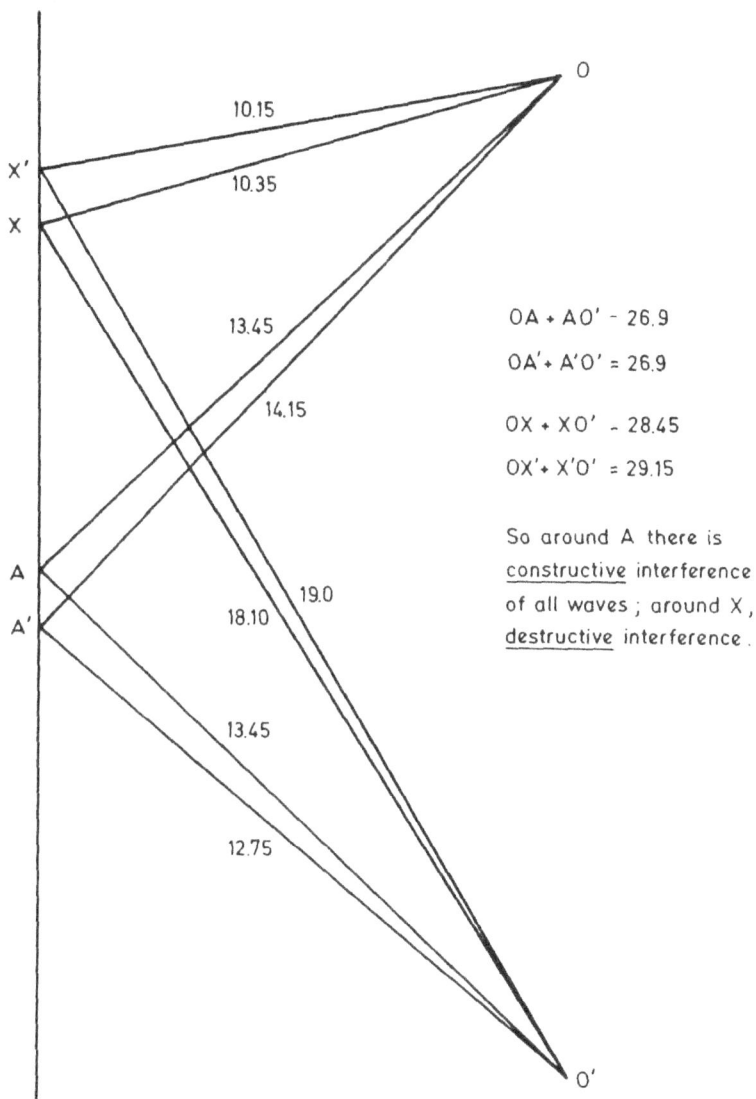

Fig. 13.15. Feynman's interpretation of the law of reflection.

For the reflection at A, the paths $O'A$ and AO are identical; in arbitrary units they are 13.45 units each. The total path is 26.90 units.

Suppose we move a small distance from A to A'. The angles are no longer equal; the paths have lengths 12.75 and 14.15 units respectively. The *total* path, however, is again 26.90 units! The same will be true on the other side of A. Thus, light reflected over a significant region around A will all arrive in phase, and there will be constructive interference, i.e. reinforcement.

Consider now the point X (which could be almost anywhere at a distance from A). The total path is now 28.45 units. Moving the same distance from X that A' is from A, to the point X', we find that the length of the path is 29.15, a difference of 0.7 units. If this distance is large compared with the wavelength of the light, the signals arriving at O from points between X and X' will arrive with all possible phases, and so will destructively interfere. Because of this cancellation, we will not see any light from X originating at O'.

The law of reflection is in fact an effect of interference; light does not go from O' to O only by being reflected at A.

This is not paradoxical; it is a well-known optical principal.

Feynman now shows us how to *increase* the intensity of light reaching our eye by painting parts of the mirror black! We simply have to imagine painting black stripes on the mirror in such a way that there will only be reflection from points for which the paths from O to O' will differ in length from that through A by integral multiples of the wavelength of the light. All the light arriving will again be in phase, but it will now be coming from many different parts of the mirror which did not contribute before.

Is this not a strange paradox? By painting large parts of the mirror black we *increase* its intensity of reflection.

How fortunate that philosophers did not discover this long ago (it was hiding there, even before Maxwell, and was studied exten-

sively by Sir William Rowan Hamilton[14] (1805-1865)). The entire
evolution of philosophy might have been changed!

This viewpoint on a classical optical problem embodies the spirit
of Feynman's own formulation of quantum theory, one that has
proven extremely fruitful. It is of course not new in substance, but
only in form. It may be taken as a prototype of another pseudo-
phenomenon popular with philosophers — the so-called "collapse
of the wave function" supposedly associated with observation. The
"wave function", in this viewpoint, extends all over space. The observer O views it as "collapsed" to the path $O'AO$. The "collapse"
is also a property of the observation! If O is moved, it "collapses"
to a different path.

When Feynman was asked at a meeting during his visit to Montreal to explain the "paradox" of the collapse of the wave function
his reply was: "there *is* no collapse of the wave function".

13.5. Energy States of Oscillating Systems and of the Hydrogen Atom

The classical view of dynamical events was that they consisted
of particles (dimensionless entities with inertia) in motion; the goal
of physical theory was to describe the "orbits" or paths of particles
from point to point. Quantum theory states that matter does *not*
consist of such particles, but rather that matter can exist in a discrete
set of non-localized states of a field-like character, rather like the
"normal modes" of a continuous classical system like a vibrating
string or drum. We do not know how to interpret intuitively the field
quantity or "wave function" ψ; we can, however, determine the states
of energy of systems. This may be done by solving "wave equations"
similar to those known in the theory of classical fields, since it can
be shown that properly bounded solutions of these equations can

---

[14] Books on physics for non-physicists never do justice to Hamilton, who was one
of the greatest physicists of the 19th century. Aside from his work on optics, his
very elegant formulation of classical mechanics forms the basis for Schrödinger's
wave mechanics. Schrödinger's wave equation is constructed from the "Hamil-
tonian function"

only be obtained for certain values of the energy, which occurs as a parameter in the equations.

When we spoke of "bounded" systems in classical physics, we imagined strings clamped at their ends, or drums clamped at their periphery. But an oscillating pendulum is also a bounded system, as is a planetary orbit. They are bounded by constraining forces, so that inadequacy of energy keeps them within spatial limits.

Classically, the energy of a particle constrained by a force to vibrate about a fixed point is well known. It consists of kinetic energy $\frac{1}{2}mv^2$, m being its mass and v its speed, and a potential energy $\frac{1}{2}kx^2$. The kinetic energy can also be written as $p^2/2m$, where p is its momentum. k is the elastic constant characterizing the restoring force, which is of magnitude kx.

The vibration takes place about a state in which $x = 0$ and $p = 0$ – the mid-point of the oscillation. Thus, the mean square value of the momentum is p^2, and that of the displacement x^2. What quantum theory designates as the uncertainty of a dynamical quantity is the root mean square of its deviation from the average value. Thus we may write

$$p^2 = (\Delta p)^2, \quad x^2 = (\Delta x)^2 .$$

The energy is then

$$E = \frac{1}{2m}(\Delta p)^2 + \frac{1}{2}k(\Delta x)^2 .$$

But we know that

$$\Delta p \Delta x \geq \frac{1}{2}\hbar .$$

Detailed calculation shows that, for the lowest state of the operator the equality holds (the oscillator is the only system for which that is true). Thus

$$(\Delta p)^2 = \frac{\hbar^2}{4(\Delta x)^2} ,$$

so that the energy becomes

$$E = \frac{\hbar^2}{8m(\Delta x)^2} + \frac{1}{2}k(\Delta x)^2 .$$

This may be written as

$$E = \left(\frac{\hbar}{2\sqrt{2m}\Delta x} - \frac{\sqrt{k}}{\sqrt{2}} \Delta x \right)^2 + \frac{1}{2}\hbar\sqrt{\frac{k}{m}} \ .$$

The energy is smallest when the first term is zero; the energy is then

$$E = \frac{1}{2}hf \ ,$$

where $f = \frac{1}{2\pi}\sqrt{\frac{k}{m}}$ is the vibration frequency. This corresponds to

$$(\Delta x)^2 = \frac{\hbar}{\sqrt{2mk}} \ .$$

We see something very important from these equations: the oscillator *may not* have zero energy, and the oscillation has a minimum amplitude. The energy is called the "zero-point energy" and is strictly a quantum phenomenon.

For the other quantum states,

$$\Delta p \Delta x = \left(n + \frac{1}{2}\right)\hbar \ ,$$

where n may be any integer. The energies are

$$E = \left(n + \frac{1}{2}\right)hf \ .$$

The energy levels are evenly spaced.

The importance of these results arises from the fact that all kinds of waves, and in particular electromagnetic waves, consist of oscillatory vibrations. The role of the quantity x in these cases is played by the *amplitude* of the normal mode. The quanta of energy describe *photons* or light quanta. These are simply waves of given frequency whose amplitudes are such that their electromagnetic energies are the Planck quanta hf, As remarked earlier, they are not particle-like, aside from the fact that their energy is only negotiable in its entirety.

One may wonder about the zero-point energy inevitably present in all electromagnetic wave modes. This appears to be an infinite well of energy. Yet it cannot be exploited in an obvious way, since it is the *lowest* energy state of the field.

Consider next the hydrogen atom, or a generalization thereof consisting of an arbitrarily heavy nucleus surrounded by a single electron. Once again, the kinetic energy of the electron is $p^2/2m$; this time its potential energy is $-Ze^2/r$, where Ze is the nuclear charge. Thus the energy is

$$E = \frac{(\Delta p)^2}{2m} - \frac{Ze^2}{r} = \frac{\hbar^2 \gamma^2}{2m(\Delta r)^2} - \frac{Ze^2}{r} ; \quad \Delta p \Delta r = \gamma \hbar .$$

Relating the average value of $\frac{1}{r}$ to Δr depends on the precise distribution of the electron wave; we may put

$$r = \beta \Delta r .$$

Leaving γ and β uncertain for the moment,

$$E = \frac{\hbar^2 \gamma^2}{2m(\Delta r)^2} - \frac{Ze^2}{\beta \Delta r} .$$

Again, algebra permits us to deduce the lowest energy state, for we may write the energy as

$$E = \left(\frac{\hbar \gamma}{\sqrt{2m}\, \Delta r} - \frac{Ze^2 \sqrt{2m}}{\beta . 2 \hbar \gamma} \right)^2 - \frac{2m Z^2 e^4}{4 \beta^2 \gamma^2 \hbar^2} ,$$

which gives an energy of

$$-\frac{m Z^2 e^4}{2 \beta^2 \gamma^2 \hbar^2} .$$

The correct value of $\beta^2\gamma^2$ is in fact 1, the energy

$$-\frac{mZ^2e^4}{2\hbar^2}$$

is called the Rydberg and is about 13.5 electron volts.

We can also calculate the spatial size of the wave function

$$\Delta r = \frac{2\hbar^2\gamma^2\beta}{2mZ^2e^2} \ .$$

If $\gamma^2\beta = \sqrt{3}$ this gives the correct result:

$$\Delta r = \frac{\hbar^2\sqrt{3}}{mZ^2e^2} \ .$$

The quantity $\hbar^2/me^2$ is known as the Bohr radius.

The higher energy states are obtained by replacing $\hbar$ in the uncertainty relation to $n\hbar$, n being any integer. We thus obtain values similar to those obtained by Bohr. The states, however, are different. The electron distributions are sketched in Figs. 13.7–13.10 (it must be remembered that we can only represent two-dimensional cross-sections of the real three-dimensional wave functions).

As in the case of the vibrating drum, the lowest energy state has a distribution with no nodes; it is spherically symmetric. Angular nodes signify angular momentum of the atom; if there are m nodes the angular momentum of the electron in the atom is $m\hbar$. The quantity designated n in the energy is one more than the sum of the number of angular nodes and radial nodes.

Of course, unlike the case of the drum, the "wave function", which in that case is simply represented by the drumhead displacement at each point, has for the atom no simple intuitive interpretation.

The problem of interpretation is thrown into a new light if we accept that the atom is *completely* described by a small number of fixed numerical quantities, primarily its energy and its angular momentum. Even the latter may not be completely specified, though its total value and its component in any one direction may be. Later considerations of relativity will reveal the existence of an intrinsic or "spin" angular momentum. The aforementioned quantities represent *total* information about the atom. There is no "orbit", no "particle" to follow along its path.

Let us return to the problem of the stability of the atom, which was undermined in Bohr's model by the acceleration of the electron in its orbit. In the quantum model, the states of the system, characterized by definite "quantum numbers", are stable. This is assured by the fact that their electron distribution – their charge distribution – is static, so that no electromagnetic radiation is emitted. The atom only has static electric fields.

We must recall, however, that all of real, experimentally verifiable physics is about *change*. A given state has no change, so the electron in one of its quantum states has no directly measurable characteristics. In that sense the "wave function" is not an element of physical reality, but simply a convenient mathematical tool which may enter into calculations of real phenomena (i.e. change). Thus the "collapse of the wave function" is not physically meaningful. It is not a physical process!

How, then, is real physics, the physics of change, described in quantum mechanics? It is described by changes of state; one may say by jumps from one state to another. Physical reality is in the quantities describing these "jumps". In charged systems like atoms and nuclei, these transitions *do* involve change in charge distributions, and consequently can give rise to electromagnetic radiation. Inability to predict precisely when each transition will take place still persists, though statistically they can be predicted with very high accuracy.

One of the early concerns was that of making the connection between classical and quantum physics. Of course quantum phenomena should not contradict classical ones in circumstances where the latter are well understood and explained within the Newtonian framework. The link is provided by the so-called "correspondence principle", which says that as one goes to higher and higher excited states (large quantum numbers) the quantum picture approaches the classical one more and more closely.

As an illustration, consider the hydrogen atom in a state of energy $-me^4/2\hbar^2 n^2$ where n is large. This will be a state with a large orbit, very weakly bound. Such a state may successively decay to states of lower n $(n \rightarrow (n-1) \rightarrow (n-2) \ldots$ etc.) emitting low-energy photons (Fig. 13.16):[15]

$$E_n - E_{n-1} = -\frac{me^4}{2\hbar^2}\left(\frac{1}{n^2} - \frac{1}{(n-1)^2}\right) \approx \frac{me^4}{\hbar^2 n^3} \ .$$

For example, if $n = 50$, $\frac{1}{2}\left(\frac{1}{50^2} - \frac{1}{49^2}\right) = -0.000082$.

$$\frac{1}{50^3} = 0.000080 \ .$$

At each step, the kinetic energy of the electron increases, the energy of the emitted photon increases, and the mean distance of the electron from the nucleus decreases. All of this is similar to what happens in a classical model, except that the changes occur in discrete jumps rather than continuously. When we get to low quantum numbers, however, quantum behaviour becomes very marked. Finally, due to the "uncertainty principle" — that is to say, the fact that the increasingly narrow localization causes a rapid rise in the kinetic energy of the electron — it stabilizes in the lowest energy state.

13.6. The Correspondence Principle

From the beginning of quantum theory, the question of reconciliation with Newtonian mechanics, which appears to work very well

---

[15] It is more complicated than that. Photons carry away a unit of angular momentum, so the states change angular momentum too. The argument made, however, is qualitatively correct.

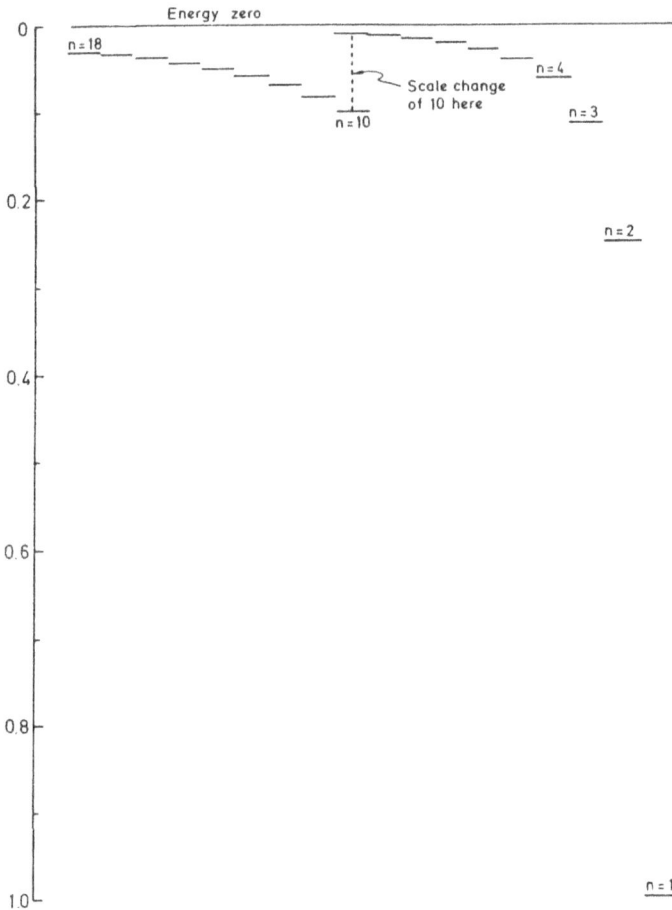

Fig. 13.16. Energy levels of the hydrogen atom.

at the macroscopic or "human" scale, was a concern. Bohr addressed this question in a paper in *Nature* in 1928, in which he enunciated the so-called "correspondence principle". According to this principle, quantum mechanics approached classical physics as "quantum numbers" became very large. This appears in the problem discussed in the preceding section, where the radiation from a hydrogen atom in a highly excited state was "almost" continuous, i.e. proceeded in very small steps. The dramatic *quantum* behaviour appeared when the quantum number n became small, and the energy "jumps" were large, and when the atom became stabilized in its lowest quantum state and no longer radiated at all.

The correspondence at high quantum numbers is even more detailed than indicated in our discussion. Each of the transitions between successive quantum states $(n \rightarrow n - 1)$ has a definite probability per unit time to take place. Thus, it is possible to calculate the *average* rate at which the atom loses radiative energy. This *average* rate, which varies slowly from transition to transition for large n transitions, is very close to the classical rate, i.e. that calculated from classical electromagnetic theory for a continuously radiating charged particle.

The transition from quantum to classical behaviour is, in this case at least, a smooth one.

13.7. The Heisenberg Viewpoint

It was not Schrödinger but Heisenberg who first enunciated the "Uncertainty Principle". His approach to quantum mechanics was quite different from Schrödinger's. Rather than using analogs with the classical theory of normal modes of vibration of oscillating systems, he took as his starting point the problem of measurement of the dynamical variables of a system. The quantum appeared to him primarily as a manifestation of the interference of the measurements of complementary characteristics of systems. For example, an attempt to measure the position of a particle changed that position in a way that was not rigorously predictable, since it gave the particle an unspecified momentum.

Starting from given quantum states, Heisenberg considered how the measurement of different dynamical variables changed those states. Thus he focused, not on wave functions but on the quantities characterizing transitions between states, that is to say, change. In this sense his approach was more explicitly physical than Schrödinger's. The array of quantities characterizing the transitions between all possible pairs of quantum states induced by the measurement of a given variable gives what is known as a *matrix* representation of that variable. It is a mathematical operator representing the physical operation of the measurement. He then demonstrated that these operators did not always commute; that is, the result of the operation A followed by the operation B was not the same as if one first performed the operation B, followed by the operation A. In mathematical terms, this can be expressed in the strange equation

$$AB \neq BA .$$

The dynamical quantities represented by A and B are known as "complementary variables"; an example is the position and momentum of a particle. The result of this "interference of measurement" as manifested in the non-commutativity was the "Uncertainty Principle".

$$\Delta A \Delta B \geq \frac{1}{2}\hbar .$$

We shall illustrate these ideas by a consideration of the properties of polarized light. But before doing so we shall examine the strengths and weaknesses of each of the approaches that we have just outlined.

The strength of the Schrödinger approach is that it clearly shows that quantum theory requires a reconsideration of the nature of matter. Any attempt to interpret it in terms of the classical concept of point particles leads to contradiction and paradox. Its weakness is that it focuses attention on the static notion of states, as characterized by non-observable wave functions, and not on change, which is the essence of the study of physics.

The Heisenberg viewpoint, on the other hand, starts from change induced by measurement, but is formulated in terms of classical concepts, that is, the evolution of the dynamical variables of particles. This leads to two unfortunate biases: one is that we are led to discuss physical systems in the terminology of classical physics, rather than recognizing from the onset that the corresponding view of matter is no longer tenable. The other is that *measurement* appears as the central issue rather than physical processes. It is precisely in this context that one is tempted into the quite illogical assumption that human intervention (some people nowadays even say human consciousness) plays a central role in the functioning of the physical world. Though the logical flaw is evident, it has been very pervasive in giving rise to spurious "philosophical" fantasies. Various considerations permit us to identify the fallacy. Let us take it as axiomatic that the physical world is real, and not a creation of our imaginations (of course, one may generate all sorts of conceptual images of the physical world in our minds, but no scientist can accept that they *are* the physical world).

First, note that quantum theory is a true and verifiable theory of physical reality; that is to say, we may make precise predictions with it which can be verified by experiment. The theory is mathematical in form and has no reference whatsoever to human consciousness, nor the fact that humans make measurements. If one believes in objective reality, it is clear that the physical universe functioned in the same way before human life appeared on earth as it has since. So much for "observer-created reality" and similar narcissistic creations of the human ego.

It is irrelevant whether "we" observe, or whether inanimate instruments do, as in the case of most experimentation. Nor does any concept of a "purpose" for instruments have a role in physical theory. When an instrument "observes" it simply takes part in a physical process. The "observer" may then be anything in the physical world, the "observation" any process in which it is involved. The concepts of "observed" and "observation" may therefore be eliminated from all philosophical discussion of physics with no loss. If they are used, it should be as a form of speech only, and with no anthropocentric connotations. This is not a denial of the special qualities of life or human consciousness. However, no iota of evidence shows that these special qualities are manifested in a special way in the physical world.

In other words, when we are involved in physical interaction with the external world, we do it as objects subject to the same physical law that governs all else.

13.8. The Theory of Polarized Light

Light quanta – photons – can be described by their energy, which is h times their frequency, and angular momentum about their direction of propagation. The latter has only two possible values, $\hbar$ or $-\hbar$. Light which is characterized as right circularly polarized consists of the first type, while that described as left circularly polarized is of the second type. Thin transparent optical devices can be constructed to let through only one or other type of photon.

Fig. 13.17.

We shall call these respectively R and L polarizers. The angular momentum of the first is clockwise about its direction of propagation, the second counter-clockwise. That quanta of light carry angular momentum is easily established. One way is to observe that transitions between atomic states take place only when those states themselves *differ* in angular momentum by one unit of h. Another is to measure the angular momentum communicated to a light-absorbing material by an intense beam of circularly polarized light. Beams of such polarized light may of course be created by passing a beam of ordinary (unpolarized) light through a filter which lets through only one of the two sorts of light.

There is another kind of light, which we discussed in Chapter 4, that is characterized by a fixed direction of its electric and magnetic fields during its propagation. For such light one defines a "direction of polarization" which is the direction of its electric field. We may choose arbitrary directions which we call x and y, for which we can have x-polarization and y-polarization. Mixing proportions of these two, we obtain light linearly polarized in arbitrary directions between these two.

Polarizers may also be constructed which only let through linearly polarized light. By turning the polarizer through 90° we may convert it from an x-linear polarizer to a y-one.

Circular polarizers, on the other hand, are undirectional devices: if one has a polarizer which lets through, from a given side, only clockwise photons, it will let through only counterclockwise photons from the other side, since what looks clockwise from one direction looks counterclockwise from the other.

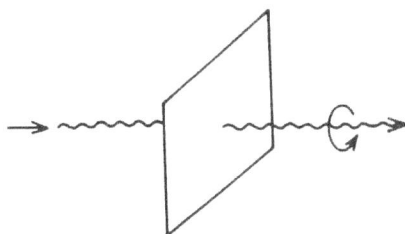

Fig. 13.18.

Can we specify for a photon the energy, the angular momentum and the direction of its electric field? Can we specify as many different properties as we please (as we wanted to do in attributing position and momentum to a "particle")?

To throw light on this question, let us define four possible operations on a beam of light:

M_+, M_- which filter out only clockwise or counterclockwise light respectively.

P_x, P_y which filter out only light with an electric field in the x-direction or y-direction respectively.

Take any light-beam and apply first the operation M_- and then the operation M_+. No light gets through!

If I want light which has a clockwise angular momentum and electric field in the x-direction I first use a filter M_+ and then a filter P_x. I write this pair of operations $P_x M_+$ where the order of the operations is from right to left.

Now, to test whether what emerges is the sort of light I am seeking, I test my resulting beam with M_- and P_y filters. Nothing gets through the latter, verifying that the outgoing light has indeed only x character. But when I test it with the M_- filter I find that some light does get through. Thus, although the very first step purported to filter out the counterclockwise $(-)$ light, some evidently remains!

In fact, the outgoing beam has exactly equal intensities of $+$ and $-$ light.

Suppose we had reversed the order of the first two operations, as represented by $M_+ P_x$ and shown in Fig. 13.19.

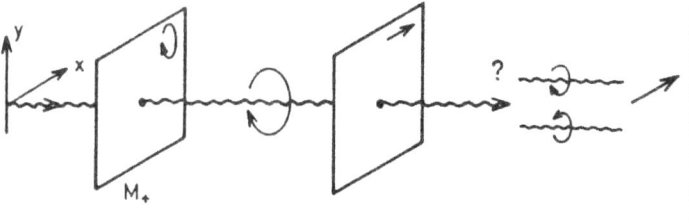

Fig. 13.19.

This time, nothing gets through the M_- filter, but, although the x-polarized light was removed at the first stage, there remain at the end equal quantities of x-polarized and y-polarized light.

Remarks

1. It is clear that, operating on light of any properties whatsoever, the results of the sequences of operations $P_x M_+$ and $M_+ P_x$ are not the same. This is a Heisenberg-type relation

$$P_x M_+ - M_+ P_x \neq 0 .$$

2. This appears to be connected with our failure to be able to create a beam of light characterized by a given angular momentum
and a given direction of the electric field; just as matter does
not appear to be of such a character that it can simultaneously
have a position and a momentum, so does light not seem to be
of such a character as to have simultaneously a definite angular
momentum and a definite direction of its electric field. Between
these two qualities is once again an "uncertainty principle".

3. That this is due to the "interference" of measurements of these
two quantities is evident from the following observation: the
sequences of measurements

$$M_+ P_x M_- \quad \text{or} \quad P_x M_+ P_y$$

do not stop a light beam completely, though

$$M_+ M_- \quad \text{or} \quad P_x P_y$$

do. Is it not strange that, adding a third filter between two
allows some light to penetrate whereas without the third filter
the light beam would be totally blocked? How can one get more
from less, much as Feynman did by judiciously applying black
paint to his reflecting mirror?

These are typical illustrations of "quantum behaviour".

The strangeness of this behaviour becomes most evident when
we imagine the strength of the light beam constantly reduced in
intensity until the passage of individual photons can be identified.
In the experiments which we have been discussing, the emerging
beam consisted of equal intensities of two sorts of radiation. But
what happens in the case of a single photon? It is not allowed to
be ambiguous; it cannot split in two. It is at *this* level that we
have to invoke probability. Each *individual* photon emerging must
have one or other type of polarization (x or y, $+$ or $-$). If twenty
are detected (each singly) the two sorts will be distributed randomly,
just like heads or tails in twenty tosses of a coin. But as the numbers

increase, the results approximate more and more closely to an equal
division.

13.9. Dirac's Unification of Quantum Theory

Professor Paul Adrien Maurice Dirac ranks without question
among the greatest physicists of our age. He was the first person
to develop a fully relativistic version of quantum mechanics, which
led to the prediction of antimatter, as well as to the demonstration
that electron spin was simply a relativistic phenomenon. For this
he was awarded the Nobel Prize in 1933. But his influence extends
far beyond that. In a book written in 1930 he developed a form of
quantum theory more general than that of Schrödinger or Heisenberg, encompassing both of them and putting them in a unified
perspective.

Using Dirac's approach it was possible to solve most of the "solv-
able" problems of the time without recourse to wave functions. This
is quite consistent with our already stated claim that the wave function does not correspond to an element of physical reality, but is
merely a dispensable mathematical tool.

The essence of his approach was to show that the whole theory
could be developed in terms of *states* of quantum systems, which
could be acted on by operators (which turned them into other states)
representing dynamical variables of the system in question. When
the operation of a variable K (position, momentum, energy, angular
momentum, etc.) on a state reproduced that state, this meant that
the state had a definite value of that variable (in technical language,
it was an eigenstate — "proper state" — for that variable). If the
operators for two variables K_1 and K_2 did not commute, i.e.

$$K_1 K_2 \neq K_2 K_1$$

states could not be found for which these two variables could be
simultaneously specified. The "uncertainty" governing the limits of
simultaneous specifications in a state were given by the value of the
variable $(K_1 K_2 - K_2 K_1)$ in that state. Once one found the maximum

set of variables or operators which commuted with each other, these would give a complete description of the state.

Most intriguing of all, by simple algebra and without any "wave function" or Schrödinger wave equation, the states of some systems could be completely determined. Their energy levels depended explicitly on the "commutators" — on $(K_1 K_2 - K_2 K_1)$ and the implied "uncertainties".

Here we have an illustration of a striking feature of the quantum theory, which we shall see repeatedly: that it is precisely the "Uncertainty Principle" which is responsible for the vast storehouse of knowledge and understanding about the world of atoms, nuclei and fundamental particles which has made the theory the most general and richly rewarding in the history of physical science.

What has often been portrayed as hiding nature from us is in fact the key to opening our eyes to a dazzling new vision of the physical world.

13.10. The Deeper Level of Quantum Theory: Theory of Fields

The foregoing discussion of the foundations of the quantum theory has been developed in a historical context; it is based on the view of the theory extant in its period of initial evolution. Regrettably popular treatments and most "philosophical" discussions of the subject remain on this level. However the contemporary particle physicist views the whole theory from a more sophisticated and much more conceptually rewarding viewpoint. In this viewpoint the wave function, viewed as a static entity representing a probability distribution for a point particle, is replaced by an operator characterizing transitions between quantum states of fields. It is usually thought impractical to develop this viewpoint in popular expositions, and perhaps it does indeed represent such a radical change of outlook from the classical one as to present an insurmountable barrier to acceptance by the layman. It seems worthwhile, however, to sketch briefly the underlying ideas, if only to provide a mise en garde against too simplistic philosophical discussion.

Perhaps a good starting point is to be found in the quantum treatment of the electromagnetic field, with which we now have a certain familiarity. The virtue of this approach is that we are accustomed here to accept the *field* viewpoint as primordial.

What, then, is the quantum status of the electric and magnetic fields themselves? The surprising answer is that they are *quantum operators* producing, from the vacuum state, electromagnetic waves whose energy comes in discrete (quantum) bundles, i.e. photons, or, when applied to other quantum states, adding such a photon to the previously existing state. To be more precise, these operators may in fact not only *create* such photons, but may also *annihilate* them. What, then, is the relation between the electric and magnetic field operators? A simple answer is that the electric field operator produces the electric field and the magnetic field operator the magnetic field. But, you will quite rightly object, one cannot exist without the other. In fact, the *electric* field operator creates an electromagnetic wave in which the electric field is in a given direction while the magnetic field direction is undetermined (or in the terminology of quantum mechanics, uncertain).

The magnetic field operator on the other hand creates an electromagnetic wave which has a fixed magnetic field, the electric field direction being undetermined. In each case the undetermined field may be in either of two opposite directions perpendicular to the one which is known. In neither case is the angular momentum determined; the field operators E and B neither commute with each other *nor* with the angular momentum operator. On the other hand, electromagnetic waves of given angular momentum exist, in which neither the electric nor the magnetic field directions are determinate. All this is very reminiscent of our earlier discussion of polarized light.

In a parallel way, the "wave function" of an electron, when thought of physically as a wave field, has quanta associated with it. The possible states of the field are again generated by an *operator*

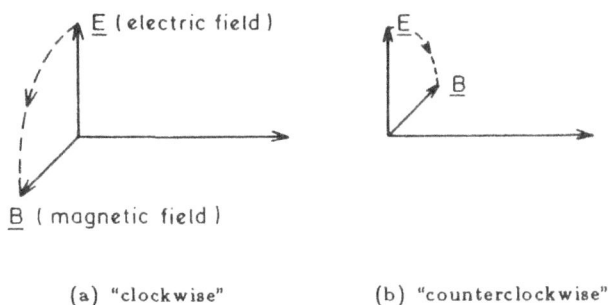

(a) "clockwise"   (b) "counterclockwise"

Fig. 13.20. Electromagnetic waves of a given angular momentum.

associated with that field, which makes a transition from the vacuum state to the state in which that field has definite values. We thus treat electrons and photons, both nonlocal entities, in the same way.

This is very gratifying if we are seeking a unified picture of the physical world. Before quantum theory, Maxwell's electromagnetic waves were one kind of physical entity; a particle, like an electron, another. Now we see that though they are different in their detailed attributes, like an electron and a photon, they appear in the basic structure of the theory in a similar way. We no longer need to be more puzzled by questions posed with respect to one than with respect to the other. This is helpful in clarifying so-called "paradoxes" in quantum theory.

13.11. Identity Among Particles

In classical physics, we assume that two particles can be identified, i.e. distinguished from each other. Even if two "identical" twins were truly identical, we would assume that if, for example, Jim were at school and Tim at home, this would be a different situation than if Tim were at school and Jim at home. But it is not unreasonable to ask whether the same is true of, for instance, electrons.

If electrons were identical material particles, we would probably not question whether it was also true for them. But if they are energy quanta of a field, can we be so sure?

Consider, for example, a violin string. As we have already seen, it has possible motions in which it vibrates with the same frequency at every point. The lowest frequency state we call the fundamental, the higher ones harmonics. We know also that there can be superposed motions of the string in which the fundamental and harmonics are simultaneously present. The amplitudes of these various waves are simply added. Or, if we throw two stones into calm water at different places, we know that these waves may interfere, that is, be superposed on each other, in a similar way. But there is no evident way, if we look at the displacement of the water at a given point, to identify what part of that displacement is associated with one wave and what part with the other.

However, if we look, not at a single point, but rather at the whole pattern of waves, we *can* disentangle them. This is precisely what our ear does in the case of sound waves.

Suppose that we jiggle a small rubber ball up and down on the surface of a pond — always at the same frequency. A fixed frequency will assure a definite wavelength, so that the crests or troughs of the wave will be equally spaced. Then imagine that we jiggle the ball more vigorously, but still with the same frequency. The crests and troughs will remain at the same places, but their amplitudes will be greater — the crests will be higher, the troughs lower. Can we still identify which component of the motion results from the original stimulus and which one results from the "extra" part resulting from the enhanced stimulus?[16]

The question is similar to the following: suppose we have an electromagnetic wave of a given frequency, and an amplitude such that it is made up of two photons. Imagine that one of the photons is absorbed in passing through a thin detector. Can we tell which one it was? Or does this question have any meaning?

Quantum mechanics tells us that it does not. There is no way of "tagging" photons. All photons of a given frequency are identical.

---

[16] An apt analogy for indistinguishable particles, suggested by my philosopher friend Storrs McCall, is the identity of two dollars (not dollar bills, but dollars on balance sheets).

How can this be established? The answer lies in statistical physics. There, entropy was defined in terms of the number of ways in which one could realize various possibilities which had equal à priori probability. An important problem, of which that of the spectrum of black-body radiation is an example, is this: how do we partition a given amount of energy among various energy states? What is the most probable way of doing it, i.e. what is the division which we can accomplish in the greatest number of ways? If we treat this problem in too general a way, we are led into fairly sophisticated mathematical problems. But we can give an illustration, which requires only simple arithmetic, which illustrates nicely the importance of the question of distinguishability or indistinguishability.

Suppose we have four units of energy to be partitioned among two states, one with one unit of energy and the other with two. The possible partitions are given by a simple table:

| | No. of photons in | |
	state 1	state 2
(a)	0	2
(b)	2	1
(c)	4	0

The possibilities are (a), (b) and (c).

If the photons are indistinguishable, each possibility can be realized in only one way. Thus the probabilities of (a), (b) and (c) are each 1/3. The average number of particles in state 1 is thus

$$\frac{1}{3}(0 + 2 + 4) = 2$$

and the average number in state 2 is

$$\frac{1}{3}(0 + 1 + 2) = 1 \ .$$

However, the *probability* of that distribution is only 1/3.

Suppose now that the photons were distinguishable. Then, though there would only be one way to realize (a) and (c), there would be three ways of getting the distribution (b). The probabilities of (a), (b) and (c) respectively would then be $\frac{1}{5}$, $\frac{3}{5}$ and $\frac{1}{5}$. The average numbers in the two states would be as before

$$\frac{1}{5} \times 0 + \frac{3}{5} \times 2 + \frac{1}{5} \times 4 = 2 \quad \text{for state 1}$$

and

$$\frac{1}{5} \times 2 + \frac{3}{5} \times 1 + \frac{1}{5} \times 0 = 1 \quad \text{for state 2 .}$$

There is an important difference: in the case of distinguishability, the probability of the 2:1 distribution between the states is 3/5 rather than 1/3! The entropy of that state is

$$-\ln \frac{3}{5} = 0.51$$

when the photons are distinguishable and

$$-\ln \frac{1}{3} = 1.10$$

when they are indistinguishable. The entropy is lower in the first case!

This sort of effect can be measured in a more realistic system. The measurement points unequivocally to indistinguishability, so this will be made a hypothesis from here on.

A more important lesson can be learned from all this, which reveals why quantum theory is so essential to the realities of the physical world. The existence of atoms and molecules which are everywhere and always identical is essential to the regularity which gives order to our world. The atom is not like a planetary system. Planetary systems can come in an infinite number of forms – a continuum. A world of planetary-system atoms could not exist. The 79 electrons in the atoms of gold are always distributed in *precisely* the same way, as are the 16 in the molecule of the oxygen we breathe. It is the precision and reproducibility of their construction which gives these materials the exact physical and chemical properties on which we depend. Nature did not have a choice between constructing the world according to Newtonian or quantum laws. Its most familiar details, as well as its surprises, are a precise reflection of quantum reality.

13.12. The Pauli Exclusion Principle

We have focused our attention, up to this point, on two particles: the electron and the photon. While we have insisted that quantum mechanics treats them in the same way, they are in an important respect different sorts of particles.

We have discussed from a quantum viewpoint the simplest of atoms, that of hydrogen. What makes it particularly simple to study is that its energy states are those of a single electron. The next simplest atom is that of helium, which consists of a doubly charged nucleus surrounded by two electrons. One complication appears in the interaction – a mutual repulsion between the two electrons. Strictly speaking, we cannot in the light of this interaction speak of the state of each of these electrons separately. Clearly, the state of each depends on that of the other. They are not independent, but correlated.

Let us then think of a two-particle state, which we shall designate $|1, 2\rangle$. The fact that the particles are not distinguishable does not prevent us from this apparent labelling, which simply reflects the fact that the system has twice as many degrees of freedom. In fact, we can formalize the indistinguishability by saying that the state $|2, 1\rangle$ is not different from the state $|1, 2\rangle$, whatever that state may be.

To see what the consequences of this are, imagine first two non-interacting particles of the same sort, which are in states $|a\rangle$ and $|b\rangle$ respectively. If we could distinguish the particles (1 and 2, say), then

$$|a1, b2\rangle$$

could represent the state in which particle 1 was in state a and particle 2 in state b. On the other hand, if we considered the super-position:

$$|a1, b2\rangle + |a2, b1\rangle \ ,$$

we would have a state in which the two particles were not distinguished. In the situation in which the two particles were in the *same* state (a), it could be designated simply as

$$|a1, a2\rangle$$

(one might first be inclined to say $2|a1, a2\rangle$ but in fact changing the multiples does not change the state).

Another possibility is the superposition

$$|a1, b2\rangle - |a2, b1\rangle \ .^\dagger$$

Exchanging 1 and 2 in this case gives

$$|a2, b1\rangle - |a1, b2\rangle$$

which is in fact the same state, so indistinguishability is maintained.

Atomic spectroscopy led Wolfgang Pauli to enunciate the principle that electrons could only exist in *antisymmetric* states (Pauli exclusion principle). There are two kinds of particles in nature. Those obeying the principle (electrons, protons, neutrons, neutrinos, μ-mesons) are known collectively as "fermions" — after Enrico

---

†If we define an "exchange operator" which interchanges particles 1 and 2: $P|1,2\rangle = |2,1\rangle$, operating with this operator twice restores the original state and permits $P|1,2\rangle = |2,1\rangle$ or $P|1,2\rangle = -|2,1\rangle$. These may be designated as "even" or "odd" states (alternatively, "symmetric" or "antisymmetric").

Fermi — whereas those which do not are known as "bosons", after the Indian physicist S. K. Bose. Examples of these particles are photons and pi-mesons, which we shall discuss subsequently.

The foregoing argument, based on the assumption that each electron has its own independent state (i.e. the particles are non-interacting) leads to the formulation of Pauli's principle in the following form: no two fermions may occupy the same state. Though this is the commonest popular formulation, our discussion shows that the principle can be formulated more generally thus: in a multi-particle system, the state is antisymmetric in exchange of any two identical particles. The first formulation is simply a particular case.

An interesting consequence of Pauli's principle has had several historical manifestations. In applying the principle one must be sure of having identified *all* of the properties of the particle. When Schrödinger formulated his "wave equation" for the hydrogen atom, it was not known that the electron had an intrinsic ("spin") angular momentum. In such a circumstance, the application of the principle would lead to the conclusion that the state of two electrons would have to be antisymmetric under a *spatial* exchange of the two particles. But this is an unjustified restriction, since it neglects spin. In fact, the *spatial* and *spin* properties of electrons are to a first approximation independent. It is then possible for a state to be spatially antisymmetric and symmetric in exchange of spins, *or* spatially symmetric and antisymmetric in spin exchange. Thus, the discovery of a new property modifies in a very fundamental way the physical properties of multi-particle states.

Similar considerations have in more recent times led to an important modification of the theory of quarks as basic components of "fundamental" particles. Because states were found by experiment which were symmetric in all already known properties, a new property, not directly measured and in fact not understood, was assigned to quarks labelled (quite misleadingly) colour. So the principle may be used to deduce the existence of previously hidden properties.

A more mundane consequence of the principle also deserves mention. The fact that bosons, and in particular photons, are not subject to it makes possible very intense light beams. The properties of electron beams are complicated by their charge and hence electric repulsion, but even in the case of neutral particles (e.g. neutrons) there is the further constraint that, since only one particle can occupy a given state, the attempt to concentrate a high density of particles requires forcing more and more of them into higher energy states. Another example of this point is the "degeneracy pressure" which may halt catastrophic collapse of stars too heavy to stabilize as white dwarfs and too light to collapse to black holes.

13.13. An Aside on "Schrödinger's Cat"

Erwin Schrödinger, though known as one of the founders of the quantum theory became, later in life, one of its most irreconcilable critics. His problem, like Einstein's, stemmed from a revulsion against the necessity of abandoning a classical framework of concepts for it. He had a particular distaste for its manner of dealing with transitions between quantum states. The mathematical theory describes these changes as taking place in a continuous way, but is interpreted to mean only that the *probability* of the transition varies continuously, while the transition itself takes place abruptly. Schrödinger is reported to have said "Had I known that we were not going to get rid of this damned quantum jumping, I would never have involved myself in this business". One cannot be entirely unsympathetic to this sentiment, which however concerns only the "interpretation" of the theory, and not its verifiable predictions.

In the context of such misgivings the "gedanken-experiment" of "Schrödinger's Cat" was born. Schrödinger's aim was to show that the standard "Copenhagen" interpretation of quantum mechanics, according to which classical characteristics of physical systems could only be discussed statistically, led to unacceptable conclusions on the macroscopic level. But over the years it has become unclear what the Copenhagen interpretation really was — how should the interpretation be interpreted? In some recent popular books purporting to explain the theory,[17] a rather naive and extreme viewpoint on this question has been presented; one might call this the "radical Copenhagen interpretation" (RCI for short). Schrödinger's conceptual "experiment" was this: we imagine a cat sealed in a box along with a radioactive source and a detector for the emitted particles. The whole might then be placed in a space vehicle in orbit around the earth. Suppose too that the detector is set to be turned on automatically for a brief period of time, during which the probability of a particle being emitted by the radioactive source is $1/2$. Finally, if a particle is detected, a poisonous gas is released in the box and kills the cat.

It is now claimed (Pagels) that "according to the strict Copenhagen interpretation", even after the interval of time has passed "we cannot speak of the cat as in a definite state — alive or dead — because as earthbound people we have not actually observed if the cat is alive or dead. A way of describing the situation is to assign a probability wave to the physical state of the dead cat and another probability wave to the physical state of the live cat. The cat-in-the-box is then correctly described as a wave superposition consisting of an equal measure of the wave for the live cat and the wave for the dead cat".

Ultimately, someone goes and looks. The cat is either alive or dead. Pagels continues: "The Copenhagen interpretation of this event is that the scientists by opening the box and performing an observation have now put the cat into a definitive quantum state — the live cat". There is more in this vein.

---

[17] See for example Heinz R. Pagels, "The Cosmic Code", P. C. W. Davies, "The Ghost in the Atom", or C. Gribbin, "In Search of Schrödinger's Cat", among others. For a critical discussion of the issue see "Three Perspectives on Schrödinger's Cat" by J. G. Loeser (*American Journal of Physics*, **52**, 1089, Dec. 1984) or the more ponderous article in the same journal (H. P. Stapp, **40**, 1098, Aug. 1972), but especially the illuminating treatment in Fritz Rohrlich's book, "From Paradox to Reality", Cambridge University Press, 1987, pp. 166-169.

This sort of reductio ad absurdum argument is unfair, however, because there is no evidence to justify attributing this sort of interpretation to the Copenhagen school. The expression "Copenhagen interpretation" has come to be so indiscriminately used that it is appropriate to quote an old friend of Bohr's, Professor Rosenfeld, who in a letter to Heisenberg speaks of the phrase "Copenhagen interpretation" which we in Copenhagen do not like at all. Indeed, the expression was invented by people wishing to suggest that there are other interpretations of the Schrödinger equation, namely their own muddled ones. Moreover, as you yourself point out, the same people apply this designation to the wildest misrepresentations of the situation".

The implication of the sort of argument we have cited is that one must draw from quantum mechanics the conclusion that it is the "observer" who *creates* physical reality. Of course, there is a sense in which everyone who does a physical experiment or observation creates a bit of "reality"; so does everything we do. This statement is so trivial that it cannot be the meaning of the proposition. That meaning must be that *the only reality there is* is that created by human observation. This is a position so extreme that it undermines the very rationale of science, whose goal is to understand how the world works.

Of course, in the cat "experiment" one could put a clock in the box, set to register the moment when the poison is released or the cat dies. Does one then say that the cat goes into the dead state only when the clock is read, or that it retroactively died at the time registered on the clock? A simpler, more natural resolution is to say that we are inevitably actors in the drama of nature, and that when we intervene, that intervention is subject to the same laws as any interaction in the physical world. Certainly one is reminded of one's surprise at Maxwell's belief in the aether when it did not appear anywhere in his equations, for there is no more a "human factor" in quantum theory than there was an "aether factor" in electromagnetic theory.

But to return to the cat: can one really describe live and dead cats, and the transition between them, by quantum wave functions? One argument at least denies the possibility; wave functions not only have amplitudes, which describe probabilities, but also internal phase relationships between the states involved. The interactions involved in the history of the cat are too complicated to permit the carrying of this sort of phase information; therefore, we cannot set up correspondence relationships between cat and radioactive nucleus.

The function of the cat in this story is in effect quite different from that of a "mirror" of the microscopic quantum system; its role is, rather, that of a measuring instrument. This sort of distinction is clearly made in Bohr's concept of measurement in quantum theory; a measuring instrument is a macroscopic object which makes an irreversible record of a microscopic (quantum) event. The cat is not as good a measuring instrument as a clock, but its function is in essence the same. So Schrödinger's concern that the macroscopic cat-in-the-box might mirror the quantum behaviour of the atomic nucleus is unfounded.

If the cat has a wave function, it must be enormously complicated and carry vast quantities of information that is irrelevant to the question at hand. On the other hand, its death carries less microscopic information than the decaying nucleus. We tend to forget that the "probability interpretation" interprets only the magnitude and not the phase of the wave function; in this sense it is incomplete. The phase does not relate to a concept (probability) but carries a message about the nature of physical reality (as Feynman's "paths" testify).

So Schrödinger cats may profitably be relegated from the serious world of science to the small talk of Feynman's "cocktail party philosophers".

Chapter 14

ATOMS, MOLECULES, CHEMISTRY AND
CONDENSED MATTER

The first concern of the new quantum mechanics was with spectroscopy, which was based on a knowledge of the states of atoms.
The hydrogen atom is the simplest. At first it seemed very simple
indeed, since the Schrödinger wave function could be solved exactly;
its energy levels were precisely as given by the simple theory of Bohr.
Its spectrum depended, however, on transitions between the energy
states. The *frequency* of these transitions depended only on the
differences of their energies, but their intensities, which concerned
the probability per unit time that the transitions would take place,
involved further questions: what transitions were actually possible
(by virtue of consistency with the principle of conservation of angular momentum, basic symmetries, etc.); how were these transitions
affected by external factors such as electric and magnetic fields, and
what were the polarizations of the emitted photons? These problems could be resolved either by calculating wave functions or using
Dirac operator methods. Once again, though, for hydrogen atoms
these were easy calculations.

All the same, the spectrum of the hydrogen atom turned out
to be not quite so simple as initially appeared. First, to understand
the observed spectra it was necessary to postulate that the electron
had an intrinsic angular momentum ("spin"). Later, when Dirac
succeeded in developing a relativistic generalization of Schrödinger's
theory, this spin was found to appear naturally. It was manifested by

a splitting or separation of one energy level into two in the presence of an external magnetic field. More important than the separation in energy, however, was the doubling of the number of states available to electrons, since this completely changed the consequences of the Pauli exclusion principle.

When a finer resolution of the spectral lines was achieved, further splittings were observed. This was attributed to the fact that electrons in states with "orbital" angular momentum about the nucleus created an internal magnetic field which also interacted with the spin; this spin-orbit coupling also affected the energy states, even though only weakly in hydrogen.

Still further complications appeared in the 1950's, quantitatively almost insignificant but important in principle, as we shall see later.

So even the hydrogen atom turned out not to be terribly simple when all complications were taken into account. Luckily, however, there were many interesting aspects of it for which such refinements were unimportant.

It is useful for further study of atoms to enumerate the states of hydrogen. They are characterized not only by the quantum number n which determines the energy, but also by the total value of their angular momentum, and the angular momentum in any one (arbitrarily chosen) direction: this angular momentum has possible values which differ by $\hbar$, and run between a minimum value $-l\hbar$ and a maximum value $l\hbar$. Thus, there are $(2l + 1)$ states all of which have the same energy. These levels are designated by letters, thus

$$l = \begin{matrix} 0 & 1 & 2 & 3 & 4 & 5 \\ s & p & d & f & g & h \end{matrix}, - - - - -$$

with multiplicity (degeneracy)

$$1 \quad 3 \quad 5 \quad 7 \quad 9 \quad 11, - - - - -$$

or, including spin

$$2 \quad 6 \quad 10 \quad 14 \quad 18 \quad 22, - - - - -$$

14.1. Multi-electron Atoms

The next atom in complexity is helium (2 electrons). The wave function is greatly complicated by the electric repulsion between the two electrons, so that their configurations are correlated; we cannot rigorously speak of a distinct state or wave function for each of them.

We can do so, though, in an approximate sense, by the following argument: let us treat each electron as moving in the *average* electric field of the other. Note, however, the curious circularity of this approach. In order to calculate the average electric field of an electron, we have to solve its wave function in the average field of the other! The mathematical problem thus posed is rather difficult, but it can be solved.

The idea behind this approach can be applied to other, more complicated problems. The present case gives us a rationale for speaking, at least in an approximate sense, of the states of individual electrons.

For helium, our experience suggests that the lowest energy state will be spherically symmetric and have no nodes; a 1s state in fact ($n = 1, l = 0$). The energy of this state is found to be about 79 electron volts (negative). The energy of a single electron around the helium nucleus is four times that in hydrogen, i.e. 54.4 electron volts. Were it not for the mutual repulsion of the electrons, the ground state energy for helium would be 108.8 ev. The repulsive energy therefore is about 30 electron volts, a significant figure.

An important qualitative observation is that the two helium electrons, being in the "same state" and thus having a symmetric wave function must, by the Pauli principle, have opposite spins (or more accurately, be in an antisymmetric spin state). Thus, the total electron angular momentum in this state is zero.

Excited states of helium may be obtained by promoting one of the two electrons to a higher energy state. In this case, of the two possibilities for the spins, the one with the spins *parallel* has the lower energy, because it forces the spatial configuration to be antisymmetric in the two electrons, i.e.

$$|2,1\rangle = -|1,2\rangle \ .$$

If the two electrons are at the same place, the electron wave function is *zero* there (anything equal to its own negative is zero). Thus, the Pauli principle has forced the electrons to maintain some distance from each other, thus lessening their mutual repulsion and lowering the energy.

Lithium has three electrons and follows helium in the periodic table. Only two electrons may now be in the same state (with their spins opposite); the third must go into a higher state. Just as in hydrogen, the next state is a spherically symmetrical state (thus with no angular momentum) and one radial node — a nodal sphere, in fact.

In hydrogen, however, this state has the same energy as a state with one angular node (a p-state) but no radial node apart from the origin at the position of the nucleus. In lithium, this state, which is called $2p$, has a *higher* energy than the preceding one, which is designated $2s$. There are in fact three different $2p$ states, whose axes of symmetry are in three different directions, each, of course, capable of accommodating two electrons with different spins.

We may schematically represent the lowest energy states of the various atoms by energy diagrams (Fig. 14.1).

In beryllium, with *four* electrons, two electrons, with opposite spins, are in $2s$ states (Fig. 14.2).

We shall not carry on in this way, but shall simply indicate on a quite arbitrary scale, the filling of the levels.

For carbon and nitrogen, we show the three equivalent $2p$-states (Fig. 14.3). Three $2p$-states permit up to three electrons with parallel spins. This means that the electron states are spatially antisymmetric; which keeps the electrons apart and therefore minimizes their energy of mutual repulsion.

Continuing the level filling for the remaining three elements of the period of the Mendeleev table, we get the representation shown in Fig. 14.4.

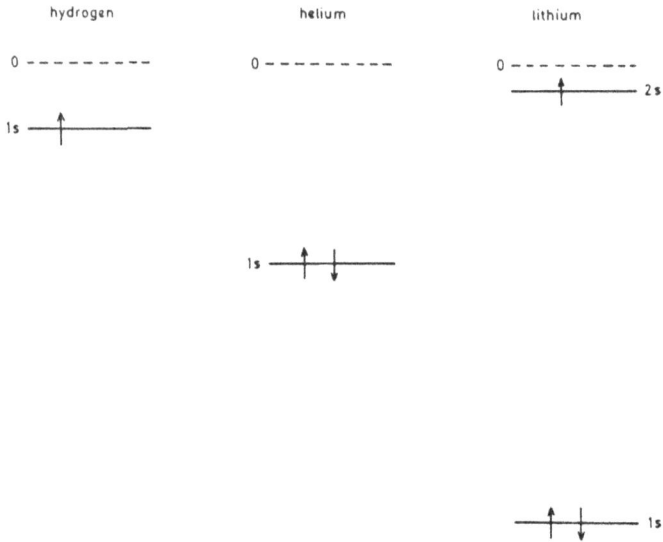

Fig. 14.1. The arrows indicate electrons occupying the states, with the direction of their "spin". The diagrams are roughly to the same energy scale.

Fig. 14.2.

After this, starting with sodium, which is chemically similar to lithium, down to argon, which, like helium and neon, is chemically inactive, there is another series of eight elements which are exactly related to their analogs in the preceding series:

Boron   Carbon   Nitrogen

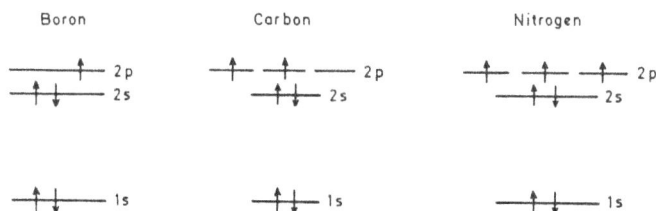

Fig. 14.3.

Oxygen   Fluorine   Neon

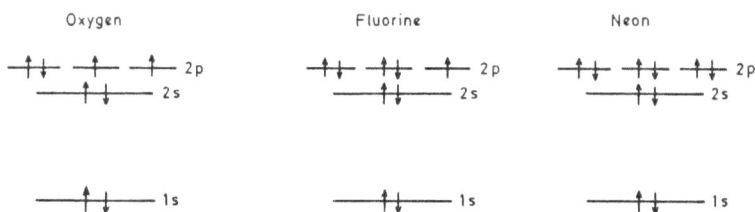

Fig. 14.4.

$$
\begin{array}{ll}
\text{Na} & \text{– Li} \\
\text{Mg} & \text{– Be} \\
\text{Al} & \text{– B} \\
\text{Si} & \text{– C} \\
\text{P} & \text{– N} \\
\text{S} & \text{– O} \\
\text{Cl} & \text{– F} \\
\text{Ar} & \text{– Ne}
\end{array}
$$

After this, the situation becomes complicated. First, there is another s-level ($4s$) for which the wave function has three radial nodes. Before the next p-levels ($4p$) is a new kind of level, called $3d$, which has two radial nodes and two angular ones (the $4p$ levels have three radial nodes and one angular one). There are five $3d$ levels (accommodating 10 electrons); the component in any given direction of the electronic angular momentum around the nucleus can be

$$-2\hbar, -\hbar, 0, \hbar, 2\hbar$$

where $\hbar = h/2\pi$. Thus, ten more elements can be inserted here, whose outer, most weakly bound, electrons are in d-states. These are known as the transition metals, and include the three known ferromagnetic (spontaneously magnetic) metals — iron, nickel and cobalt.

We shall not pursue the question of atomic states beyond this point. In fact, most essential principles of the subject can be well illustrated by considering only the first period of the periodic table of Mendeleev, that is, the first ten elements up to neon. This set includes the elements most important for our everyday life: hydrogen and oxygen which give us water, nitrogen and oxygen in the air, silicon and oxygen in the rocks and soil, and carbon which is necessary for life.

The following table of ionization energies provides clues to the role of these elements. The question of greatest interest is, how and why do the elements form the compounds they do? Most importantly, what is it about carbon which permits it to form a far wider range of molecules than any other element, so many that a whole branch of chemistry, *organic chemistry*, is devoted to it?

First ionization energies (in electron volts) of the elements of the first period of the Mendeleev table are:

Table 14.1.

H	13.595	C	11.256
He	24.481	N	14.53
Li	5.39	O	13.614
Be	9.32	F	17.418
B	8.296	Ne	21.559

(from "Handbook of Chemistry and Physics", CRC Press, 58th edition, 1975).

Note that helium and neon are the most tightly bound, and at the same time the most chemically inert. The jump between He and Li is because, with Li, we start to fill a new "shell" much less tightly bound than the first one. Li is highly active chemically.

If an electron were removed from a Li atom to form a Li ion, it would be chemically identical to He (the same levels would be filled). Similarly, if fluorine were to take on another electron to form an F^- ion, it would be chemically identical to neon. Thus, at first sight one might not expect a crystal of Li^+ and F^- to be bound, even though it might be energetically favourable to form ions. The crystal exists for another reason, although the ions would not be bound *chemically*, the crystal would be held together by electric forces.

Before proceeding to the question of the chemical combination of atoms into molecules, let us consider the strange quantum mechanics of angular momentum.

The strangeness is manifested in the following statement: the angular momentum about any given direction must be an integral multiple of the constant $\hbar$. A second important fact is that if there is angular momentum about one direction, there must be more angular momentum about other directions, though what other direction is not determined. An alternative statement is that we can never say exactly what the direction of the angular momentum is. This is analogous to the earlier statement that we can never say where an electron with a given momentum is. Yet if we pick any particular direction, it can be determined exactly how much angular momentum it has about that direction.

Once again, things are happening in what appears to be a paradoxical manner. But again, too, the apparent paradox is merely a consequence of trying to reason about the electron as though it were a classical point particle, which it definitely is not.

It does seem curious, all the same. Let us pick out some direction; we are then able to see how much of its rotation might be about that direction; it might be $\hbar$, or $2\hbar$, etc. We can then say nothing about the direction of the rest of it, though the total rotational energy is known.

Consider a direction just slightly different from the original one. The particle may now be in a state in which it is the angular momentum about this direction which is well determined; but when it is in this state, the angular momentum about the original direction becomes uncertain.

We are trying to attribute physical reality to the wave function, but we have no justification for this. Physics is about change; that which is unchanging may not be observed. Let us reformulate the problem. The hydrogen atom can only change by interacting with something else. So suppose that it is brought into interaction with something else, e.g. another atom. How could angular momentum be transferred to this second atom? An integral number of units of angular momentum in any direction can be transferred. What does happen depends on the precise details of the interaction, though it is not rigorously predictable, any more than the time of a radioactive decay is. It seems less surprising if we realize that the interaction with the other atom changes the angular momentum of the original one, since it exerts a force (or torque) about its nucleus. While the atoms are interacting, there is no definite angular momentum about either nucleus. If the atoms subsequently separate so that there is no interaction between them, their individual internal angular momenta will perhaps be changed in a manner which is *statistically* predictable, but not predictable in any one case.

14.2. Binding of Atoms into Molecules

Let us look at the problem of the binding of atoms into molecules first in the case of the simplest known molecule, that of hydrogen. How and why do two hydrogen atoms stick together to form H_2, whereas helium does not appear in any but atomic form?

There are two ways of looking at this problem, which seem at first glance quite different. Each one gives some useful insights.

One method is to think of the molecule as being formed

gradually from the union of two atoms. The atoms interact and modify each other in such a way that the energy of the system is reduced.

The other method is more similar to that used for atoms. We envisage two nuclei and electrons moving in the electric field of the two together in a hierarchy of states determined by solving Schrödinger's wave equation.

What can we deduce from the first point of view? As the atoms come together, the electric field of each affects the electrons of the other. If we call the identical atoms A and B, then the nucleus of A will attract the electron of B and vice versa. Of course, the interaction of the atoms will change ("polarize") each of them, pulling the charge of each atom toward the other. The way one describes this is to mix some p-state in with the s-state in which the atom exists before the interaction.

Spin comes into the picture, but not primarily because of magnetic effects. Rather, if the spins of the electrons on the two atoms are the same ("parallel"), the Pauli exclusion principle keeps the electrons apart, so that each is on the average further from the opposite nucleus. The binding is thus decreased; in fact, the two atoms no longer attract each other, but repel.

If, on the other hand, the spins of the two electrons are *opposite*, the Pauli principle does not prohibit the overlap of the two electron distributions and the attraction is maintained.

But, you may reasonably object, the electron distributions can overlap, thus causing an electrostatic repulsion. So how do we know that the attractive forces dominate? Detailed calculation shows that they do, but there is another indication which points to the same conclusion. Imagine that the two atoms approach until finally their nuclei coincide. The problem would then be that of two electrons surrounding a 2-proton nucleus. This is the problem of the helium atom, where everything is strongly bound together, *provided* that the two electron spins are opposite. If they are equal (parallel) one of the electrons would have to go into an excited state at considerably

higher energy. Careful calculation shows that the hydrogen atoms
in this case repel rather than attract, so that no hydrogen molecule
is formed. The limiting case serves as a guide to what will happen
as the atoms approach each other (Fig. 14.5).

Fig. 14.5.

The "valence bond" which holds the atoms together is formed
only when the spins are opposite.

This conclusion is reinforced when we try to construct the
molecule by starting with the two bare hydrogen nuclei and add
the electrons in the field which they create (Fig. 14.6).

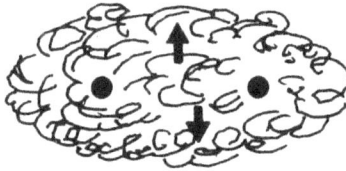

Fig. 14.6.

The wave functions will no longer be spherically symmetric, but
will instead have a symmetry about the axis joining the two nuclei.
The lowest state will again be one with no nodes and no angular
momentum about that axis. It will become more and more like the
1s (lowest energy) state of the helium atom as the two nuclei are
brought close together. The two electrons will both be in this lowest
energy ("ground") state.

14.3. Do a Hydrogen Atom and a Helium Atom Cohere?

What does the "combined atom" limit suggest? If the two nuclei were to be brought together, a lithium atom would be formed. Of the three electrons, two could be accommodated in the lowest (1s) state, but the third would have to be in the much higher energy 2s state. If we think of building the "molecule" by adding three electrons to the bare nuclei, once again two of the electrons, with their opposite spins, could occupy the lowest state, but the third would have to occupy a higher energy state. The total energy would be lower if the atoms were totally separated, since then all three electrons could be in the lowest (1s) state of the respective atoms. Thus, no molecule, no bound state, of the two atoms would be formed. The two electrons which occupy the lowest state and have opposite spins will form a bond. But the third electron will constitute no bond. As the atoms approach, its energy will be forced up; it will tend to stay near the helium nucleus, but will be pushed further away from the hydrogen nucleus by the action of the Pauli principle.

The argument for the mutual repulsion of two helium atoms is even stronger. Again, when the atoms are separated, all four electrons can be found in the lowest energy states of their respective atoms. When they are brought together, however, two of the electrons are forced into much higher energy states. In the case of the complete union of the two nuclei, these will be the 2s states. When the nuclei are separated, it will be difficult to say, without detailed calculation, whether the excited states will be the ones with a "radial" node, symmetric about the two nuclei, or "p-like" states, with a nodal plane halfway between the two nuclei. In neither case will a valence bond be formed.

A common feature of the above three cases can be identified. Pairs of electrons from the interacting atoms form a "valence bond" with their spins opposite. The introduction of another electron, whose spin must be the same as that of the electron on the other atom, gives rise to a repulsive force, and militates against the bonding of the two atoms. Stated somewhat differently, two atoms which have electrons unpaired as to spin can bond to each other when the spins of the two electrons are opposite. If all the electrons on one or both of the interacting atoms already have only paired spins, electron repulsions due to the Pauli principle dominate and the atoms will not bond together into a molecule.

These are general principles for atoms of any complexity, and form the basis of a general theory of the valence bond.

14.4. Valence and Valence Bonds

Let us, with the aid of these principles, consider the chemical properties of the other atoms in the first period of the periodic table. Though the basic ideas are simple, their manifestations turn out to be more varied and interesting than might be imagined in advance.

Lithium is a quite simple case. Two of its electrons are in 1s states, the third in 2s. The former do not enter significantly into the problem of molecule formation. Their spins being opposite, they form a chemically repulsive core to the atom. But their wave functions are much more compactly concentrated around the nucleus than that of the 2s electron (in hydrogen the 2s state has a radius four times that of the 1s state). The latter can therefore interact to form a valence bond without the inner electrons getting close enough to the other atom to become involved). Lithium then forms compounds as hydrogen does; lithium hydride (LiH) is a quite strongly bound molecule.

The next element in the table is beryllium. A Be atom has four electrons, two in 1s states and two in 2s, with opposing spins. This appears at first sight reminiscent of helium, which is of course chemically inert. Beryllium is not chemically inert; it forms the hydride BeH_2. The difference from helium is that just above the 2s states in energy are 2p states. There is therefore an excited state of the atom Be with the electrons as shown in Fig. 14.7. By promoting one electron from a 2s to a 2p state the electrons can have parallel spins, which reduces the repulsive energy between them.

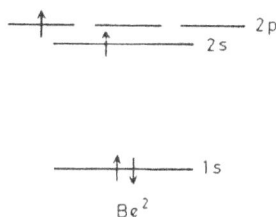

2p

2s

1s

Be^2

Fig. 14.7.

The energy of Be in this state is only slightly higher than that of the ground state (i.e. lowest energy state). Suppose now that the beryllium atom interacts with two hydrogen atoms, each having an electron with a spin which balances off one of the Be electrons. It would seem possible to bind both hydrogen atoms to the beryllium one thus creating BeH_2. We can push the argument even further than this.

If we recall that each beryllium electron can lower its energy, for a given distance between the nuclei, by adjusting its wave function to be closer to a hydrogen nucleus, we see that if one Be electron is in an s-state and one in a p-state this does not represent the most favourable situation energetically. Imagine that the p-wave function has its axis of symmetry and of elongation in the x-direction. Suppose now that one electron is in a state which is half $2s$-state and half $2p_x$ state. We will call this a *hybrid* state; the amplitudes of the two component states will simply be added. This situation is portrayed in Fig. 14.8 which shows, in a cross-section cut by a plane passing through the two nuclei, sections of surfaces of constant amplitude. For comparison, contour diagrams for the p-state wave functions alone are also shown (Fig. 14.9).

The $2p_x$ state wave function has an amplitude with opposite signs on opposite sides of the y-z plane.

As shown in Fig. 14.10, if the phase of the s-wave is positive everywhere, the wave function will be enhanced on the right. If the signs are reversed for the p-state, the enhancement will take place on the left. The existence of these two combinations make it possible

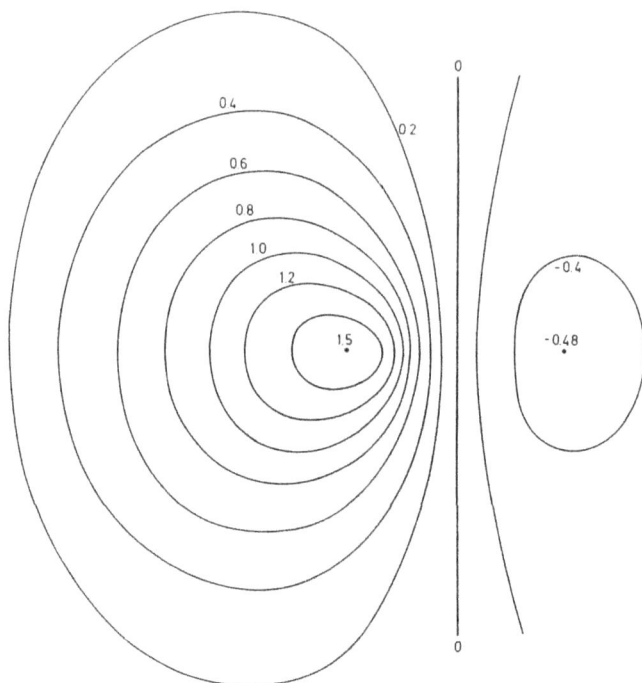

Fig. 14.8. Hybrid states. Hybrid wave function contours. The function is positive to the left of the line 0-0, negative to the right.

for the hydrogen atoms to attach themselves at the two ends of the x-axis (as shown by the circles). The molecule will then be a linear one, which can be represented schematically by

$$H - Be - H \ .$$

14.5. Boron

Consider boron, which has 5 electrons (Fig. 14.11).

The three p-state wave functions may be taken to be symmetric about the x-, y- and z-axes respectively (Fig. 14.12). Suppose those associated with the x- and y-axes — the horizontal axes — are occupied.

"p - state" wave functions

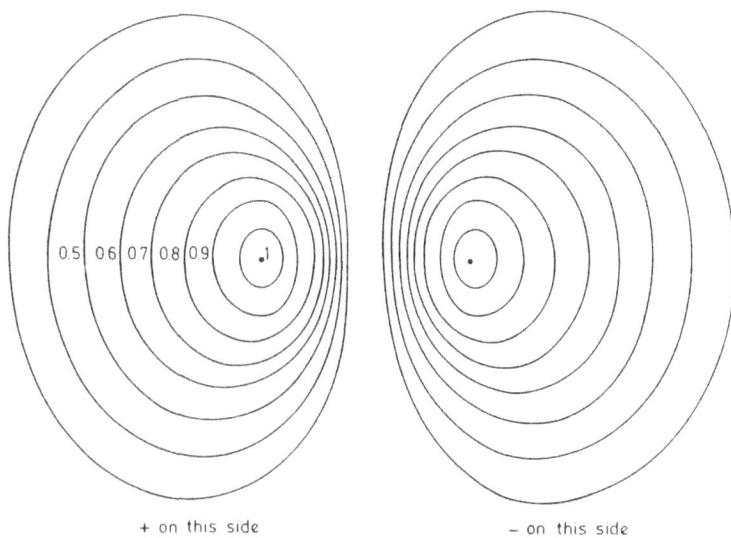

+ on this side   − on this side

Fig. 14.9.

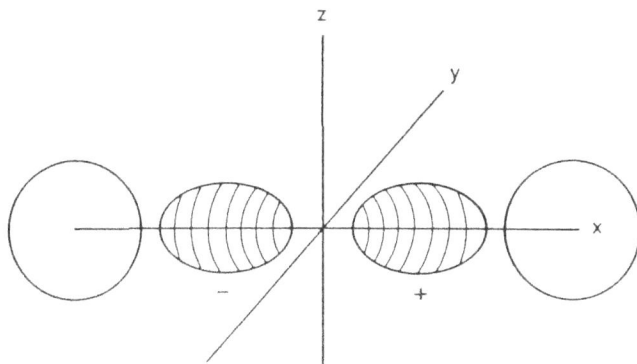

Fig. 14.10.

We have designated the electron states which are superpositions of *s*- and *p*-states as "hybrid states". Be had two such hybrid states whose extensions were in opposite directions to the *x*-axis. Boron

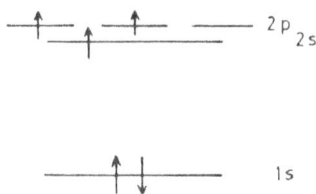

Fig. 14.11.

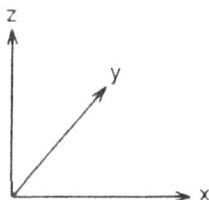

Fig. 14.12.

had no states with extension along the z-axis; different combinations are extended in different directions in the horizontal (x-y) plane. Three different combinations can be made to accommodate the three electrons. At this point symmetry considerations come into play. From the viewpoint of physics, there are no preferred directions in the horizontal plane. The three hybrid states must thus have axes of extension (and of symmetry) making equal angles with each other, as shown in Fig. 14.13.

In the compound BH_3 if the boron atom is at O, the three hydrogen atoms will be at A, B and C. Boron is trivalent, i.e. it can form three valence bonds with other atoms.

14.6. Carbon and Organic Chemistry

We can now understand what makes carbon such a special element.

Carbon can have *four* unpaired spins, and thus can form *four* valence bonds (Fig. 14.14).

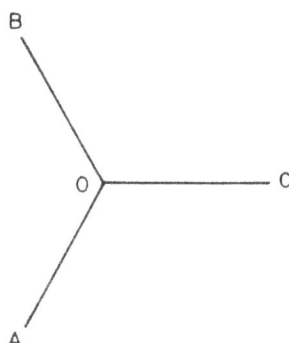

Fig. 14.13.

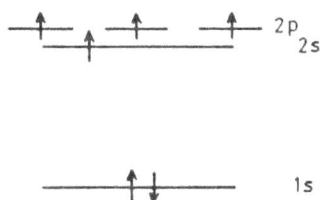

Fig. 14.14.

One possibility, which is suggested by the situation in boron, is that the three valence bonds involving planar hybrids are maintained while the fourth electron is in a pure p_z-state, with its axis perpendicular to the plane of the other bonds. The virtue of this configuration is its ability to "grow" without limit in the plane. Consider, for instance, the "benzene ring" C_6H_6 (Fig. 14.15).

Here the carbon atoms are fitted together in a ring, each with a trigonal valence configuration, two of its bonds being with neighbouring carbon atoms and one with a hydrogen atom.

What is happening to the fourth electron from each carbon atom? Firstly, we may imagine that these electrons, all in $2p_z$ states

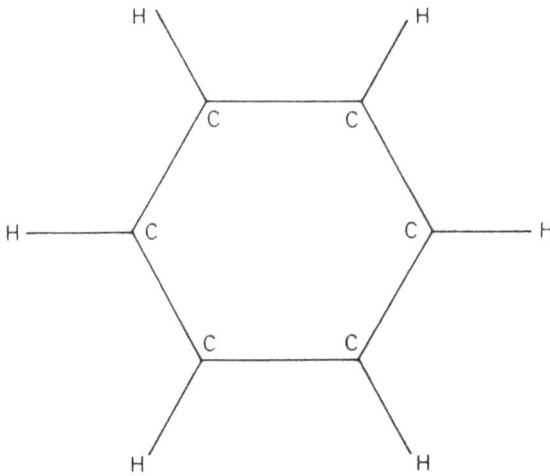

Fig. 14.15.

centred on one or other of the C-atoms, form second bonds in pairs between themselves. Of course, this may happen in more than one way. Starting with the upper left hand C-atom, it may bond with the carbon atom on its right, or with that down to its left. But there is absolutely no way of choosing between these possibilities, so each must occur with equal probability. Thus, we must superpose the wave functions of the two possible states with equal weight.

From a second viewpoint, these six "orphan" electrons are free to wander from atom to atom. Their states extend around the whole ring, forming a sort of wave. The energy of the state will, as usual, depend on the number of nodes of the wave. The possibilities are:

    (i) no nodes, same amplitude on each atom.

    (ii) one node, phases opposite for opposite atoms on the ring (one wavelength in the ring).

    (iii) one whole wave in half the ring (thus, electrons on opposite atoms in phase).

    (iv) reversal of phase form each atom to the next; three waves in the ring.

Each state has a higher energy than the preceding one,

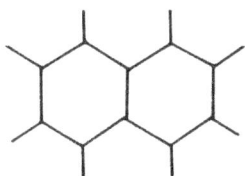

(a) Napthaline : $C_6 H_8$

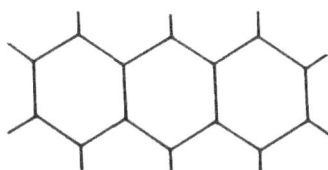

(b) Anthracine : $C_{14} H_{10}$

and something more complicated:

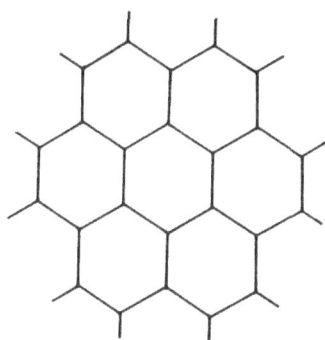

(c) Coronene : $C_{24} H_{12}$

Fig. 14.16.

Other things can be "stuck on" in place of hydrogen atoms, e.g. OH, NH_2 ..., to make countless other compounds.

The limitless possibilities are due to the fact that a 2-dimensional space can be filled with fitted hexagons. A *huge* pattern of hexagons like this is a LAYER OF GRAPHITE. Graphite is formed of such layers, stacked on each other, and only weakly interacting by mutually polarizing (Van der Waals) forces.

Graphite is a good lubricant because the layers slide over each other easily (Fig. 14.17).

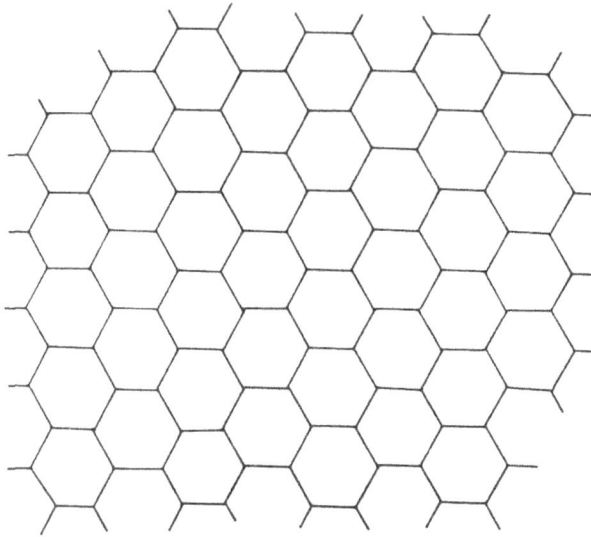

Fig. 14.17. Part of graphite layer.

Why does carbon tend to "waste" its fourth valence electron on simply forming weak supplements to already strong bonds? Solid carbon, for example, occurs in two forms: graphite, which is common, and diamond, which is extremely rare. Diamond is a much more "efficient" form in that the atoms form four completely equivalent bonds. As in the two-dimensional case, it is possible to surmise the three-dimensional diamond structure from symmetry considerations alone. How does one construct, with maximum three-dimensional symmetry, four hybrid bonds?

The answer is shown in Fig. 14.18. We construct a cube, with a carbon atom at its centre. We then put atoms at opposite diagonal corners of the top face. Finally, we put two more at opposite corners of the bottom face, directly below the empty corners on the top.

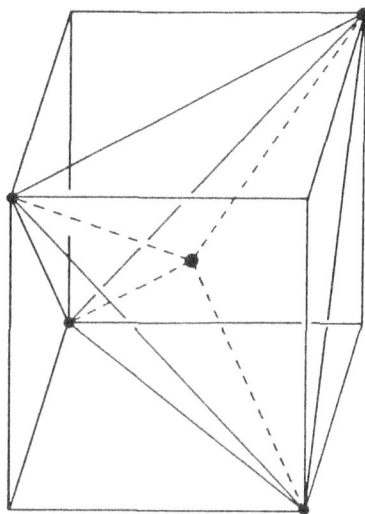

Fig. 14.18. Diamond structure (one cell).

We then see that:

(i) the lines joining the central atom to its nearest neighbours form equal angles with each other.

(ii) all faces are equivalent.

If hydrogen rather than other carbon atoms are placed at the corners, a carbon atom being at the centre, we have a molecule of methane.

But our cube may also be used as the basis for generating the diamond crystal. The way in which the cubes are assembled is shown in Fig. 14.19. The cube portrayed is made up of *four* cubes like the preceding one; each of these has an atom marked *b* at its centre. Interlocking cubes may be found; in fact, each atom is at the centre of one, with valence bonds projecting out toward its corners.

This is the structure of a diamond crystal. Many well-known semiconductors on which modern electronic technology is founded also have this structure. Silicon and germanium, which lie directly below carbon in the periodic table, are examples.

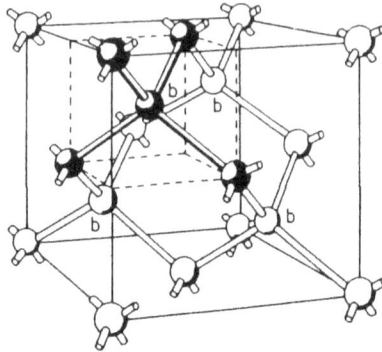

Fig. 14.19. Arrangement of atoms in the diamond structure, showing each atom with four nearest neighbours in a tetrahedral arrangement. (With permission from J. S. Blakemore, *Semiconductor Statistics*, copyright 1962, Pergamon Press PLC.)

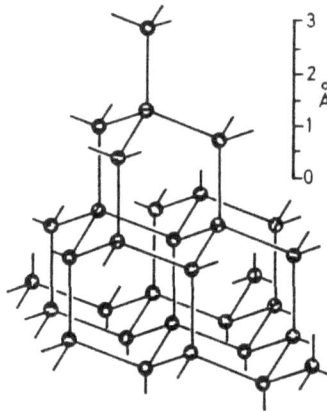

Fig. 14.20. The crystal structure of diamond. Each carbon atom has four nearest neighbours located at the vertices of a tetrahedron (nearest neighbours are joined by solid lines).

Thus, carbon atoms can fill, not only a two-dimensional surface, but also three-dimensional space. This is what gives carbon its special position in the chemistry of nature.

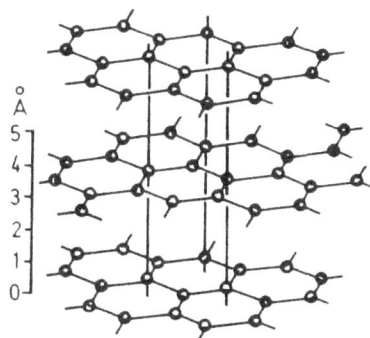

Fig. 14.21. The crystal structure of graphite.

Many multi-ring compounds can be formed; this is why organic chemistry is such an exceptionally rich subject. Several interesting questions may be posed:

1. Why is the graphite form of carbon so common in nature, whereas diamonds are extremely rare, despite the fact that the diamond form has lower energy?

The answer is to be found not so much in terms of energy as of entropy. The entropy of diamond is much lower than that of graphite. Thus, it takes a much higher temperature (and/or pressure) to make the diamond form.

Similar considerations hold for large numbers of molecules formed from chains of rings.

2. Why is there no comparable profusion of compounds for chemically related elements, such as silicon and germanium, which occur commonly with the diamond structure?

The difference is found in the energies involved. As we move from the states $2s$, $2p$ for carbon to $3s$, $3p$ for silicon and $4s$, $4p$ for germanium, the gap between the s and p states decreases (as in fact does the whole scale of energies, as manifested by the $1/n^2$ law). It is correspondingly easier to form the diamond bonding configuration in these substances, whereas the trigonal bond configuration will be less stable.

14.7. The Elements Beyond Carbon

For the remaining elements in the first period, the situation is simpler but less interesting.

(a) *Nitrogen*

Since in carbon we created the maximum number of states of unpaired electrons (four) when we come to nitrogen, we must add another electron in one of these states to create a pair, thus reducing the valence by 1. Since the $2s$ state is lower in energy than the $2p$ states, this is the obvious place to put the extra electron (presumably we could pair off electrons in one of the p-states, and make trigonal hybrids with the other three electrons, but this is not found to happen). We are left then with electrons with unpaired spins in each of the three spin states. The axes of elongation of the wave functions of these states are along three perpendicular directions (x, y and z). At first sight we might expect to find the NH_3 complex to have the hydrogen atoms joined to the nitrogen atom in three perpendicular directions. Actually, the molecule opens out somewhat, since there are repulsive forces between the H-atoms due to the fact that their electrons are subject to a repulsive force, being already involved in paired electron configurations.

Actually, the picture is not very different if we think of all electrons being in diamond-type hybrid states, one of which is occupied by electrons with paired spins. The remaining three bonds should then make the same angles with each other as in diamond, which is $109°28'$. The actual bond angle in NH_3 is $107°$.

The important thing to remember is that the sort of argument we have been making is qualitative and suggestive rather than rigorous.

The "valence-bond" viewpoint and the "molecular-orbital" viewpoint often provide quite different explanations of an observation. For example, in the case of the hydrogen molecule, two valence electrons are never found on one atom. In the molecular orbital viewpoint, since the two electrons are taken to be in the same state around the two nuclei, they are as likely to be both close to one nucleus as not. Neither of these situations is realistic; the truth lies somewhere between the two. It is possible to "improve" on each model, and so obtain a better fit to reality. But all such "improvements" tend to break down the conceptual differences between them. It disappears completely when each model is progressively modified until it provides agreement with reality.

(b) *Oxygen*

With eight electrons to be put into ten states, we can only leave two electrons unpaired, presumably in two of the p-states. Because of the repulsion of the attached molecules (hydrogen in H_2O) the angle between the remaining bonds will be a bit larger than $90°$.

This does not offer much promise for explaining the beautiful hexagonal patterns of snowflakes! Ice, however, is not held together by valence bonds of the sort we have been discussing. The basic building blocks are in fact not atoms but H_2O molecules. The nature of the bonds holding them together, the "hydrogen bonds", is quite complex and beyond the scope of this discussion.

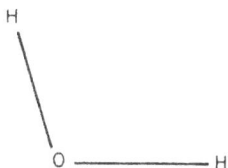

Fig. 14.22.

(c) *Fluorine*

Fluorine can only have one electron in a p-state in which it has no partner of opposite spin. It is most strongly bound to atoms with one excess unpaired electron outside closed shells of electrons in spin-paired states: hydrogen, sodium, potassium (the alkali metals).

This is because such an atom tends to give up its extra electron to fill the last of the states in fluorine. In this way the filling of the states in sodium is precisely like that in Ne. This is also true of fluorine with its newly captured electron. There ought therefore to be no chemical bonding between sodium and fluorine atoms. In fact evidence shows that there is not. Instead, they are held together purely by electric forces. The sodium atom, having lost an electron, becomes a positively charged ion. The fluorine atom, having gained one, becomes a negatively charged ion. These oppositely charged ions are then held together by Coulomb forces.

Since Cl sits just below F in the periodic table, and hence has the same electronic configuration in the $3s$-$3p$ shell that fluorine has in the $2s$-$2p$ one, the crystal of ordinary salt (NaCl) is bound in the same manner as NaF. The fact that the ions Na and Cl are both chemically equivalent to inert gases (neon in the one case, xenon in the other) makes it easier to understand why a compound made from a poisonous gas (Cl) and a violently alkaline metal (Na) can be such an innocuous component of our daily diet.

At the end of the period in the Mendeleev table lies neon, with all s and p states doubly filled and thus no power of chemical binding to anything. It is chemically equivalent to helium, though physically quite different because of its greater mass (five times that of helium).

Logically, our next subject of study should be the atomic nucleus. But the study of the nucleus will lead us into a new realm of the physics of particles — that concerned with the so-called antiparticles of Dirac. We shall therefore deal with this problem first.

Chapter 15

PARTICLES AND ANTIPARTICLES

We have now come to one of the most interesting, and important, topics in modern physics. We are again indebted for it to one of the greatest theorists of modern times, P. A. M. Dirac (deceased 1984).

The problem which led Dirac to the discovery was finding a relativistic quantum theory for the electron.

From the technical viewpoint, which is too sophisticated for discussion at the popular level, Dirac's theory was extremely imaginative and ingenious and fraught with various unexpected discoveries. It was a triumph of the search to find the most elegant possible mathematical description of nature. Dirac was wont to say that it was more important that a theory be mathematically beautiful than that it agree with experiment. The physical picture which emerged contained several surprises: it portrayed the free electron as undergoing a constant "Zitterbewegung", or "jittering motion" between states with velocities c and $-c$ (c being again the velocity of light); it predicted the existence of an intrinsic angular momentum ("spin"), measured in half-units of the Planck constant $\hbar$; and it predicted the existence of states of negative energy.

The intrinsic electron spin had already been postulated empirically by Goudsmit (1926), so that the discovery of a fundamental theory for it was a great triumph. But the prediction of negative-energy states (of arbitrarily great magnitude as well!) appeared at first sight ridiculous — an indication of a flaw in the theory. However, even before the advent of quantum theory it was known that in relativity the relation between energy and momentum of a particle of rest-mass m_0 was

$$E^2 = c^2 p^2 + (m_0 c^2)^2 \; ,$$

p being the momentum and E the energy of the particle. Taken literally, this meant that the energy could be either

$$E = \sqrt{c^2 p^2 + (m_0 c^2)^2}$$

or

$$E = -\sqrt{c^2 p^2 + (m_0 c^2)^2} \; .$$

Thus, free electrons could have energy states greater than $m_0 c^2$, or less than $-m_0 c^2$. It was, however, assumed that the latter were spurious, mathematical solutions only with no physical reality. Quantum mechanics, however, allowed no such escape. The negative energy states could not be ignored without losing the relativistic invariance of the theory. One way of looking at the question is to invoke the fundamental uncertainty principle

$$\Delta E \Delta t \gtrsim \hbar \; .$$

For time intervals less than $\hbar/2m_0 c^2$, the uncertainty of the energy was greater than the gap between the positive and negative energy states. Thus, the negative states might, for sufficiently short periods of time, manifest themselves at positive energy! These intervals of time were of the order of 10^{-21} second, a sufficient time for a photon of light to go about 3×10^{-11} cm, a distance many times greater than the diameter of the nucleus and therefore not negligible! (This constitutes one of the great *positive* triumphs of the "uncertainty principle". Many others will appear as we continue.)

Dirac, understanding that these negative energy states could not simply be ignored, confronted the question of whether they were inconsistent with the established facts of physics. He argued that they were not, provided that all of these negative energy states were assumed to be permanently occupied by electrons.

Since the negative energy states are infinite in number, this appeared to imply that the "vacuum" contained infinite negative energy. But Dirac understood that nature is only concerned with energy *change* or *transfer* — that a never-changing pool of energy, whether *positive* or negative, could never manifest itself in nature.

That these states had to be occupied was of course clear. Otherwise, electrons of positive energy such as those that fill the world around us could make radiative transitions to the negative energy states, with the emission of photons with energy greater than $2m_0c^2$. By making transitions to even lower energy states, ultimately infinite amounts of radiative energy could be emitted, accompanied by a constant disappearance of electrons. On the other hand, if the negative energy states were already filled, the Pauli principle would prevent radiative decay into them.

Consider, however, the reverse situation. Presumably, a photon of energy greater than $2m_0c^2$ could cause the *excitation* of an electron from a negative energy state to a positive one. We would have created a new "visible" electron — but what else?

When the "negative energy sea" is completely filled, and a constant electric field is applied, no electron in that sea can move, even though the field must evidently be applying a force to it. This is because no state is available for it to move into; any attempt it might make to change its state is frustrated by the fact that all other available states are already filled. It is therefore immobilized.

Now imagine that one electron is *removed* from the negative energy sea. The charge of the vacuum is now changed; it is as though a positively charged entity appeared, distributed in exactly the same manner as the electron in the negative-energy state. More than that: suppose an external field were trying to accelerate (i.e. increase the energy of) all electrons. An electron in a state just below the one vacated could now move up to occupy the vacated state; its former state would then in turn be vacated. This would enable an electron of still lower energy to move up to fill it. The positive charge would then be moving, due to the field, in the opposite direction to that of electrons in the field. Their motion would be a reflection of the motion of electrons, so it would appear that this positive charge was moving exactly as an electron would, only in the opposite direction.

We call this an anti-electron, or a positron. It is simply a hole, an unfilled state, in the negative energy sea. According to this picture, no new matter has been created. An electron has been moved to a state where it can be influenced. It has left a hole behind, which gives a bit of freedom to its neighbours, but the hole moves like an oppositely charged electron.

Feynman, in a technical paper (not a popular one) written in 1949 gave a very vivid analogy, illustrated in the following series of diagrams (Fig. 15.1). Imagine that you are in a helicopter, looking down on a crowded, one-lane highway. Cars are bumper to bumper as far as you can see. They are all trying to go forward (to the left, say) but are inhibited because there is nowhere for them to go.

Now imagine that one car turns off, leaving a gap where it had been. The car behind pulls up, leaving the gap one space further back. The following car pulls into that space, etc. From above you see a hole, a gap, moving backward. The force to the left produces the movement of that hole to the right. All the movement of the hole corresponds *exactly* to the movement of cars, so it has the dynamics of a car.

While Dirac originally developed his relativistic theory as a theory of *electrons*, it is in fact applicable to all fermions (i.e. particles obeying the Pauli exclusion principle, e.g. protons, neutrons, neutrinos, μ-mesons, etc.).

The verification of the existence of the positron was made by C. D. Anderson, at the California Institute of Technology, in 1932 and was published in Physical Review in 1933. Anderson was completely unaware, at the time when his observation was made, of Dirac's theory and hence of the theoretical prediction of antiparti-

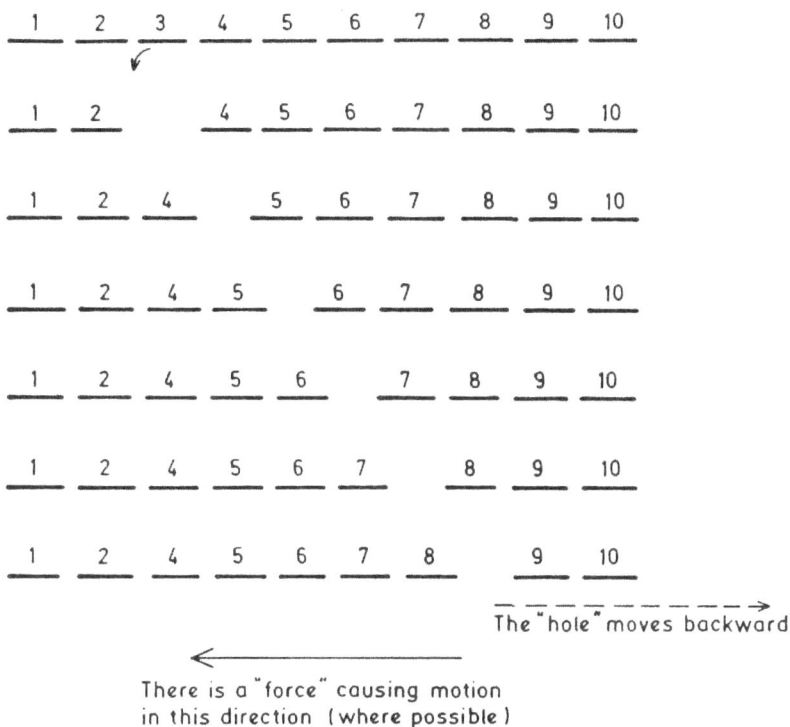

Fig. 15.1.

cles. The discovery was made in the course of studies of cosmic rays in a cloud chamber. A magnetic field was used to analyze the cosmic ray particles, because it was capable of distinguishing between particles of different charge. This is because charged particles have curved trajectories in a magnetic field, and the direction of the curvature about the direction of magnetic flux is in opposite directions for positively and negatively charged particles.

Actually, Anderson was not the first person to observe positrons in cosmic rays. The Italian physicist Bruno Rossi had done so earlier, but the result was entirely unexpected, so that when he sent a report of his observation to a scientific journal for publication, the paper was rejected. P. M. S. Blackett and G. Occhialini at the Cavendish Laboratory, also in 1933, published cloud chamber photographs showing electron and positron tracks simultaneously, i.e. tracks which curved in opposite directions and appeared to originate at the same point.

Occhialini, a dynamic and voluble Italian, was twice in the shadow of a Nobel prize. Blackett obtained his in 1948 for his cosmic ray researches, though not specifically for the positron, which had already earned Anderson his Nobel prize in 1936. In any case, in 1933 Occhialini was only a struggling young assistant. But by 1946 Occhialini was working in Bristol with C. F. Powell on the photographic detection of cosmic-ray particles when they discovered the pi-meson, which had already been predicted theoretically by Yukawa in 1935. It was Occhialini who took plates coated with a newly developed photographic emulsion to the observatory of the Pic du Midi in the French Pyrenées. In the process of careful examination of these developed plates, the meson track was found.

Powell, however, as the senior scientist directing the cosmic ray group, was awarded the Nobel prize in 1950.

I had the good fortune myself to be in the laboratory in Bristol in 1946 to work with Professor N. F. Mott, and remember the excitement generated by the Powell-Occhialini group. I might have known Occhialini better had I shared his favourite recreation, which was cave-crawling. Those with less enthusiasm spent their weekends instead on long hikes in the English countryside. I preferred this latter activity.

There is always an element of arbitrariness in the attribution of life's rewards; in any case, the most significant reward that any scientist can enjoy from his/her work is simply the satisfaction derived from doing it.

15.1. Consequences of Antiparticles, Feynman Diagrams

We shall show how to represent some fundamental physical processes by diagrams, due originally to Richard Feynman. For us, they represent simple pedagogical devices. For the theoretical physicist, however, they are "blueprints" for calculation. Every line, every point is a coded instruction for calculation which follows well-defined rules. This is what made Feynman's diagrams the cornerstone of modern theoretical physics.

Consider first Fig. 15.2:

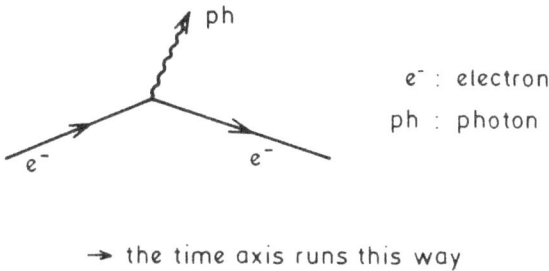

e^- : electron

ph : photon

→ the time axis runs this way

Fig. 15.2.

This portrays an electron emitting a photon and going into a lower energy state. Momentum and energy are conserved.

A similar diagram is Fig. 15.3, which portrays an electron absorbing a photon and going into a state of higher energy.

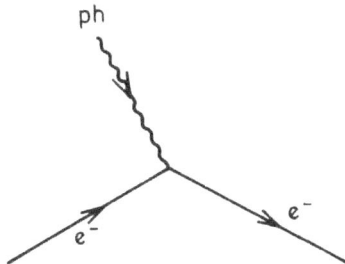

Fig. 15.3.

Suppose that the electron which absorbs the photon is a member of the "negative energy sea". If we interpret this process in terms of positrons, we may represent it as in Fig. 15.4.

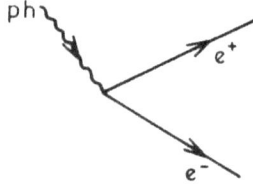

Fig. 15.4.

We have eliminated the "phantom" electron coming from the left, and indicated, instead, a positron, as well as an electron, emerging on the right.

An electron "before the event" has been replaced by a positron "after the event".

Let us now think back to our traffic model, and the diagram with which we represented it. From top to bottom, we follow the path of a *positron* from left to right. The *positron* moves freely.

From bottom to top, however, it is precisely the path of a free *electron* (represented by the empty circle). This sort of consideration, formalized mathematically, led Feynman to state that a *positron* behaved like an *electron moving backward in time*.

This led him to represent the process sketched in the last diagram in the form shown in Fig. 15.5.

Aside from being twisted about, this looks like the diagram shown earlier for absorption of a photon by an electron. That is of course exactly what it is, except that the electron goes *forward* in time in the negative energy sea!

What Feynman was able to show was that mathematically the processes shown in Figs. 15.3 and 15.5 are identical; the same calculation holds for each.

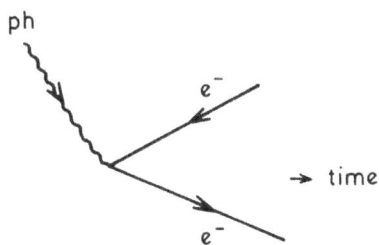

Fig. 15.5.

Similarly, the following diagram (Fig. 15.6) is equivalent to Fig. 15.2. Figure 15.2 represents an emission of a photon by an electron, leaving a positive energy electron. Figure 15.6 is similar except that the final electron is in a state of the negative electron sea. Alternatively, one may replace the "after the event" electron in the negative electron sea as a "before the event" positron, or an electron-moving-backward-in-time.

It is now interesting to put Figs. 15.5 and 15.6 together to form Fig. 15.7.

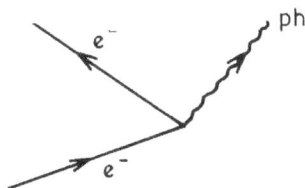

Fig. 15.6.

Fig. 15.7.

What this portrays is an initial photon creating an electron-positron pair, which subsequently annihilate each other to recreate the photon. A new feature has appeared. Because the electron-positron pair exists only temporarily, its energy cannot be sharply defined; thus, in the intermediate state, energy is not conserved. This means that the original photon need no longer have an energy greater than $2m_0c^2$. If the photon energy is E_{ph}, we need only "borrow" an amount of energy, $2m_0c^2 - E_{ph}$, to create the pair temporarily. For how long? The uncertainty principle

$$\Delta E \Delta t \gtrsim \hbar$$

gives us the answer: for a time

$$\Delta t \lesssim \frac{\hbar}{2m_0c^2 + E_{ph}} \ .$$

We therefore envisage that all photons *spontaneously* create virtual (temporary) electron-positron pairs.

Another process can also be envisaged: an electron emits a photon and absorbs it again. This is portrayed by combining Figs. 15.2 and 15.3 to give Fig. 15.8.

Fig. 15.8.

Photons of any energy, even very high, can be created virtually.

We are showing that electrons and photons are inextricably bound up with each other in nature. So are photons and proton-antiproton pairs. Part of the *physical* photon *is* the complex of virtual electron-positron pairs it can create, just as part of any electron *is* the complex of virtual photons it can create. A photon, or an electron, without these "interactions" is fiction, not physical reality.

In other words, charges and electromagnetic fields are inextricably tied up with each other. The physical entities do not exist separately.

The picture can now be generalized in very complex ways. Consider, for example, Fig. 15.9 which shows how electrons can produce electron-positron pairs. Another example is Fig. 15.10.

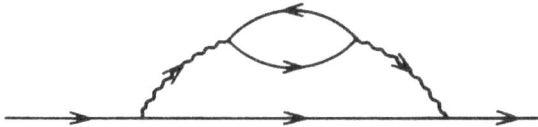

Fig. 15.9.

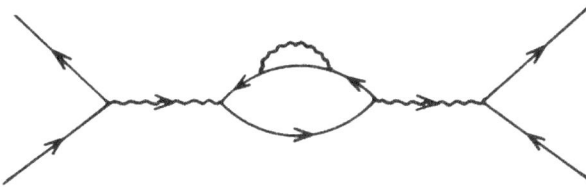

Fig. 15.10.

The possibilities are endless!

A particularly interesting one is to imagine the situation near an atomic nucleus, which induces a strong electric field. This field can be described in terms of virtual photons of various wavelengths and energies.

Consider one of these photons. It may produce a virtual electron-positron pair (or more).

The members of this pair now find themselves in the electric field of the nucleus. This field will tend to *polarize* them, that is,

nucleus

Fig. 15.11.

to pull them in different directions. We may electrically polarize (change the charge distribution in) the vacuum! This polarization of the vacuum may in turn affect the energy of the atomic electrons surrounding the nucleus. Other subtle effects are also predicted.

Problems of this sort were brought to light by the discovery of very small, but very important, spectroscopic effects by Willis Lamb and P. Kusch in 1955. That these effects were found to be important made physicists aware of the importance of vacuum in quantum electrodynamics. In this sense they generated a revolution whose effects are still felt to this day. Two Nobel prizes followed from work on these problems: one to Lamb and Kusch in 1955 for the key experiments, another to Feynman, Schwinger and Tomonaga in 1965 for unravelling the associated theoretical problems.

A disturbing aspect of these problems was that when it was attempted to resolve such vacuum effects, they were found to give rise to infinite energies (the calculated real mass of the electron, for example, was found to be infinite). Of course this was nonsense. The problem was circumvented in an ingenious but troublesome way. "Corrections" to the electron mass and charge can be calculated in terms of the "free" mass and charge m and e. These corrections (which contain infinite terms) are replaced by the *experimental* values of these quantities. This makes it possible to replace, in energy calculations, the "conventional" parameters m and e by their experimental values.

Dubious though the procedure (called "renormalization") may appear, it was outstandingly successful; Feynman, in fact, has pointed out that nowhere else in physics has such an accurate agreement between theory and experiment been obtained. So what appears fundamentally to be a weak theory is in fact the most precise theory known to physicists. Yet there remains something fundamental which we do not understand. Feynman has, only half-jokingly, called the device "a swindle". Dirac never ceased to contend that it was mathematically unacceptable. Yet it permeates all fundamental particle physics done today.

Actually, the problem is an old one. Even in classical electricity the energy in the electric field around a point charge is infinite. Thus the mass of the electron should be infinite, by virtue of Einstein's $E = mc^2$. Generally, we simply ignore this infinity. Yet, as one physicist wryly remarked "the fact that something is calculated to be infinite is not sufficient justification for assuming that it is zero".

The problem may be restated in another way. The basic assumptions with which our calculations start are quite unrealistic. We assume a simple point-like electron, charged but in the first instance non-interacting. We assume an electromagnetic field divorced from matter. Starting with these idealized entities, we try to reconstruct a theory of the *real* electron which appears in experiments, as well as the real photon. Our difficulties are probably due to the fact that we start so far from reality. What the calculations appear to be telling us is that to carry out our program successfully, we must start with "idealized" electrons and photons with infinite properties. The gulf between our idealizations and the realities of nature is too great to be crossed with a finite bridge.

Unfortunately we do not know, up to now, any better place to start. In the true pragmatic spirit, the fact that we have a theory that "works" leaves us little incentive to try.

Chapter 16

THE ATOMIC NUCLEUS
Its Constituents and Structure

The atomic nucleus consists of two components, neutrons and protons, collectively called nucleons. These particles have very similar though not identical masses. The masses can be measured in energy units Mev/c^2 (million electron volts divided by the square of the speed of light). The neutron rest energy is 939.50 Mev. It has no charge. The rest-energy of the proton is 938.21 Mev. Its charge is equal and opposite to that of the electron. Strangely enough, although the neutron is overall electrically neutral, it has a magnetic moment, negative in value, of -1.9128 nuclear magnetons. The nuclear magneton is $e\hbar/M_p c$ where $\hbar$ is the modified Planck constant and M_p, the proton mass. The magnetic moment of the proton is 2.7928 magnetons. Both of these values appear anomalous, and suggest a substructure for the two particles, as we shall see later.

Some important properties of nuclei have been established experimentally:

1. All nuclei, with the possible exception of the very lightest, have approximately the same density. If there are A nucleons in the nucleus, the volume of the nucleus is proportional to A.

2. When we speak of the volume of the nucleus, we imply that it has a fairly sharply defined radius. This appears to indicate that the range of the nuclear forces is quite short, and that the forces are strong.

3. Studies of nuclear reactions enable us to determine the bind-

ing energies of nucleons in the nucleus. These binding energies are found to be roughly independent of the size of the nucleus, that is, of A.

4. The total internal energy of the nucleus may be determined from the difference between its mass and the sums of the masses of its constituents. It is found that this does not rise systematically with the nuclear mass or nucleon number. This fact, and the constancy of the nuclear density, suggests that individual nucleons do not interact with all the other nucleons in the nucleus; their total interaction energy in the nucleus remains roughly constant. This can be attributed to the short range of the nuclear forces, which only permit nucleons to interact with their nearest neighbours.

5. All experimental data on scattering and nuclear structure are consistent with the assumption that *in a given state* the neutron-neutron, proton-proton and neutron-proton nuclear forces are the same. This is usually referred to as the "charge independence of the nuclear force". A consequence of this is that the individual nucleons move in a fairly constant potential throughout the nucleus. As a result, not too complicated nuclear models can be developed.

The foregoing observations make it possible to estimate the range of the nuclear force. This range can also be determined from experiments on the scattering of single nucleons on each other.

16.1. Yukawa, Pi-mesons and Nuclear Forces

The preceding considerations, along with the nearly constant density of nuclear matter, enable us to estimate the range of the nuclear force between pairs of nucleons to be about 1.4×10^{-13} cm. This estimate is confirmed by experiments on neutron-proton or proton-proton scattering.

Using the notion that the electromagnetic force between charged particles (e.g. electrons) was mediated by the exchange of photons, Yukawa, in 1935, proposed a model of the nuclear force by postulating a new field whose quanta would, in a similar manner, mediate the interaction between nucleons. What is particularly striking

about this idea is that it is yet another example of an important discovery arising from the use of the uncertainty principle. Once again,
that principle is seen not to obscure our vision of the physical world
nor to hide reality from us, but rather to point the way to a deep
understanding of important physical phenomena.

What property of the field quanta determines the range of the
force in question? In the electric case, the field quantum, the photon, has no rest mass. Thus, virtual photons of all energies down
to zero can be emitted. Now a virtual photon of frequency f has
energy hf and so, by the uncertainty principle relating simultaneous
uncertainties of energy and time, can only exist for a time of the
order of $1/(2\pi f)$. It can therefore contribute to the force out to a
distance of $c/(2\pi f) \approx$ photon wavelength. But since photons of all
wavelengths exist, the electromagnetic force has infinite range. A
force may have a lesser range only if the minimum energy of the
field quanta is sufficiently high that they have not enough time to
go beyond the range of that force before being reabsorbed. This can
be accomplished if the field quanta have an appropriate rest-mass.
If that mass is m, the lifetime of the virtual quanta is $\hbar/mc^2$ and
their range $\hbar/mc$.

What must then be the mass of Yukawa's field quantum to give
a force of range 1.4×10^{-13} cm? A simple calculation shows it to be
278 times the electron mass. Because this was midway between the
electron mass and the nucleon mass, Yukawa called this particle the
"meson".

So the uncertainty principle has led us to predict a new particle
of a given mass, starting with a knowledge of the range of the nuclear
force. We may carry the prediction further. Since the nuclear force
seems to be the same between neutron and proton, proton and proton
and neutron and neutron, we must allow for the possibility that the
mesons (which will henceforth be called pi-mesons or pions) may
exist with positive or negative charge or be neutral.

The way in which the interactions take place may be represented
by diagrams, just as in the case of the electromagnetic interaction.

For example, the interaction of two charged particles (such as electrons or protons) is represented by a diagram in which the two particles exchange a photon. (Actually, not one but many photons are exchanged.) In nucleonic interaction, pi-mesons are exchanged. If the two nucleons are of the same type, they can interact only through the exchange of neutral mesons, as in Fig. 16.1.

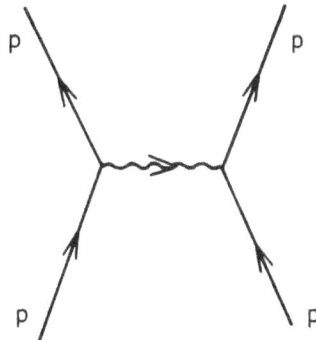

Fig. 16.1.

The mesons may of course be transmitted in either direction.

In the interaction of a neutron and a proton the exchange can take place in three ways: with neutral, positively charged or negatively charged mesons, as shown in Fig. 16.2.

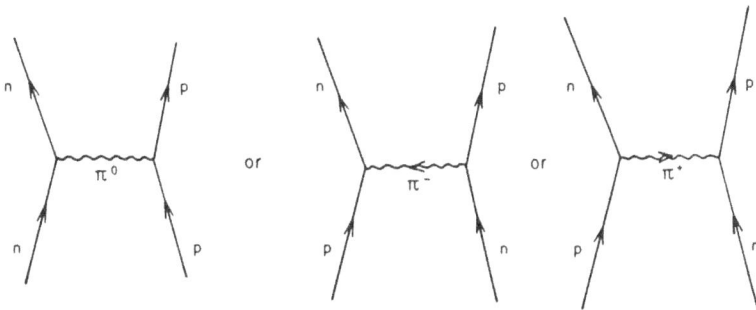

Fig. 16.2.

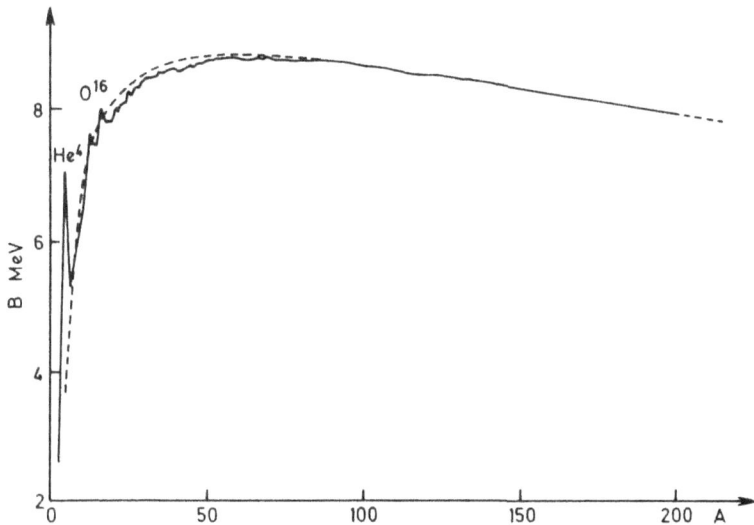

Fig. 16.3. Graph of the mean binding energy per nucleon (B in Mev) plotted against A for stable nuclei. The dotted line is a theoretical curve.

We have deduced the range of the nuclear force, but what about its strength? The strength of the electric force depends on the charges of the interacting particles. That of the nuclear force must similarly depend on a nucleonic charge. To get some idea of the strength of the force, we can look at the binding energy per nucleonic particle (nucleon) in the nucleus. The binding energy can be determined from the mass of the nucleus; it is the square of the speed of light multiplied by the amount by which the mass of the nucleus is less than the sum of the masses of its constituents. The binding energy per nucleon varies somewhat from nucleus to nucleus (Fig. 16.3), which plots this quantity against the mass numbers of the stable nuclei of the successive elements. It is on the average between 8 and 9 million electron volts (Mev), about a million times stronger than the binding energy of the least tightly bound electrons in an atom. (This does not give us direct information about the relative strengths of the forces between two particles, because of the difference in the laws of force, and particularly, the difference in their ranges.) More direct information may be derived from experiments on the scattering of nucleon by nucleon. In any case, nuclear forces are much stronger than electric ones; quantitatively, by a factor of a hundred or so.

A charged particle is surrounded by an electromagnetic field, and this field has energy. Should not this energy be considered as part of the intrinsic energy of the particle, that is, of its mass energy? Unfortunately, calculation of this field energy shows that it is infinite! We shall see later that this infinity is somewhat artificial. What is evident, however, is that if we consider the electron, say, first as a point particle dissociated from any field, the mass which we must attribute to it is not the mass of the true electron, since the electron constantly carries with it an electromagnetic field the quanta of which, the photons, carry some of its energy.

If this is the situation for an electron because of its electric field, it is much more so for a nucleon, for it creates around itself a much stronger meson field. This meson field is a much more important part of the nucleon than the electromagnetic field is of the electron.

Our argument about the range of nuclear forces suggests another consequence of the strong coupling of the nucleons to the meson field. Suppose a proton "emits" a positive pi-meson that is not absorbed by a neutron in its proximity, but is rather reabsorbed by itself. This is the process which we have already called virtual emission; the meson does not have an independent existence, but is merely a part of the nature of the true nucleon. By the same argument as above, based on the uncertainty principle between energy and time, this meson may, during its existence, only travel a distance $\hbar/mc$ before being reabsorbed. Thus, the meson is surrounded by a cloud of pi-meson quanta extending this distance.

It also appears that once a proton has emitted a positive pi-meson, it is itself left with no charge (albeit temporarily); it becomes, until the meson is reabsorbed, a neutron!

But we must be careful about using this sort of image. In the above statement, we must be aware that the neutron and the proton referred to are not the real neutron and proton, but rather idealized entities deprived of the accompanying meson fields. In terms of these quantities the real proton has the spatial extension of the "meson cloud" and its total charge characterizes the real, experimentally observed proton.

Similar observations can be made about the neutron. The hypothetical neutron can emit a negative pi-meson (pion) whose range is that of the nuclear force, becoming, in the process, a hypothetical proton! (Actually it is not a matter of emitting and reabsorbing one pion since pions are constantly being emitted and absorbed.) Therefore, both the neutron and the proton have a structure, with their charge distributed over a range of $\hbar/mc$, m being the meson mass – about 1.5×10^{-13} cm.

This description of the physical nucleons is verified experimentally by scattering experiments designed, like Rutherford's classical experiment, to probe the charge distribution of the particles. The results are shown in Fig. 16.4.

The structure for the proton is as expected — a central charge and an extended positive charge distribution. That for the neutron is more curious. The positive central charge reflects the hypothetical proton expected after negative pion emission. The negative charge between about 0.2 and 1.25 fermis (one fermi = 10^{-13} cm) reflects the charge of the negative pion. But why is there positive charge beyond that? It demonstrates that the situation is even more complicated. We have, for instance, neglected electromagnetic forces inside the structure postulated, as well as the fact that the hypothetical pion, too, is complicated by its interaction with nucleons.

What we learn from all this is not only that particles are more complicated than we first imagined, but that they form a tangled web of interaction in which they all become enmeshed in a multitude of complex interactions. For example, the neutron and proton, first considered as distinct particles, now simply appear as two interchangeable manifestations of a single particle. This sort of family

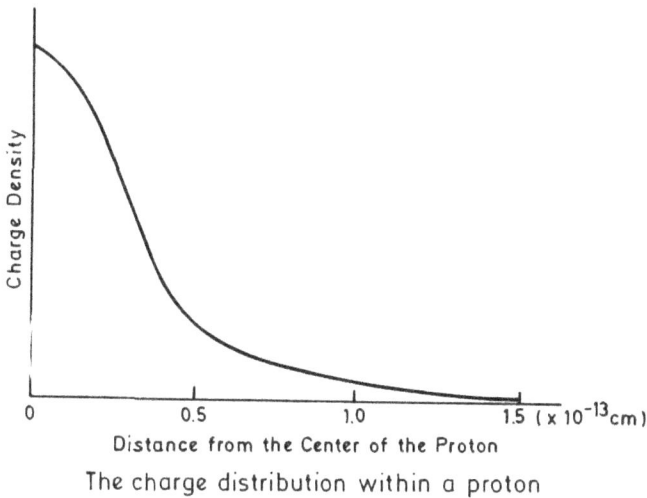

Distance from the Center of the Proton

The charge distribution within a proton

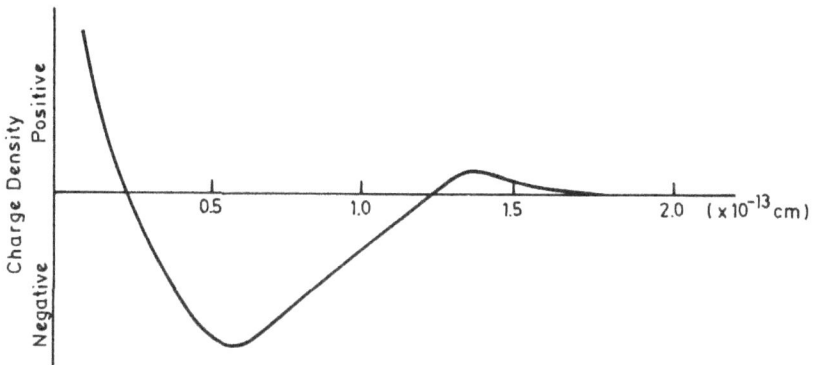

The charge distribution within a neutron

Fig. 16.4. (With permission of Macmillan Publishing Co. from *The Fabric of Reality* by S. K. Kim. Copyright © 1975 by Sung Kyu Kim.)

linkage will become a dominant theme in the theory of fundamental particles. There we shall meet other members of the family – "excited states" of the nucleon family.

Similarly, the three charge versions of the pion are all states of "the pion", again a single particle.

We shall see subsequently that there are excited states of this particle as well — states of higher energy (higher mass), with otherwise similar properties. This is a sort of third level of spectroscopy: first there is that of the atoms, then that of the nuclei, and finally that of nuclear constituents.

Just as we started our discussion of atomic states by considering the simplest of atoms, hydrogen, so we shall deal first with the simplest nucleus (excepting that of ordinary hydrogen, which is just a single proton); that nucleus is the deuteron, "heavy hydrogen", which consists of one proton and one neutron.

We already know that the nuclear force is charge-independent, i.e. it is the same between two neutrons or between two protons as between a neutron and a proton. This, however, applies only to the nuclear force; in the case of two protons the electric force acts as well. Secondly, the force is only the same when the state (other than that of charge) is the same. It is necessary to mention these two points because the nuclear force is very complicated, depending in fact on every property of the interacting particles (as, for example, their spin).

As for the deuteron, we expect it to be in the lowest possible spatial energy state; thus, as in the atomic case, a state whose wave function is spherically symmetric and has no nodes, and thus no relative angular momentum. But what is its spin? Do the spins of the neutron and the proton add, or do they cancel each other? Does the system have rotational angular momentum 1 (times $\hbar$) or zero? The former is found to be the case; thus, in a magnetic field, it is seen to be in a triplet state; its angular momentum may be $\hbar$, 0 or $-\hbar$. Evidently, the nuclear binding force is stronger in this state than when the spin is zero.

The force between neutron and proton may also be studied by observing the way they scatter each other when they collide. If in an experiment the relative spins of the colliding particles are random, there should be a different pattern of scattering in those cases where the spins are similarly oriented than in that in which they are opposite. This is indeed found to be the case, and it is even possible to deduce that, for the case of opposite spins, the force is not quite strong enough to hold the particles together. Even when the spins are aligned with each other, the system is quite weakly bound. The graph of nuclear binding energies (Fig. 16.3) shows that the binding energy is only about 1 Mev, though that for most nuclei is 8–9 Mev per particle.

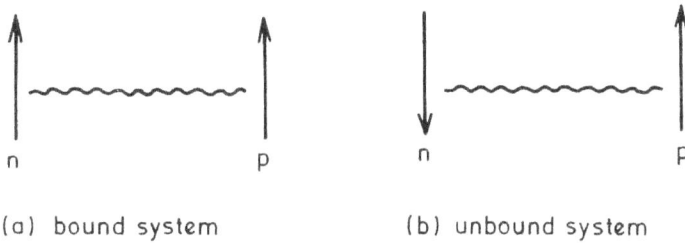

(a) bound system   (b) unbound system

Fig. 16.5. Nuclear binding energies.

Since the charge is independent of the nuclear force, what consequences does this have for a system of two neutrons? Can the di-neutron exist? It cannot since, by the Pauli exclusion principle, it would have to exist in the lowest spatial configuration with its spins parallel. In that case, the particles would be in the same state with regard to spatial configuration, spins and particle type (charge). It is possible to have a state with the spins opposed, because then the particles differ only in this respect while being symmetric in all others. But when the spins are opposed, the force is not strong enough to hold the particles together. A stable di-neutron therefore does not exist.

It follows that no stable system of two protons exist, because there is a further repulsive electric force.

Let us now describe some qualitative features to be expected of heavier nuclei.

16.2. The Nuclear Shell Model

One simple viewpoint is suggested by the earlier remark that the force acting on each nucleon, at least in a heavier nucleus, is more or less constant within that nucleus (except, of course, near its boundary). We can imagine a series of energy states in this common potential, and, as in the case of atoms, imagine filling these states, starting with that of the lowest energy and going on to successively higher ones. This approach is known as the "shell model". The key to the success of the shell model was to incorporate into it the phenomenon of spin-orbit coupling, that is, the interaction between the orbital angular momentum of the nucleon and its intrinsic spin. This leads to a series of "shells" of states, just as appeared in atomic spectroscopy, though the shells have a more complex structure than in atoms. We shall use this model only to show some simple features of light nuclei. We shall confine ourselves to the first shell; the general principles emerge from consideration of these simple cases.

The simplest nucleus, aside from the single proton which is the hydrogen nucleus, is the deuteron, the deuterium nucleus or heavy hydrogen. This nucleus consists of one neutron and one proton in the lowest level ($1s$), with the particle spins aligned, as shown in Fig. 16.6.

H^2 (deuterium)

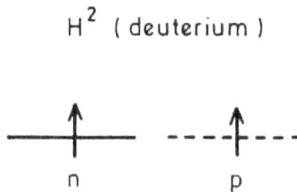

Fig. 16.6.

The next two nuclei, with three nucleons, are H^3 (tritium) and He^3 (helium-3). In the former, there are two neutrons (with spins aligned) and one proton in the lowest level (whose energy is not the same as that in the deuteron). He^3, with two protons and one neutron, has one significant difference. Here, because of the electric repulsion between the two protons, the lowest proton level lies above the lowest neutron level. The difference is slight, because the electrostatic force is weak compared to the nuclear force. As we proceed to progressively heavier nuclei, however, the energy difference between similar neutron and proton levels increases. When there are Z protons, for example, there is a repulsive interaction between $Z(Z - 1)/2$ pairs of particles, while nuclear interactions only take place between a small number of neighbouring nucleons; consequently, electric forces will ultimately dominate over nuclear ones.

The schema for the filling of levels in the 3-nucleon nuclei are shown in Fig. 16.7.

With four nucleons, there is only one stable nucleus, in which two particles of each type, of opposing spin, fill the lowest available levels (Fig. 16.8).

Fig. 16.7.

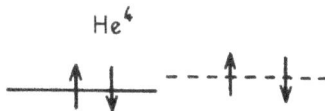

Fig. 16.8.

Thus, at helium, the lowest shell is filled. A similar situation exists in the helium atom, whose two electrons fill the $1s$ atomic state.

The next nucleus is that with six nucleons, since there is no stable nucleus of five nucleons. Li^6 has three neutrons and three protons; in each case two are in the lowest $(1s)$ state and the third is in an excited $(2s)$ state. The unpaired neutron and proton have their spins aligned, just as in the deuteron. Both $1s$ and $2s$ states of the proton are, again, higher in energy than the analogous neutron states (Fig. 16.9). The most common lithium nucleus, however, is Li^7 rather than Li^6.

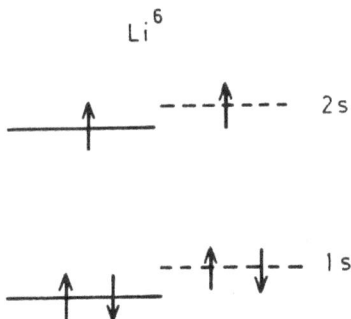

Fig. 16.9.

These two lithium nuclei, which may exist in atoms having the same chemical properties, are known as isotopes. Li^7 differs from Li^6 in having two neutrons, necessarily of opposing spins, in the $2s$ state (Fig. 16.10).

Beryllium-8 (Be^8), which would correspond to the filling of the $2s$ shell, is not stable against decay into two helium nuclei; the common form of beryllium is thus Be^9. This requires one neutron in the $2p$ state (Fig. 16.11).

We shall not continue developing this picture of nuclei, since the principles are clear from the foregoing simple examples. Suffice it to say that the shell model is very successful in predicting nuclear

Li^7

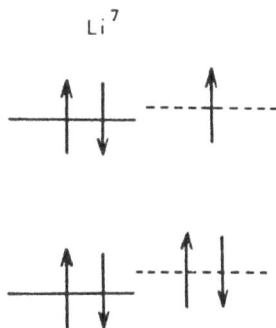

Fig. 16.10.

Be^9

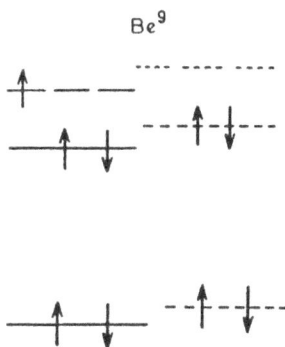

Fig. 16.11.

properties. It is, however, not found to be sufficient as a basis for nuclear spectroscopy; especially for heavier nuclei excited levels appear which it does not describe or predict. We are therefore led to a different model, the "liquid drop" or "collective" model.

16.3. The Collective Model

In a number of respects, a nucleus is rather similar to a liquid drop. In each case, the interparticle forces are of short range, so that each particle interacts only with close neighbours. This means that, to a first approximation, the total cohesive energy is proportional to the number of particles. Also the interparticle force becomes repulsive at short distances, so that the particles have a constant density. Thus, the binding energy is, at least in the interior of the nucleus or drop, proportional to the volume.

Suppose that our nucleus has A particles, of which Z are protons and $N = A - Z$ are neutrons. To a first approximation, then, its binding energy is proportional to A.

The nucleons near the surface of the nucleus, however, are less strongly bound. The force toward the inside of the nucleus is normal; but the outward force is reduced since the nucleon density is lower there. This gives rise to a phenomenon similar to that of surface tension in a liquid drop. The deficiency of binding energy (which appears as a negative contribution to the binding energy) should then be proportional to the surface area. For a spherical nucleus, its volume is $4\pi R^3/3$, and this times the density (number of nucleons per unit volume) is equal to A, the total number of nucleons. Thus A is proportional to R^3, so that R, the nuclear radius, is proportional to the cube root of $A(A^{1/3})$. The nuclear surface area is in turn $4\pi R^2$, which is then proportional to $A^{2/3}$. So far we have

volume effect proportional to A ,

surface effect proportional to $A^{2/3}$.

But there is another quantum effect, which we encountered above, due to the fact that each type of nuclear particle tends to fill the lowest energy states available to it. Ignoring the electrostatic force for the moment, a consequence is that the lowest energy state of the nucleus is that in which the number of neutrons and protons in the nucleus is equal. This implies a term in the energy proportional to $(N - Z)^2$:

Quantum effect proportional to $(N - Z)^2 = (A - 2Z)^2$.

Finally, we must take into account the repulsive electric forces. Since the number of pairs of mutually repelling particles is $\frac{1}{2}Z(Z-1)$, we have a corresponding contribution to the energy:

Electrostatic contribution proportional to $\dfrac{1}{2}Z(Z-1)$.

If we put all these contributions to the energy together, we can estimate the energy of a hypothetical nucleus made up of arbitrary numbers of neutrons and protons. In particular, for a given total number of nucleons, we can determine for what proportions the energy of the nucleus is lowest. This then enables us to determine the most stable nucleus of this sort.

The picture of the nucleus as a charged liquid drop may be developed into a dynamic model, for which energy levels may be calculated. The first calculation of this sort was made by Marcus Fierz in 1940. The problem was taken up again much later by A. Bohr and B. Mottelson in Copenhagen and Rainwater at Columbia, and was used to predict collective energy levels of nuclei. This work was awarded the Nobel prize in 1975.

Subsequently, the shell model and the collective model were combined into a unified model, in which all particles were viewed as being in quantum states in the distorted "drop" defined by the collective motion, these quantum states themselves in turn defining that collective motion. A calculation of this sort, in which each nucleon moves in a potential created by the totality of nucleons, is said to be *self-consistent*. Normally one must solve such problems by a succession of steps: one first makes a guess at the potential in the nucleus, then calculates the individual nucleon states, whose potential may then be calculated from the nuclear force law, etc.

We should also mention that the theory of fission was developed in 1939 by N. Bohr and J. A. Wheeler, using a distorted liquid drop model. The idea is clear. As one goes from lighter to heavier nuclei, the increasing repulsive electric (Coulomb) force results in a progressive decrease in stability. At the same time, collective excitations are characterized by distortions of nuclear shape, which can ultimately lead to the breaking of the nucleus into two parts.

16.4. How Do Nuclei Decay?

When nuclei are unstable, they undergo radioactive decay, giving off their energy in the form of radiation. This takes place according to the law of radioactive decay which we have discussed earlier. Excited states of a given nucleus normally undergo gamma-decay; that is, they radiate a photon whose energy is equal to the excitation energy. There is, however, another process which takes place, as suggested in the previous section, in which the ground state of a nucleus of a given constitution of neutrons and protons is itself not stable against decay to another nucleus of the same total number of nucleons. The common process in this case is a form of beta-decay.

16.5. Beta-decay and the Weak Interaction

Beta-decay is a form of nuclear transformation in which a nucleus emits an electron, a neutron in the nucleus being converted into a proton in the process. It is a very weak interaction, in the sense that it takes place very slowly. We must bear in mind, however, that this slowness is to be measured on the nuclear time-scale, which is $\hbar/Mc^2$, M being the nucleon mass. This corresponds to about 2×10^{-14} second. Radioactive nuclei exist whose half-lives are measured in seconds, minutes, hours, days, years ... He^6 has a half-life of slightly less than a second — already long! — but the following are examples of very long ones:

Ni^{59}	80,000 years
Al^{26}	740,000 years
Zr^{93}	1.5 million years
Pd^{107}	7.0 million years
Lu^{176}	30 billion years
Rb^{87}	500 billion years

The electrons which are emitted have a wide range of energies for different nuclei, ranging generally from a few KeV (thousand electron volts) to more than 15 million (Mev). More important is the fact that the electrons are not emitted with one specific energy, but with a continuous spectrum of energies running from zero up to a maximum value. The beta-decay spectrum (i.e. frequency of electrons as a function of energy) is given in Fig. 16.12.

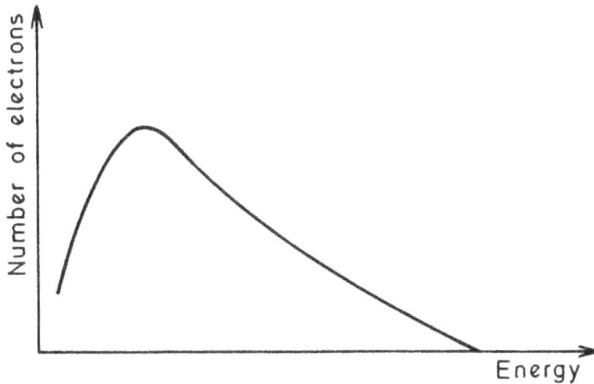

Fig. 16.12.

How can this fact be reconciled with the conservation of energy? And what about the conservation of angular momentum? Both the initial and final nuclear states have angular momenta which may be either integral or half-integral units of $\hbar$ or of $\frac{1}{2}\hbar$, as the number of nucleons is an even or odd number. But the electron has an intrinsic angular momentum of $\frac{1}{2}\hbar$, and normally has no orbital angular momentum about the nucleus. So there is a problem with both energy and angular momentum conservation.

Wolfgang Pauli first suggested in 1930 that both difficulties would be eliminated if one assumed that another particle (unobserved) were emitted along with the electron. The constraints on such a particle could be defined:

- it must be uncharged,
- it must have a half-integral angular momentum (in units of $\hbar$),
- it must have no rest-mass. For the energies (Mc^2) of the initial and product nuclei can be measured, and it was found that the maximum energy (including the rest-energy) of the emitted electron was the difference of the energies of the two nuclei; that is, the electron could be emitted with all the available energy.

It was obvious, then, that the extra particle in beta-decay would be very hard to observe: not only did it have no charge and no mass, so that it had no appreciable electromagnetic interaction with its surroundings; it also had the speed of light and interacted very weakly with the other particles in the decay. It was christened by Enrico Fermi the neutrino — "the little neutral one".

The theory of beta-decay was proposed by Fermi in the mid-thirties; it was again based on loose analogy with electromagnetic theory. One of its important aspects, we now know, is that it permits us to maintain an important conservation law, the conservation of number of leptons, leptons being the family of light fermions of which the neutrino is the latest member. In fact, at that time, the importance of the principle was less well understood than it is now. Dirac's theory naturally implies that all fermions have antiparticles, so that there must be an antineutrino. What was unclear, however, was how the neutrino and its antiparticle might differ, since they seemed to share all of their essential properties due to the lack of charge. But we have understood since 1957 that they are different, that the particle emitted with the electron is an antineutrino, and that lepton number is conserved.

This understanding, in the light of Dirac's theory, yields the following picture of beta-decay: a neutron interacts with a neutrino in its negative-energy "sea"; the neutrino is converted into an electron of positive energy (a real electron) and the neutron into a proton (Fig. 16.13). The enclosure around the dotted neutrino line indicates its negative-energy character. If the neutrino were a "real" (positive-energy) one, the reaction would be written as

$$\nu + n \rightarrow e^- + p \,.$$

In antiparticle terms, the neutrino on the left becomes an antineutrino on the right to give what is "observed" (Fig. 16.14):

$$n \rightarrow e^- + p + \bar{\nu} \,.$$

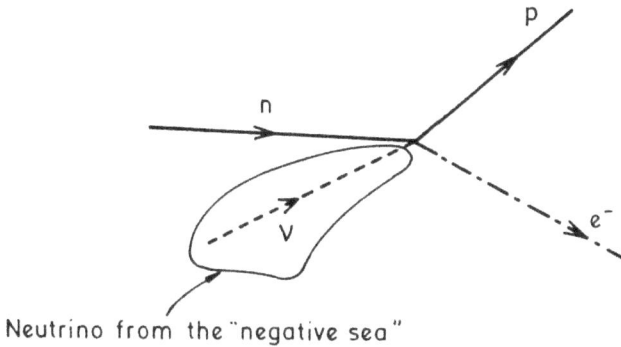

Neutrino from the "negative sea"

Fig. 16.13.

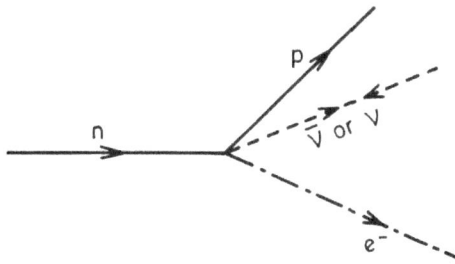

Fig. 16.14.

Our earlier discussion of antiparticles now permits us to generate a number of equivalent processes. One is the time-reversed process of the original one (Fig. 16.15):

$$e^- + p \rightarrow \nu + n \ ,$$

which will appear later in a process known as "internal conversion", in which an atomic electron interacts with a proton in the nucleus to produce a neutron and a neutrino. If, on the other hand, the same process involved an electron in the negative-energy sea, the process would appear as a proton decay (Fig. 16.16):

$$p \rightarrow n + \nu + e^+ \ .$$

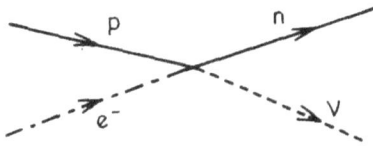

Fig. 16.15.

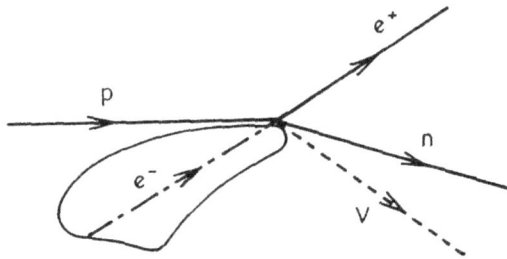

Proton decay

Fig. 16.16.

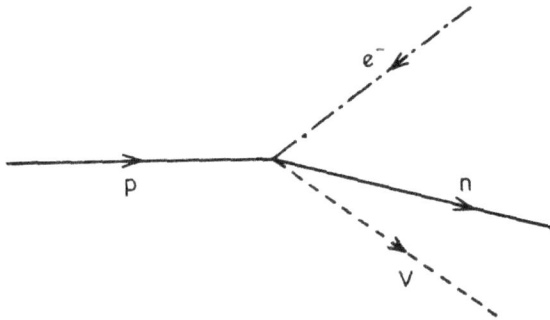

Fig. 16.17.

This describes a process by which a nucleus may emit a positron and a neutrino, one of its protons being in the process converted into a neutron, a reaction also commonly observed. There are two ways to describe this by diagrams, the second of which is as shown in Fig. 16.17.

If we are to maintain the parallel with electromagnetism or the meson theory, we should introduce a mediating field as the agent of the weak beta-decay force. This field should be coupled to the neutron-proton system on the one hand and the electron-neutrino one on the other.

Figure 16.18 shows a process involving a hypothetical positively charged W^+-particle. This particle was introduced as a theoretical assumption some years ago, but its existence was not verified until 1983. Certain of its properties could be deduced in advance; for example, the beta-decay interaction was known to be of such a short range that it was at first assumed to be operative only at a sharply defined point. Thus, by a now familiar argument, it must be very massive, so that its quanta can only have a very short range. The mass was found experimentally to be approximately 81 GeV/c^2, greater than that of an iron nucleus; the corresponding range of the force is about 10^{-3} fermis.

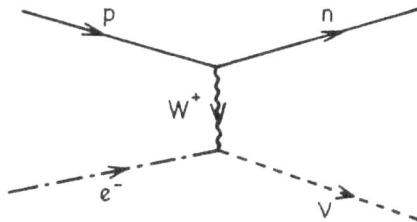

Fig. 16.18.

One does not expect only a positively charged W-meson; a negatively charged one could produce the same effect if the reaction went the other way (Fig. 16.19).

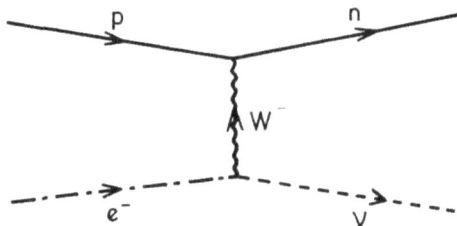

Fig. 16.19.

By analogy with the meson case, a third, neutral version should also exist. This would interact with particle-antiparticle pairs. This particle will be considered later.

If, in the reaction $e^- + p \to \nu + n$, the electron on the left comes from the negative-energy sea, the reaction takes the form

$$p \to n + e^+ + \nu \ .$$

Representing the positron as an "electron going backward in time", this process can be represented by the alternative diagram shown in Fig. 16.20.

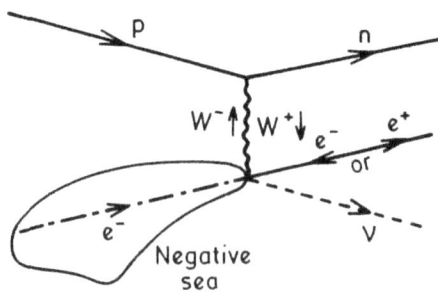

Fig. 16.20.

The strengths of the various forces that we have so far introduced may be represented in relative terms by:

strong: 1,   electromagnetic: $1/137$,
weak: 10^{-13},   gravitational: 10^{-42} .

The foregoing summary of the beta-decay theory lays the groundwork for a consideration of the role of beta-decay in permitting nuclei of a given nucleon number A to transmute to the nucleus of lowest energy of this A; it is, therefore, the ultimate key to the stability or instability of most nuclei.

16.6. Some Examples of Beta-decay Processes

If we consider "adjacent" nuclei of A nucleons (e.g. nuclei having $(N+1)$ neutrons and $(Z-1)$ protons or $(N-1)$ neutrons and $(Z+1)$ protons are adjacent to a nucleus of N neutrons and Z protons), one of the following processes may be possible without violating the law of energy conservation:

$$n \rightarrow p + e^- + \bar{\nu}$$

or

$$p \rightarrow n + e^+ + \nu .$$

The first involves the transformation of a neutron into a proton, the second the reverse. Whether one or the other of these processes takes place depends on the relative energies of the adjacent states. Let us consider several examples:

(a) Decay of Br^{78} ($N = 43, Z = 35$) into Se^{78} ($N = 44, Z = 34$):

Binding energy of $Br^{78} = 50.42$ Mev .

Binding energy of $Se^{78} = 53.91$ Mev .

The decay releases 3.49 Mev, which is more than enough to produce a positron (0.51 Mev). The rest of the energy is divided between the positron and the neutrino in the form of kinetic energy.

(b) Decay of Si^{31} ($N = 17, Z = 14$) into P^{31} ($N = 16, Z = 15$):

$$\text{Binding energy of } Si^{31} = 13.83 \text{ Mev .}$$

$$\text{Binding energy of } P^{31} = 15.31 \text{ Mev .}$$

The energy released on decay is 1.50 Mev, which is ample to produce an electron.

Note that the first of these reactions is characterized by the emission of a neutrino, and the second by that of an antineutrino.

(c) Consider now the case of Cd^{113}, whose adjacent nuclei are In^{113} and Ag^{113}. While the binding energy of Cd^{113} is 55.65 Mev, that of Ag^{113} is only 53.75, so decay cannot take place in that direction; that is to say, positron decay is excluded. On the other hand, the cadmium nucleus has $N = 65$ and $Z = 48$, while In^{113} ($N = 64, Z = 49$) has a binding energy of 55.85 Mev. Although there is a favourable energy difference, electron beta-decay is not possible either, since the energy difference of 0.20 Mev is not sufficient to create an electron. Cd^{113} is therefore stable.

There is still another possibility for proton-rich nuclei. Consider:

(d) Mo^{93} ($N = 51, Z = 42$) with binding energy 59.83 Mev, Nb^{93} ($N = 52, Z = 41$) with binding energy 60.32 Mev. The energy available for decay is 0.49 Mev, not quite enough to create a positron. Another method of decay is:

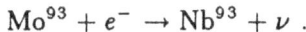

$$Mo^{93} + e^- \rightarrow Nb^{93} + \nu .$$

The energy on the left hand side of the reaction is $-59.83 + 0.51 = -59.32$ Mev while that on the right is -60.32 Mev, which makes 1.0 Mev available for the reaction! But where does the electron come from? It is in the innermost shell of the atom whose nucleus is undergoing the decay! Because that electron is itself bound to the nucleus, its energy is a bit less than 0.51 Mev (by about 0.024 Mev).

Since only a neutrino is emitted, it must appear with a definite energy, not a spectrum of energies as is the case for ordinary beta-decay.

The process involved here is called K-capture, since the electrons in the innermost atomic shell are called K-mesons.

Finally, a comment about fission. Since the proportion of neutrons to protons is higher in heavy nuclei than in lighter ones, the products of the fission process must be quite unstable, having a considerable excess of neutrons. This may be so drastic that free neutrons will be emitted. Even so, the decay products must undergo a series of beta-decays to attain the nucleonic proportions necessary for stability. There are many different and competing fission decay modes; let us just take one as an example. Consider the absorption of a neutron by a U^{235} nucleus. This will create an unstable U^{236} nucleus which may decay into Kr^{93} ($N = 57, Z = 36$), Ba^{140} ($N = 84, Z = 56$) and three neutrons. The krypton and barium nuclei attain stability by the following sequences of decay:

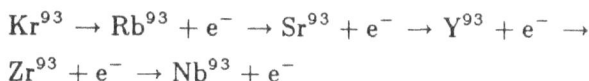

$$Kr^{93} \rightarrow Rb^{93} + e^- \rightarrow Sr^{93} + e^- \rightarrow Y^{93} + e^- \rightarrow$$
$$Zr^{93} + e^- \rightarrow Nb^{93} + e^-$$

and

$$Ba^{140} \rightarrow La^{140} + e^- \rightarrow Ce^{140} + e^- \ .$$

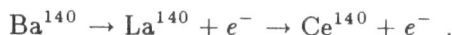

The three neutrons may now be absorbed by other U^{235} nuclei, and so forth. This is a chain reaction. From the above decay schemes it appears that fission reactions, such as those that take place in a nuclear reactor, not only emit neutrons, but also a large amount of quite intense beta-radiation.

Chapter 17

THE PHYSICS OF FUNDAMENTAL PARTICLES

The world of atoms, molecules and nuclei can be constructed from a small number of basic elements:

- the electron
- the photon
- the neutron and proton
- the pi-meson
- the neutrino

To investigate the properties of these particles, and of systems constructed from them (nuclei, atoms, molecules, etc.), a succession of generations of high-energy accelerators has been built. At the same time, we have looked to the natural world (radioactive materials, cosmic rays) for further clues. The results of these efforts have been the creation, or observation, of new kinds of particles, which have proliferated as the energy employed has been increased.

Another way of defining the process is to say that we have been exploring the physical world at constantly diminishing scales of length. To investigate the behaviour of matter at a certain scale of distance, we must use particles whose wavelengths become equal to or less than that distance. The shorter the wavelength, the greater must be the energy of the particle used as a probe. Table 17.1 gives some idea of the energies needed to investigate various scales for different particles. The last figure in the right-hand column is approaching the limits of present accelerators.

At the higher energies now available, many new "particles" are found. This should not be surprising, if we take the viewpoint that particles fall into "families" (like the neutron and proton), and the new particles are *excited states* of some sort of the well-known ones. For atoms or molecules, excited energy states become plentiful at increasing energies of excitation. Many of the new "particles" are only excited states of some new sort of entities, and "particle physics" a new level of spectroscopy!

Table 17.1.

	10^{-8} cm	10^{-13} cm	10^{-15} cm
Photon	1.25×10^4 ev	1.25×10^9 ev	1.25×10^{11} ev
Electron	150 ev	$\approx 1.25 \times 10^9$ ev	$\approx 1.25 \times 10^{11}$ ev
Proton		$\approx 6 \times 10^8$ ev	$\approx 1.25 \times 10^{11}$ ev

Even in the limited list of particles given above, we can already distinguish different sorts. The neutron, proton and pi-meson interact through the strong nuclear force; all such particles are called *hadrons*. The electron and the photon interact through the electromagnetic force; the electron, neutrino, neutron and proton interact through the weak force.

The pi-mesons and the photon are field quanta which mediate the strong and electromagnetic interactions respectively. A W-particle has been postulated to mediate the weak β-decay force.

The electron, neutrino and nucleon are fermions; the photon and pi-meson are bosons.

Heavy fermions (mass greater than or equal to that of nucleons) are called *baryons*; light ones are called *leptons*.

Underlying the notion of families of particles are considerations of *symmetry*; families are defined by common symmetry characteristics.

Three of the most important symmetries are the following:

(1) Symmetry between particles and their corresponding anti-particles.

This symmetry relates to the question: do the laws of physics remain the same if particles are replaced by their antiparticles, or vice versa?

(2) Time reversal symmetry: are the laws unchanged under reversal of the direction of time?

(3) "Mirror" symmetry (parity): are the laws unchanged when a system is reflected in a centre of symmetry, i.e. is it unchanged when every point with given coordinates (x, y, z) is replaced by its mirror image point $(-x, -y, -z)$? If systems *have* mirror symmetry and if they are characterized by a field $f(x, y, z)$, we may define an operator P which changes $f(x, y, z)$ into $f(-x, -y, -z)$:

$$Pf(x, y, z) = f(-x, -y, -z) \,.$$

But then $P^2 f(x, y, z) = f(x, y, z)$, where P^2 means that we apply the operator twice. There are then only two possibilities:

$$Pf(x, y, z) = f(-x, -y, -z)$$

or

$$Pf(x, y, z) = -f(-x, -y, -z) \,.$$

In the first case we say the field has *even* parity, in the second, *odd* parity.

Until 1956, it had been generally assumed that conservation of parity governed all particle interactions. This was first thrown into doubt with the discovery of two particles, designated respectively as θ- and τ-mesons, which seemed in almost every respect identical; they differed only in that the θ-particle (+ or −) decayed into two pi-mesons

$$\theta^\pm \rightarrow \pi^\pm + \pi^0$$

and the τ into three

$$\tau^\pm = \pi^\pm + \pi^+ + \pi^-$$

with very different lifetimes. That for the former process was 7×10^{-11} second; for the latter, 4×10^{-8} second. It was T. D. Lee and C. N. Yang who suggested that there might be only one particle with two decay modes. The problem was that the π-meson was known to have odd parity, so that the two decay modes implied different parities for the particles. This in turn threw doubt on the hypothesis of parity conservation.

Spurred by this puzzle, Lee and Yang undertook an exhaustive search for evidence of parity conservation in nuclear β-decay processes. No such evidence was found.

A very ingenious experiment was devised by Mme. C. S. Wu at the U. S. National Bureau of Standards to test parity violation explicitly. The idea was the following: a sample of Co^{60}, which undergoes β-decay, was subjected to a strong magnetic field which aligned the nuclear spins and the direction of emission of the β-decay electrons was observed. If parity were conserved in β-decay, it would be expected that the electrons would be emitted equally in the direction of the field and in the opposite direction. If they were emitted preferentially in one direction, it would demonstrate that parity was *not* preserved. The situation is illustrated in Fig. 17.1. On the left of the imagined mirror, the nuclear spins are aligned in the upward direction; thus, seen from above, they are spinning in a counter-clockwise direction. The experiment showed that the β-decay electrons were emitted preferentially in the direction *opposite* to the direction of spin alignment.

In the configuration shown, what is seen in the mirror image of the experiment? The electrons now seem to be spinning in the clockwise direction, thus, their spins point *downward*. But that is also the direction of electron emission. Thus, in the mirror image, the electrons are emitted in the direction of the spin alignment, contrary to the situation in the experiment itself. That is to say, parity is *not* conserved. The non-conservation of parity would disappear if the electrons were emitted equally in the two directions.

Physics: Imagination & Reality

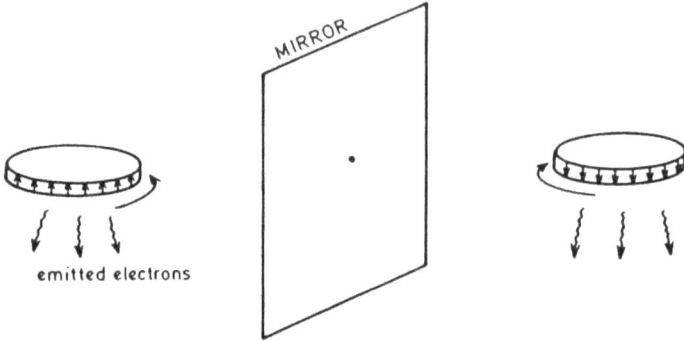

Fig. 17.1.

No evidence exists of non-conservation of parity in *strong* interactions; it was established, on the other hand, that in *weak* interactions not only was parity not conserved, but that the violation of conservation was as great as it could be. It is, then, not unreasonable to suspect that weak interactions are weak *because* parity is not conserved.

The non-conservation of parity in weak interactions turned out to have interesting consequences for the neutrino. Zero-mass particles may spin about their direction of motion in either a clockwise or a counter-clockwise direction, somewhat like a rifled bullet. It had formerly been thought that the neutrino spin could be in either direction. Analysis of β-decay experiments like that of Wu *et. al.* indicated otherwise. If the electrons in the Wu experiment were emitted preferentially in one direction, it followed that the antineutrinos were emitted in the *opposite* direction.

Non-conservation of parity for neutrinos is illustrated in Fig. 17.2. The experiment indicated that anti-neutrinos spun *clockwise* about their direction of motion. Since, as Fig. 17.2 shows, mirror symmetry reverses the direction of propagation when the spin is conserved, neutrinos must spin counter-clockwise about their direction of propagation.

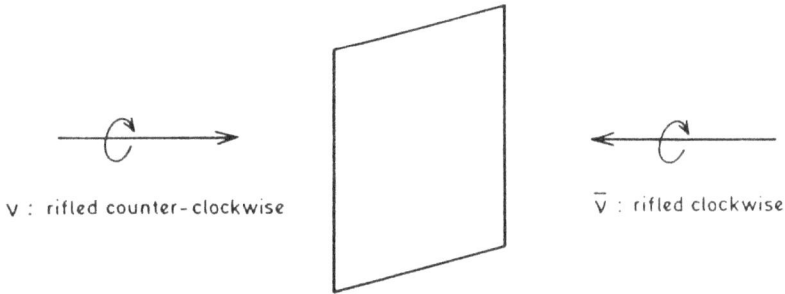

v : rifled counter-clockwise   v̄ : rifled clockwise

Fig. 17.2.

17.1. Non-conservation of Parity and Strange Particles

Experiments undertaken at the Brookhaven cosmotron in 1952 added a new dimension to the problem of the curious parity "anomaly" posed by the tau and theta mesons. One event showed the result of the collision of a π^--meson with a nuclear proton to form two uncharged particles. That there were two neutral particles was deduced from the fact that, although the track of the meson ended abruptly, two other pairs of particle tracks appeared to emanate from that end point a bit further on. A similar event pictured in a bubble chamber photograph from the Lawrence Radiation Laboratory in 1957 is shown in Fig. 17.3. The sketch in the upper left corner provides a guide to the interpretation of the event. Dotted lines have been added to sketch the paths of the presumed neutral particles. The identity of the particles making observed tracks can be established from their mass and their charge as indicated by the direction of bending of the tracks in the applied magnetic field. Their masses may be deduced from conservation of momentum and the range and density of those tracks.

The existence and properties of the neutral particles marked Λ^0 and K^0 are deduced; they decay at the points from which the new pairs of particle tracks emerge. On the left, a neutral particle designated as K^0 decays into a pair consisting of a positive and a negative π-meson. To the right, a Λ-type track of oppositely charged

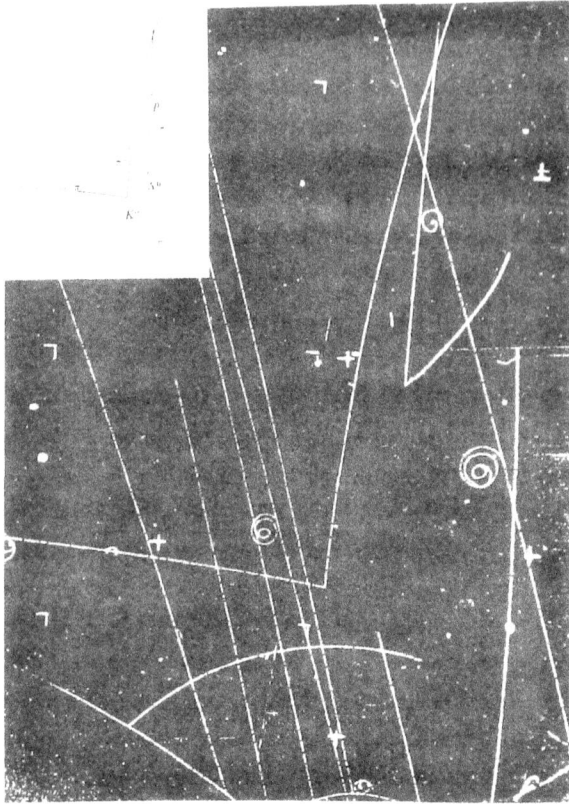

Fig. 17.3. (From E. Segre, *Nuclei and Particles*, ⓒ 1977, Addison-Wesley Publishing Co., Inc., Reading, Massachusetts. Reprinted with permission.)

particles, a proton and a π-meson, are evidence of the decay of the other neutral particle, the so-called Λ^0. Summarizing, then, one has the reactions

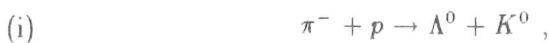

(i) $$\pi^- + p \rightarrow \Lambda^0 + K^0 \ ,$$

while the particles produced decay by the processes are

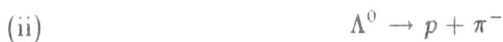

(ii) $$\Lambda^0 \rightarrow p + \pi^-$$

and

(iii)
$$K^0 \to \pi^+ + \pi^- \ .$$

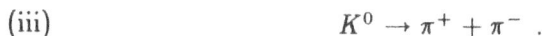

The K^0 appears to be none other than a neutral version of the θ-meson of the θ-τ puzzle which we have already discussed.

The most interesting and puzzling feature of the process (i) is the following: the large number of events of this type found make it clear that they take place through the strong interaction. On the other hand, the decay (ii), which involves the same two particles as the original reaction, takes place very slowly, on the time scale expected of weak interactions. But this is only the reverse of the original reaction, and so might be expected to produce the Λ^0, though without the associated K^0. Why is the interaction strong when K^0 is involved whereas it appears weak when Λ^0 is involved alone? (Of course, if the reverse reaction to the Λ^0-decay *is* weak, we would not really expect to see it, because it would be $\approx 10^{-13}$ times rarer than the associated production.)

All this is *strange* behaviour, which Gell-Mann "explained" by postulating a new property of particles. He in fact called it "strangeness" S (one fears that nowadays, in the age of "charm" and the rest, it might be called "subtlety"). The chief characteristic of this quantity was that it had to be conserved for interactions to be strong, but did not need to be conserved in weak interactions. Gell-Mann then attributed to the K^0-particles a strangeness of 1, and to the Λ^0 a strangeness of -1 (it would have made no difference had these been reversed). Particles already familiar, such as nucleons and π-mesons, were assigned strangeness zero; in other words they were not strange.

So now the situation is the following:

(a) In the original reaction

$$\pi^- + p \to \Lambda^0 + K^0 \ ,$$

the strangeness of the incident particles is

$$0 + 0 = 0 \ ,$$

while that of the outgoing particles is

$$-1 + 1 = 0 \ .$$

Strangeness is conserved: the interaction is *strong*.

(b) Consider next the decay of the Λ^0:

$$\Lambda^0 \rightarrow p + \pi^- \ ,$$
$$S = -1 \rightarrow 0 + 0 \ ,$$

so that strangeness is not conserved, and the process proceeds *weakly*.

Then we have the decay of the K^0:

$$K^0 \rightarrow \pi^+ + \pi^- \ ,$$
$$S = 1 \rightarrow 0 + 0 \ ,$$

again, a weak interaction.

Note that, just as Goudsmit introduced a *new characteristic* (spin) for the electron to account for its otherwise inexplicable behaviour in spectroscopy, so Gell-Mann deduced the existence of a new characteristic of hadrons to explain the new facts of the high-energy reaction. The only difference here is that spin, a variety of angular momentum, is related to a familiar classical dynamical concept. Strangeness, on the other hand, was an entirely new non-classical (quantum) property.

This marked a new departure in physics, which forced the abandonment of classical intuition as a guide to quantum phenomena. It is, for this reason, of the utmost importance. Having taken the first step, we have since gone further and further down this path, continually abandoning old intuitions to a reliance on purely mathematical structures.

We have entered a new wonderland, cutting our links to the old and familiar, and have not yet learned how to reconcile its new reality with our old familiar one. We are sailing on unfamiliar new waters, learning how to navigate without a compass, using instead new instruments which we have not yet learned to read with confidence.

Returning from this brief but important digression, let us follow the story of strange new particles somewhat further. Shortly after the discovery of the Λ^0-baryon and K^0-meson, several related particles appeared:

 - a group of three baryons, the Σ^+, Σ^0 and Σ^-.
 - a charged variant of the K-meson, the K^+.

Once again associated production was copious and the decays were weak. For instance, the π^0-meson reacting with protons gave

$$\pi^0 + p \to \Sigma^+ + K^0$$

and also

$$\to \Sigma^0 + K^+ \ .$$

Once again, the K-particles were assigned strangeness 1, the Σ-particles strangeness -1, so that strangeness was conserved. But the Σ^+-meson slowly decayed into a proton and a π^0-meson

$$\Sigma^+ \to p + \pi^0 \ .$$

Since in this reaction strangeness was not conserved the Σ^0, like the Λ^0, decayed by the reaction

$$\Sigma^0 \to p + \pi^- \ ,$$

again with violation of strangeness conservation. The Σ^0 seemed very similar to the Λ^0, but had slightly higher mass, approximately that of Σ^+ and Σ^-.

Stranger things were to come. π-mesons colliding with Λ or Σ particles produced still somewhat heavier particles, either neutral or negatively charged: these are called Ξ^0 and Ξ^-:

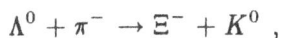

$$\Lambda^0 + \pi^- \to \Xi^- + K^0 \ ,$$
$$S = -1, 0 \to -2, 1 \ .$$

The strangeness -2 was assigned to Ξ^- to explain again strong associated production. Similar strong reactions are

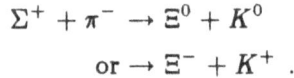

$$\Sigma^+ + \pi^- \to \Xi^0 + K^0$$
$$\text{or} \to \Xi^- + K^+ \ .$$

Typical (weak) decay processes are

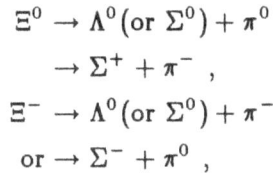

$$\Xi^0 \to \Lambda^0 (\text{or } \Sigma^0) + \pi^0$$
$$\to \Sigma^+ + \pi^- \ ,$$
$$\Xi^- \to \Lambda^0 (\text{or } \Sigma^0) + \pi^-$$
$$\text{or} \to \Sigma^- + \pi^0 \ ,$$

all of which have a strangeness of -2 on the left and -1 on the right.

It may at first sight appear curiously arbitrary that (for example) the Σ-particles come in all charged variations, a bit like π-mesons, while the so-called "cascade particles", the Ξ's, do not have a positively charged number. Closer observation, however, reveals a pattern.

17.2. An Observation about Strangeness

The neutron-proton family, which is not "strange", is biased toward positive charge. The Σ family, which has strangeness -1, has no charge bias. The cascade particles are biased to *negative* charge, and are doubly strange. There appears to be a pattern in this; if Q is the average charge of the family, it is related to strangeness by the simple relation

$$S = 2Q - 1 \ .$$

Q for the neutron-proton family is $\frac{1}{2} = \frac{1}{2}(1+0)$ and $S = 0$. For the Σ-particles, $Q = 0$ and $S = -1$. Finally, for the Ξ's, $Q = -1/2$ and $S = -2$.

Gratifying though this is in showing the element of a pattern, we may wonder why the situation is different for mesons. The π-meson family, which has strangeness zero, also has $Q = 0$. The K-mesons, K^0 and K^+, have average charge $\frac{1}{2}$ and strangeness 1. For these particles

$$S = 2Q \ .$$

One suspects that the difference in the S-Q relationship in the two cases has something to do with spin, and thus with the fact that the baryons are fermions and the mesons, bosons.

17.3. The Eight-fold Way

In molecular spectroscopy, one of the convenient properties for characterizing the properties of molecules was *symmetry*. The branch of mathematics dealing with the theory of symmetry is known as "group theory", and was first developed by the young French mathematician of the late 18th century, Evariste Galois.[18] In the same spirit, Murray Gell-Mann organized these new particles according to their symmetries; the results can be neatly demonstrated (Fig. 17.4).

In Fig. 17.4, particles of a "family" are represented on horizontal lines of the hexagon; each horizontal row contains particles of the same strangeness, the strangeness increasing from bottom to top. Particles of the same *charge* lie along sloping lines from bottom left to upper right, the charge increasing by 1 with each successive displacement to the right.

The masses, which are approximately equal for members of a given family, are shown. They increase with strangeness.

The whole group of eight particles (hence "eightfold way") have an approximate symmetry; thus mass differences reflect the breaking of that symmetry.

---

[18] Galois, a very unconventional character, was involved in radical politics and had difficulty gaining recognition from the conservative mathematical community. He was killed in a duel in 1832 at the age of 21, after feverishly writing up his novel mathematical ideas. Leopold Infeld has written a semi-fictional biography of Galois, whom he admired greatly. It is called "Whom the Gods Love", and makes fascinating reading. Infeld was working on the book while directing my doctoral thesis.

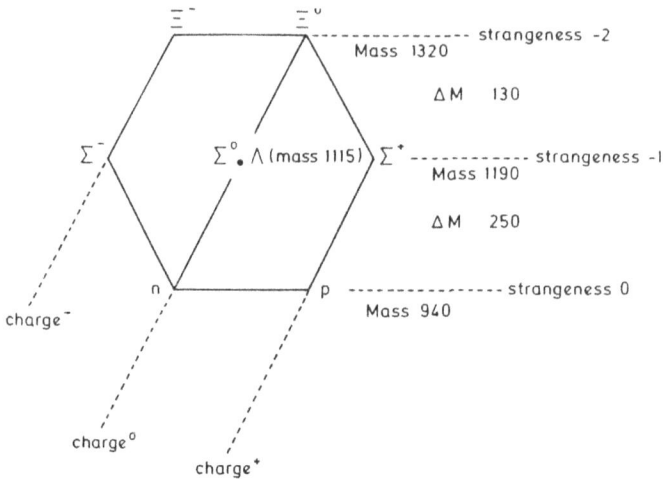

Fig. 17.4.

A similar family of mesons is shown in Fig. 17.5.

We have spoken of five mesons. The three needed to complete the set of eight are the $\overline{K}^0$, which is the antiparticle of the K^0, and the K^-, which is the antiparticle of the K^+. (These have opposite strangeness to the K^0 and K^+, i.e. $S = -1$.) Finally, there is an η^0-meson which is, for this family, the analog of the Λ^0 in the baryon family.

17.4. The Strange Story of the K^0, $\overline{K}^0$ Mesons

We have already suggested a connection between the K^0-meson and the θ- and τ-mesons which led Lee and Yang to challenge the conservation of parity. What precisely is the connection?

There is a clue in the decay processes. Recall that θ and τ had different *parity*. The θ decayed quite rapidly into two pions, the τ considerably more slowly into three. The K^0 has strangeness 1, the $\overline{K}^0$ strangeness -1, so neither of these can be the θ-meson, which decays rapidly into non-strange pions. Furthermore, θ, which we shall henceforth call K_1, and τ, which we shall call K_2, have

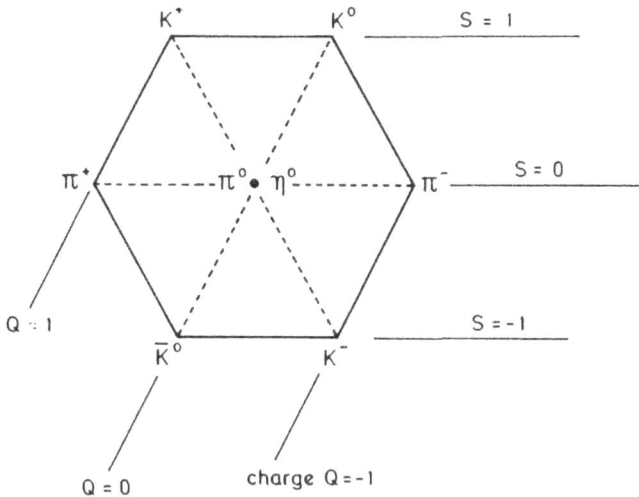

Fig. 17.5.

different parities, while K^0 and $\overline{K}^0$ have the same parity.

The situation may be better understood if we consider it as parallel to what we studied earlier with respect to polarized light. Light can be characterized by two incompatible characteristics: the direction of its electric field and its angular momentum about its direction of propagation. K-mesons may be characterized by *parity* or by *strangeness*, but not by both simultaneously. In fact, K_1 and K_2 have definite (but opposite) parities, but their strangeness is indeterminate. K^0 and $\overline{K}^0$, on the other hand, have definite (but different) strangeness, but their *parity* is completely indeterminate. Here is the "uncertainty principle" manifesting itself with respect to quite non-classical characteristics; this is perhaps not too disturbing because we do not have strong prejudices about them.

As in the case of polarized light, states characterized by one variable may be written as equally weighted superpositions of the two states of the other variable. This has some quite unexpected and striking consequences, which represent typical quantum behaviour.

Consider, for example, the production of the K^0 meson by the reaction

$$\pi^- + p \rightarrow K^0 + \Lambda^0$$

in which strangeness is conserved. We must now think of K^0, whose parity is quite uncertain, as consisting of an equal combination of K_1 and K_2. K_1 dies out quite quickly; its lifetime is 7×10^{-11} second. Therefore, what is originally an equal combination becomes, with time, simply K_2. If we have an initial beam of K^0's moving at nearly the speed of light, the K_1 component will only go on the average about 2 cm before decaying, while the K_2, with lifetime 4×10^{-8} second, will on the average go on for about 12 m.

A beam of K^0's, as initially produced, cannot be used to generate the reaction

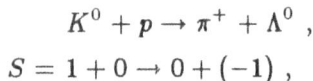

$$K^0 + p \rightarrow \pi^+ + \Lambda^0 \ ,$$
$$S = 1 + 0 \rightarrow 0 + (-1) \ ,$$

which drastically violates conservation of strangeness. But the reaction

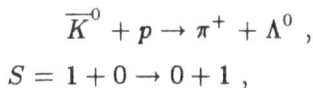

$$\overline{K}^0 + p \rightarrow \pi^+ + \Lambda^0 \ ,$$
$$S = 1 + 0 \rightarrow 0 + 1 \ ,$$

is the inverse of the initial reaction, and is strongly favoured. Thus, while a collision of a K^0 on a proton cannot produce a Λ^0 particle, a similar collision with a $\overline{K}^0$ can.

Our original beam of K^0 particles gradually changes, by decay of its K_1 component, into a beam of K_2 particles. These now have well-defined parity, but indeterminate strangeness; that is, the K_2 beam consists of equal parts of K^0- and $\overline{K}^0$-particles. The latter, colliding with target protons, can now make Λ^0-particles (along with π^+'s) — something which could not happen at the beginning of the beam. If a proton target were one centimetre from the point of the

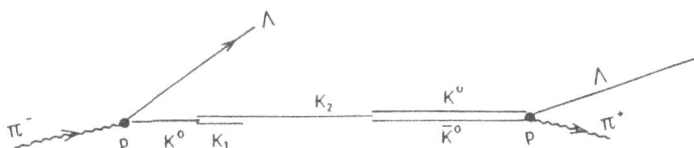

Fig. 17.6.

original reaction, very few Λ^0 particles would be produced. But at one metre, they would be produced copiously (Fig. 17.6).

A π^--particle colliding with protons (both with strangeness $S = 0$) has ultimately produced, with the intervention of another proton, two π^+-mesons, one π^-, a K^0 and two Λ^0's:

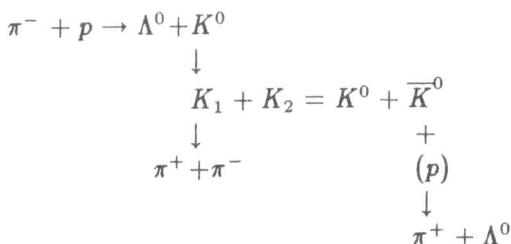

$$\pi^- + p \rightarrow \Lambda^0 + K^0$$
$$\downarrow$$
$$K_1 + K_2 = K^0 + \overline{K}^0$$
$$\downarrow \qquad\qquad +$$
$$\pi^+ + \pi^- \qquad\qquad (p)$$
$$\downarrow$$
$$\pi^+ + \Lambda^0$$

Initially we had no strange particles, and now we have produced particles of total strangeness -1, all by strong interactions!

17.5. The Story of the Ω^-

The eightfold family of baryons discussed above is not the only one consistent with Gell-Mann's symmetry principles. In addition to the neutron and proton, there is another set of non-strange baryons known as Δ-particles $(\Delta^-, \Delta^0, \Delta^+, \Delta^{++})$. The lopsided charge distribution is necessary if the group of particles is to have no strangeness; the average charge must be $1/2$. The particles Δ^0, Δ^+ are very much like excited states of the neutron and proton, but Δ^- and Δ^{++} are new. Without these latter particles, one could form a new hexagon of eight particles, since excited states $\Sigma^-, \Sigma^0, \Sigma^+$ as well as Ξ^- and Ξ^0 were found with properties generally similar to the first generation of such particles. But the two additional particles

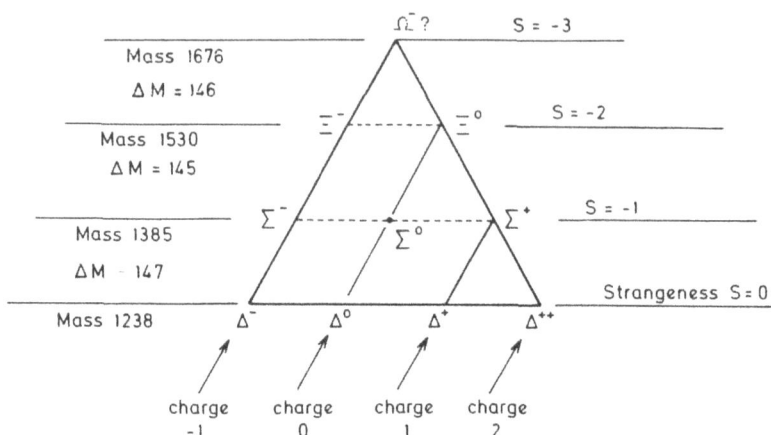

Fig. 17.7. If we cut off triangles from the three corners of this diagram we again have the hexagon, with particles resembling those of the "first generation". The spin of 3/2, which characterizes all these particles, permits the additional members of the "family"

suggest a family of a different sort, based on a triangular scheme. Such a scheme, predicted by theory, is shown in Fig. 17.7.

In 1961, Gell-Mann and independently Yuval Ne'eman postulated it as shown. The only problem was that the "summit" particle was missing.

Note the characteristics shared with the previous eightfold scheme. Particles on a given horizontal line had the same strangeness, zero on the bottom line, -1 on the next, -2 on the second. Thus, if a member existed to complete the family, it would be expected to have strangeness -3. This would require it to have a charge of -1.

Another property distinguishing this family of particles from the earlier one is that whereas they all had spin $\frac{1}{2}\hbar$, the new ones had spin $\frac{3}{2}\hbar$. That would also be expected of the missing particle.

Finally, with each step in increasing strangeness, the particle mass increased by about the same amount: 147 with the first step, 145 with the second. A further guess, then, about the missing particle, which we shall henceforth call Ω^-, is that its mass is 146 units greater than that of the Ξ^- and Ξ^0, i.e. 1676.

Thus, the search was launched for a hitherto unknown particle with a set of very precise properties. The hunt lasted for two years and involved scanning tens of thousands of photographs of high-energy events. It was eventually found.

Events of this sort convince theorists that they are on the right track. The way was now cleared to pursue the track further.

17.6. The Invention of the Quark

Among all the schemes for families of particles, the simplest of all had been bypassed. It was the simple triangle (Fig. 17.8), from superpositions of which our hexagon and triangle of ten can be constructed.

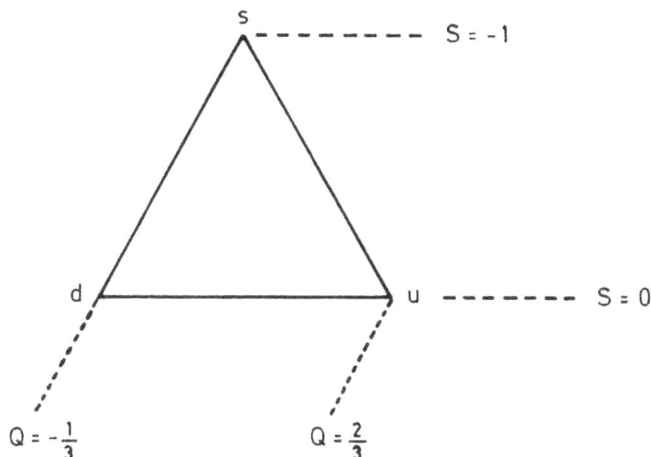

Fig. 17.8.

Consistent with the previous diagrams, one would expect two "ordinary" (non-strange) particles and one strange one. Call the two ordinary ones d and u (this originally stood for "down" and "up") though we have not rotated the diagram through 120° clockwise.

We will call the top (strange) one s. Gell-Mann borrowed a name for them from James Joyce: *quarks*. His idea was to construct all other particles from them, thus radically reducing the number of truly "fundamental" particles from a few dozens down to three.

There seemed to be some justification for considering nucleons, for instance, as being constituted of quarks. In experiments reminiscent of those with which Rutherford had established the existence of the atomic nucleus, high energy collisions of electrons with nucleons were found to give rise to too many wide-angle scatterings to be consistent with a continuous distribution of charge in the nucleus. This suggested that nucleons were made up of several particles much smaller than themselves, which were first given the name "partons". Suppose that these were in fact Gell-Mann's quarks. Since the Ω^- particle had strangeness -3, it would have to consist of three s-quarks. But since its charge was -1, each quark would have to have charge $-\frac{1}{3}$.

Likewise, the highest charge found on any baryon was 2 (the Δ^{++}). If this were also made up of three quarks, each would have to have a charge $\frac{2}{3}$; it would then be constructed of three u-quarks. By similar arguments, the d-quark would also have to have charge $-\frac{1}{3}$.

So all known baryons seemed to be constructed of combinations of three quarks:

n	udd	Δ^-	ddd
p	uud	Δ^0	udd
Λ^0	uds	Δ^+	uud
Σ^-	dds	Δ^{++}	uuu
Σ^0	uds	Ω^-	sss
Σ^+	uus		
Ξ^-	dss		
Ξ^0	uss		

If quarks are fermions (as they must be), the distinction between the proton and the Δ^+, which are constituted of the same quarks, is based on the relative alignment of the quark spins: in the proton, two will point one way and one the other, leaving a net spin of $\frac{1}{2}$. In the Δ^+, all quark spins will be aligned. The same consideration distinguishes the lighter and heavier Σ and Ξ particles, the heavier ones (those with higher energy) having their spins aligned.

What is *less* evident is the distinction betwee Λ^0 and Σ^0, which are also composed of the same quarks.

Before raising other questions, let us look at the families of mesons. Since the quarks may be said to have baryon number $\frac{1}{3}$ (three of them make a baryon) and since a meson has baryon number 0, we must assume that mesons are constructed of a quark and an antiquark; the baryon number is then $\frac{1}{3} - \frac{1}{3} = 0$. Table 17.2 shows how this can be done.

Table 17.2.

$\pi^+ (\rho^+)$	$\bar{d}u$
$\pi^- (\rho^-)$	$d\bar{u}$
$\pi^0 (\rho^0)$	$u\bar{u} - d\bar{d}$
$\eta^0 (\omega)$	$u\bar{u} + d\bar{d}$
K^+	$u\bar{s}$
K^0	$d\bar{s}$
K^-	$s\bar{u}$
$\overline{K}^0$	$s\bar{d}$
$\eta^1 (\phi)$	$s\bar{s}$

The particles indicated in brackets are excited states of quark and antiquark, having spins aligned (total spin 1) rather than opposite (total spin 0).

Would not the particles $\pi^0 (\rho^0)$, $\eta^0 (\omega)$, and $\eta^1 (\phi)$ which are constituted of a quark and its own antiquark decay by mutual annihilation into two photons, just as an electron and a positron do? Indeed they do; these mesons are unstable and *do* decay with quite a short lifetime ($\approx 10^{-16}$ to 10^{-19} second). Thus, they bear some resemblance to positronium, which is a bound state of electron and positron. This shows that these mesons are not fundamental particles themselves, but are composite structures.

Let us now discuss some problems arising from the outline which we have given of a quark model for hadronic (strongly interacting) particles.

A very obvious one relates to the Pauli exclusion principle, since the quarks are fermions. Consider for example the following particles: $\Delta^0, \Delta^+, \Delta^-$ and Ω^-. These particles all have spin $\frac{3}{2}$, so that the spins of the three particles of which they are constituted are all parallel. Thus, they are states symmetric in exchange of the spins of identical quarks, -3 s-quarks in the case of Ω^-, 3 d-quarks for the Δ^-, 2 d-quarks for the Δ^0, 2 u-quarks for the Δ^+. We also know that spatially symmetric states (with no nodes) are generally the lowest energy states of a system. How, then, can we reconcile our model with the Pauli principle, which requires that no two identical particles can be in the same state?

The answer is found in postulating still another characteristic for the quarks. This characteristic has rather unfortunately been named "colour" though it has nothing whatsoever to do with colour as normally understood. Nor have we any intuitive idea of what this new property signifies; we know only that it is a property which does not appear in "classical" systems. Once again, it seems to have a peculiarly quantum character.

Whatever the new characteristic is, it must come in three forms; these are designated red, green and blue. Those hadrons which are made of three otherwise identical quarks ($\Delta^-, \Delta^{++}, \Omega^-$) must then, in order to conform to the Pauli principle, consist of a red quark, a green quark and a blue quark.

We are left with another difficulty, which can be illustrated by considering the Δ^+ particle. All that would appear to be necessary is that its two u-quarks have different colours: red and green, say, or red and blue, or blue and green. But there would be no restriction on the colour of the d-quark. If the u-quarks were red and blue, the d-quark could be either of these colours, or green. But this would imply not one particle, but three. The door now seems to be open to far too many particles.

This difficulty can be overcome by imposing the condition that all baryons consist of one red, one green and one blue quark. Since these three primary colours combine to make white, we can formulate our rule to state that all particles appearing in nature must be *white*. This also happens to cover the case of the mesons. A red quark and an anti-red quark will have no net colour, just as an electron and positron have no net charge.

If the rule is taken to be general, it has other consequences: first, that lone quarks should not be found in nature, nor should systems of two bound quarks. Indeed, up to now, neither has ever been seen.

It seems curious to be left with a situation in which the most fundamental constituents of matter cannot be isolated or identified. We cannot even say, within a baryon, which quark has which colour! We must consider the baryon as a superposition of states in which the colours are distributed in every possible way.

We may take some consolation in the fact that the situation is not without precedent. Magnets have north and south poles, but if you try to saw a magnet in two to isolate the separate poles, you instead end up with two new magnets, each of which has, again, a north and a south pole!

Another way of stating this is to say that magnetic lines of force have two ends. Faraday, following Boscovich, was attracted to the idea that particles of matter could be *defined* as the terminal points of lines of force; the lines of force, then, become the ultimate reality. A similar notion is involved in currently fashionable "string theories" of fundamental particles.

The basic question of the force which binds quarks together, in either baryons or mesons, still remains. It is certainly none of the already discussed forces. Yet the π-mesons, heretofore considered to be the agents of the strong force, no longer play that role, being themselves composite particles whose constituents are held together by the new force that we must now evoke. The model which we will

use for this new force is electromagnetism. In electromagnetic phenomena, the mediating field is produced by charge. We now assume
that the characteristic of quarks which determines their interaction
is COLOUR; it is a new sort of charge, which comes in *three* versions
— red, blue, green — whereas the electromagnetic charge comes in
two — positive and negative. A theory can be built which closely
parallels electromagnetism, though it is more complicated in detail.
This is known as quantum *chromodynamics*, to emphasize the parallel with quantum electrodynamics. The quanta of the field, which
we can call the *colour field*, and which now becomes an agent of the
strong interaction, are known as gluons.

Note that the three *different* colour charges require a multiplicity of different gluons. In electromagnetism, we had the unique
photon. In the meson theory of the strong force, we needed *three*
mesons, since mesons can, but need not, transfer *charge* between nucleons. In the present theory, gluons must be able to transfer colour
charge. A quark may turn from red to green by emitting what we
will call a green → red (gr) gluon. This, absorbed by a green quark,
will turn it into a red one; in this way the quarks will have exchanged
colours. An antiquark will carry anti-colour. In our naive picture of
an infinite sea of negative-energy particles, a red quark in a negative-
energy state may be promoted to a positive-energy state; this is the
process of creation of a meson.

The number of gluons needed to mediate the interactions between all possible pairs of mesons would at first sight seem to be
9:

$$
\begin{array}{ccc}
r-r, & r-b, & r-g, \\
b-r, & b-b, & . \quad b-g, \\
g-r, & g-b, & g-g.
\end{array}
$$

In fact, only eight are necessary, since the $r-r, b-b$ and $g-g$ gluons
are not independent of each other. This stems from the constraint
of colourlessness imposed on ordinary matter.

An interesting feature of quantum chromodynamics is that the agents of the strong force, the gluons, themselves carry colour charge and therefore are subject to the colour force. Note that this is a new feature of the theory as compared to electrodynamics or the old meson theory of nuclear forces. The photon, the agent of electromagnetic forces, is not charged, and so is not itself subject to electromagnetic force. Similarly, pre-quark mesons did not carry strong charge, that is, a meson did not generate a meson force. With gluons the situation is different; two gluons can interact with each other by the strong (gluon) force. The gluons, after all, carry colour charge from one quark to another; therefore, they have such charge.

We have met another situation of this sort, in the theory of gravity. Since all energy produces gravitational force, the energy of a gravitational field produces gravitational force. In quantum terms, the quantum of the gravitational field, the graviton, by virtue of its energy can produce more gravitons. The theory is nonlinear, as is that of the gluon field.

Since gluons carry colour charge, the general principle of the colourlessness of distinct quanta of matter assures that gluons, like quarks, will never be seen in isolation. More and more, we are being forced to see the core of the world of particles through a glass, darkly. Is there some deep significance in this? Must we learn to live with a world of inference, and no longer a world of identifiable objects? The world of fundamental particles is becoming increasingly alien, whose patterns of behaviour elude our intuition and our capacity to understand in terms of familiar experience.

17.7. A New Picture Of Nuclear Processes

In our earlier discussion of the meson theory of nuclear forces, we drew diagrams to illustrate the processes of interaction of nucleons, e.g. Fig. 17.9 representing a mode of interaction of neutron and proton.

Consider Fig. 17.10, which describes a process involving quarks. Recall the interpretation of antiparticles as particles travelling

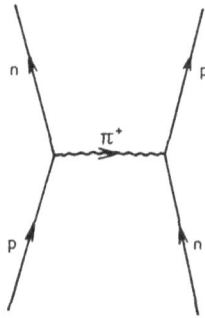

Fig. 17.9.

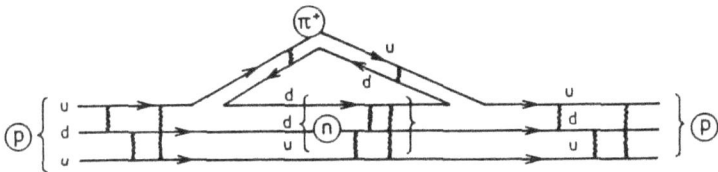

Fig. 17.10.

backward in time. The three parallel lines on the left represents a bound system of 2 u-quarks and a d-quark; thus, a proton. All lines should bear a colour, the colours of the three quarks all being different but otherwise arbitrary. The d-quark loop in the centre represents creation of a quark-antiquark pair (a "virtual" process). The $\bar{d}$ can combine with a u-quark to make a π^+ meson. The d-quark replaces the u creating temporarily a udd nucleon — a neutron. Ultimately the d and $\bar{d}$ annihilate each other, the meson disappears, and we again have a proton. This sort of process renders possible an understanding of the "structure" of a proton, and in particular its magnetic moment.

　　To complete the diagram, the various quark lines should be joined by gluon lines, which constantly permit the quarks to ex-

change colour and hold the quarks together in the colourless particles
— proton, neutron, π-meson.

Figure 17.11 shows the interaction of a neutron and a proton
through the exchange of a π^+-meson. Symbolically, we represent
gluon interactions between quarks by wavy lines. Only one gluon exchange is shown for each interaction; actually many such exchanges
may exist. They are, in fact, going on continually.

The rationale of the associated production of strange particles
now becomes clear; a strange quark (or anti-quark) must be carried
from one particle to the other.

What is represented here is the process, as shown in Fig. 17.12,

$$\pi^- + p \rightarrow \Lambda^0 + K^0 .$$

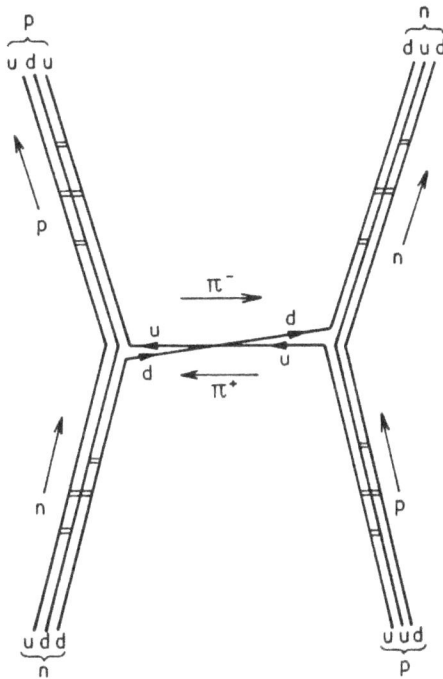

Fig 17.11.

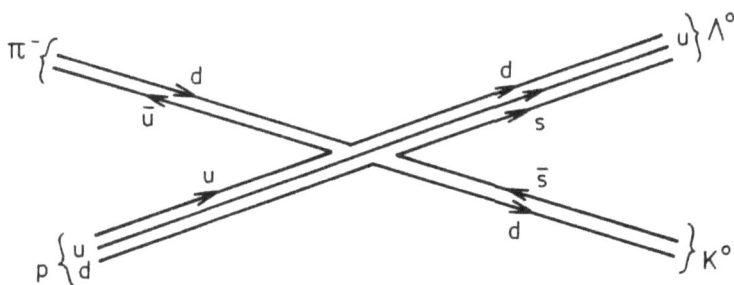

Fig. 17.12.

Note that although the introduction of quarks has significantly decreased the *number* of fundamental particles, it has complicated the description of processes such as those discussed here. Is there then a net simplification? We have decreased the number of entities in our theory, but find the number of relations between them increased. Einstein observed that the more fundamental a theory, the more complex is the explanation of specific phenomena. One cannot simplify nature. The goal of science is only to find simplicity and economy in its underlying *principles*.

17.8. Nature of the Force between Quarks

The force between quarks is quite different from any we have met up to now. Gluons have zero mass, and the force is of long range. Rather than decreasing with distance, however, as the electromagnetic force does, it remains more or less constant as the quarks are separated. Thus, the energy stored in the force is virtually unlimited. This is presumably why quarks cannot be separated. Once a certain separation is achieved, enough energy has been stored to create quark-antiquark pairs, i.e. mesons. Expending increasing amounts of energy in trying to dissociate them (by high energy particle bombardment, for instance) will only produce increasing numbers of mesons; the baryons themselves will remain intact.

One can think of the situation in terms of Faraday lines of force. In electricity or magnetism, lines of force radiated outward from a source of field; because a fixed number of force lines pass through a surface which increases in area as the square of the distance from the source, these forces obey an inverse square law. With the gluon force the lines of force do not radiate out equally in all directions from one source A, returning by increasingly circuitous paths to another, B (Fig. 17.13); rather, they tend to be bunched close to the line joining them. The number of lines of force crossing a unit area of surface between A and B does not therefore change much as the particles are separated; the force remains fairly constant.

Fig. 17.13.

Why do lines of force bunch together? This can be attributed to the mutual force between gluons, provided that this force is sufficiently attractive.

While this picture is highly oversimplified, it gives some idea of how a force law of the sort invoked can be conceived. The lines of force are a sort of string between the particles.

The most interesting feature of such a force is that it becomes increasingly weak as the particles come closer together. Under electric or gravitational forces, particles become free at large separations. Quarks on the other hand, approach freedom when they are close together. For this reason, the energy of a composite particle should be of the same order of magnitude as the sum of the energies associated with the masses of the quarks of which it is constituted.

17.9. Other Quarks

There is a gratifying elegance about the scheme we have outlined, based on the "magic number" three. However, things were not destined to remain so simple.

The first example of particles not falling within the purview of the preceding theory was found in 1974 in Stanford and simultaneously in Frascati, Italy. These particles had a lifetime $(10^{-23}$ second) which was about 1000 times longer than would be expected for decay due to strong interactions at that energy.

It soon became evident that a new quark was needed — one carrying another new property, which Glashow called "charm".

Nor was that the end. Experiments by Lederman and others in 1977 suggested still another quark, with yet another characteristic or "quantum number". This quark is now generally known as the b-quark (for "bottom", though the associated characteristic was first, unfortunately, called "beauty").

There now appear to be five quarks — an unhappy number — each appearing with its three colours. Theory suggests another, the t-quark.

17.10. Weak Interactions and Unification

The foregoing considerations have dealt with the *hadrons*, which are involved in the strong nuclear interactions.

The electron and the neutrino belong to a different family, the leptons. The charged members of this family couple to the *electromagnetic* field, but *all* are sources of the so-called "weak" field, which, as we have seen, mediate the process of beta-decay. The basic process is that of the β-decay of the neutron:

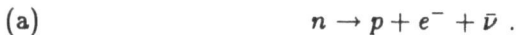

(a) $$n \rightarrow p + e^- + \bar{\nu} .$$

The appearance of the anti-neutrino conforms to the principle of *conservation of leptons*. The process may be envisaged, using the same concept as that used to discuss the electron-positron pair, as the promotion of a neutrino from the negative-energy sea, along with its conversion to an electron. This cannot be due to the absorption of a *photon*, which cannot interact with the neutral neutrino. One therefore postulates a new particle, which will be designated as W. A *negative* W-particle enters into the basic reaction:

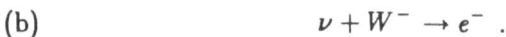

(b) $$\nu + W^- \rightarrow e^- .$$

A positive W (W^+) can interact similarly with anti-particles:

(c) $$W^+ \to e^+ + \nu \; ;$$

thus again a particle on one side of a reaction becomes its antiparticle on the other.

If the neutrino in (b) is in the negative-energy sea, it may be written:

(d) $$W^- \to e^- + \bar{\nu} \; ,$$

the $\bar{\nu}$ being the hole left in that sea.

The W^- also interacts with a neutron and a proton:

(e) $$n \to p + W^- \; .$$

Putting together (d) and (e),

$$n \to p + W^- \to p + e^- + \bar{\nu} \; .$$

This is represented by Fig. 17.14,

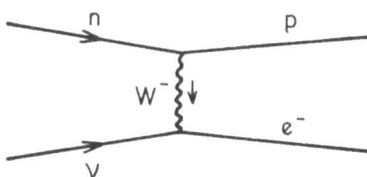

Fig. 17.14.

where the dotted line represents the negative-sea neutrino. This is equivalent to an antineutrino *emerging* from the reaction as shown in Fig. 17.15. We have thus made a parallel between electromagnetic and weak processes.

What properties can we deduce for the W-particles? Because β-decay is an extraordinarily slow (weak) process, the W-particle must

Physics: Imagination & Reality

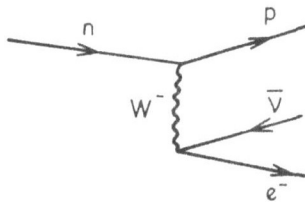

Fig. 17.15.

couple weakly to nucleons as well as to leptons. β-decay experiments also indicate that the interaction is almost a local one; that is, the W-field has a very short range. Thus, the electron and anti-neutrino are produced very close to the point where the neutron decays into a proton. An accurate determination of the range can be deduced from neutrino scattering experiments (just as neutron-proton scattering enabled us to determine the range of the nuclear interaction). In this way the W-particle rest energy has been estimated as 81 GeV (1 GeV $= 10^9$ eV). This is more than the total energy of nuclei up to around germanium in the periodic table! We can "borrow" such energy, according to the uncertainty principle, only over distances of the order of 0.5×10^{-16} cm, thousands of times smaller than an atomic nucleus.

Because of this high rest energy it was possible only recently to produce W-mesons in the laboratory. This was accomplished in 1983 at CERN in Geneva, where energies of up to 500 GeV were obtained using colliding proton-antiproton beams.

Because the neutron and proton consist of quarks, W-mesons are presumed to be coupled in the first instance not to nucleons but to quarks. Recalling that the neutron has the quark composition *udd* while the proton is *uud*, the β-process appears to involve changing a d-quark (charge $- 1/3$) to a u-quark (charge $2/3$). This leads us to write, instead of (a) above, the reaction

$$q_d \rightarrow q_u + e^- + \bar{\nu}$$

(q_d is the d-quark and q_u the u-quark). In terms of W-particles this implies the weak interaction of quarks

$$\boxed{q_d \rightarrow q_u + W^-}.$$

Where, then, does the *strange* quark fit in? It also has charge $-1/3$, so we could write equally well

$$q_s \rightarrow q_u + W^- .$$

This would imply a β-decay process for *strange* particles. Given that the Λ^0 particle, for example, has the composition (uds) while p is (udu), it follows that the Λ^0 should be able to undergo the β-decay

$$\Lambda^0 \rightarrow p + W^- \rightarrow p + e^- + \bar{\nu} ,$$

which can in fact be observed (albeit rarely).

The charmed quark, like the t- and b-quarks, can be incorporated into these considerations. The charmed quark q_c has charge $2/3$, like the q_u. The t- and b-quarks (q_t, q_b) have charge $-1/3$. We shall see shortly how the various quarks (and leptons) couple to the W-field.

We have so far talked only of the multiplicity of quarks, but in fact there are also other "families" of leptons. The so-called μ-meson, whose mass is about 207 times that of the electron, is in every other respect like the electron. It is like a sort of excited state of the electron. Another neutrino is associated with the μ-meson, *not the same as that associated with the electron*; we call this the μ-neutrino ν_μ. This neutrino cannot be involved in any weak interaction of electrons (β-decay or its inverse).

But this is not the end. A third electron-like particle, called the τ,[19] has a mass ≈ 3480 times that of the electron. Although still called a lepton, its mass is greater than that of the neutron or proton. It, once again, has its own associated neutrino, the ν_τ.

---

[19] This is a different particle from the one mentioned earlier in connection with the "$\tau - \theta$ puzzle", which is now called a K-meson.

These various leptons are connected through the W-field; the decay of the μ^--meson takes place as follows:

$$\mu^- \rightarrow W^- + \nu_\mu \rightarrow e^- + \bar{\nu}_e + \nu_\mu \ .$$

This is the μ-meson decay involved in the famous time-dilation experiment of relativity. The $\bar{\tau}$ can decay into

$$\tau^- \rightarrow \bar{\mu} + \bar{\nu}_\mu + \nu_\tau$$

or

$$\tau^- \rightarrow e^- + \bar{\nu}_e + \nu_\tau \ .$$

The fact that there are six fundamental leptons, and the desire to incorporate all fundamental particles into a single theoretical framework ("grand unification") was responsible for the postulate of the sixth, not yet observed quark, the t-quark.

Quarks and leptons can be grouped in families of four:

$$\text{(i)} \begin{pmatrix} u & \nu_e \\ d & e \end{pmatrix} \qquad \text{(ii)} \begin{pmatrix} c & \nu_\mu \\ s & \mu \end{pmatrix} \qquad \text{(iii)} \begin{pmatrix} (t) & \nu_\tau \\ b & \tau \end{pmatrix} \ .$$

The first family provides all the fermions needed to construct the real world of our everyday experience. The others appear only at high energies, whether in high energy laboratories or in cosmic rays. Most importantly, they would have played a vital role in the initial stages of the Big Bang, which is why high-energy physicists have now invaded the domain of early cosmology. Their members are unstable.

In the foregoing remarks we hint at the unification of various forces. What stage have physicists reached in this exercise? One solid accomplishment is the unification of the weak and electromagnetic interactions based on the work of Steven Weinberg and Abdus Salam. In this theory the photon and another massive neutral boson, called Z^0, appear as combinations of a charge-preserving weakly interacting W-particle, called the W^0, and another field B^0 which can annihilate to produce the pairs $\nu_e \bar{\nu}_e$ and $e\bar{e}$. Both W^0 and B^0

are expected to have non-zero mass, that of the W^0 being the same as the mass of W^+ and W^-.

In speaking of "combinations" we should again think of our discussion of polarization. W^0 and B^0 might be likened to the two linearly polarized photons, while the combinations γ (photon) and Z^0 would be analogous to circularly polarized photons. But the γ and Z^0 have vastly different rest masses; that of the photon is zero, while that of the Z^0 is somewhat greater than that of the $W^\pm$. Evidence of the existence of the Z^0 particle was found at CERN in 1983. The most important feature of the Z^0 particle is that it permits reactions converting lepton pairs (e.g. neutrino-antineutrino) into quark-antiquark pairs, or vice versa.

Can one go further, and bring the quarks and their strong interactions into the unification process? Theories that do this are called "grand unified theories", of which there are several, none yet definitive. As for gravitation, the problem is still open. Einstein spent the last 30 years of his life trying to unify electromagnetism and gravitation, without success. Most recent efforts to incorporate gravitation into an ultra-grand unification scheme have led to many difficulties and no definitive results. It is always possible that gravitation is not one more field like the others, but plays some special role in the architecture of the universe.

As for grand unification, one might ask: how can one unify fields whose strength varies over 13 or 14 orders of magnitude? The answer is that the theories predict that the coupling constants of the particles to their field quanta vary with energy, becoming equal at about 10^{15} GeV. At this energy one is probing distances of 10^{-29} cm. At a *very* early stage of the Big Bang, when the temperature was high enough to produce such energies, all interactions would have been effectively of the same strength; all sorts of particles would then have interacted freely. Accelerators being currently planned, and perhaps realizable by 1995, may reach 1 TeV $= 10^3$ GeV, but this is still a factor of $\approx 10^{12}$ short of the "unification energy". Such an accelerator would be 30 km in circumference, and be buried 100 m underground.

Are we approaching the final solutions which Richard Feynman thinks will be the triumph of our "heroic age"? In the past, every apparent approach to finality has ended in the opening up of new levels of subtlety in the structure of the physical world, and a sudden realization that, far from knowing almost everything, we had lacked an essential key to the structure of the physical world. There is no assurance that more stunning new surprises are not still in store for us.

It is not difficult, in any case, to find unsatisfactory, unaesthetic aspects in our current vision. Freeman Dyson has remarked on the lack of the simplicity of insight which characterized Einstein's theory of relativity.

Perhaps the most obviously unsatisfying aspect of our current model of the physical world is that we have no prescription for determining the basic energy levels, the masses of what we understand to be fundamental particles. Our world is clearly a quantum one, and one of the basic problems of quantum mechanics is to determine energy levels of systems. From what starting point can we set out to determine the masses of elementary particles? We appear to be confronting the difficulty of explaining a system in terms of itself, possibly in violation of some cosmic Gödel's theorem. Is the alternative not to admit the existence of a deeper level of reality from which all our models must spring?

Author's postscript: Lest this question be interpreted as an opening for the introduction of religion, that is *not* at all the author's intention. Religion may be seen as a way of terminating what appears to be an endless spiral. The risk is that one may terminate it too soon.

Chapter 18

THE INFLATIONARY EARLY UNIVERSE

The particle theories of unification and of grand unification are subject to test only at energies enormously higher than any conceivable on earth. It is hardly surprising, then, that the particle physicists, in looking for some source of evidence to verify their theories, turn to the very early universe, where, if we go back far enough, almost boundless energies are thought to have existed. They have recently come up with a model of the *very* early universe which, they claim, explains some puzzling features of the standard big bang model. This "inflationary model", as it is called, has no consequences for what happened after the first second of the universe; there, nothing is changed. What happened earlier, they claim, provides answers to some outstanding problems of the features of the present universe. What are these "problems"?

18.1. The "Homogeneity Problem"

This relates to the earlier observation of the extraordinary homogeneity of the cosmic background radiation. A simple way of disposing of this problem would be to say that the universe was simply created that way. The inflationists, however, are not satisfied with that answer. The theory states that, at the very beginning, there was a horizon, a distance from any point beyond which one could not "see" (i.e. from which one could not receive information propagated at or below the speed of light). This is illustrated in Fig. 18.1. If we look at the perimeter of the horizon, in the fourth

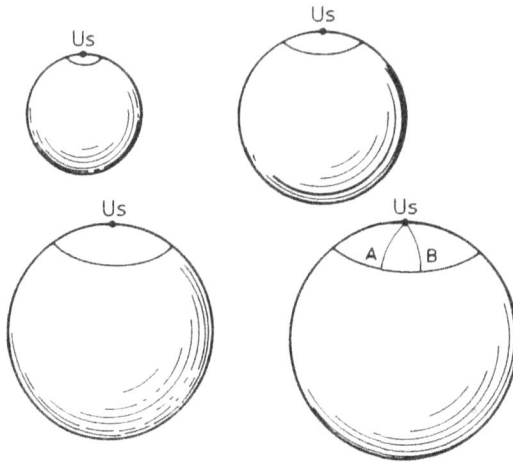

Fig. 18.1. Growth of the "horizon" with expansion of the universe. "Us" represents the point for which the horizon is determined: it can be anywhere. The
"circle of longitude" represents the horizon.

frame, signals reaching A since the beginning could not have come
from further than a point B; AB is the distance a light signal could
have propagated since the initial instant. Signals from further away
could have had no influence on what happened at A. There would
have been no mechanism for producing equilibrium all around the
horizon. Why then should one have isotropy for all directions from
A? Figure 18.2 shows the same thing in a different way. Here P
and Q are two points on our horizon. Everything that could have
affected P and Q at time *t* lies within the light cones terminating at
these points. Obviously, their histories were quite independent, yet
they have identical radiation intensity.

There is of course absolutely nothing to explain if the *creation*
was perfectly homogenous and isotropic.

18.2. The "Smoothness Problem"

To account for the formation of galaxies, galactic clusters and
superclusters, inhomogeneities (fluctuations?) are necessary; matter

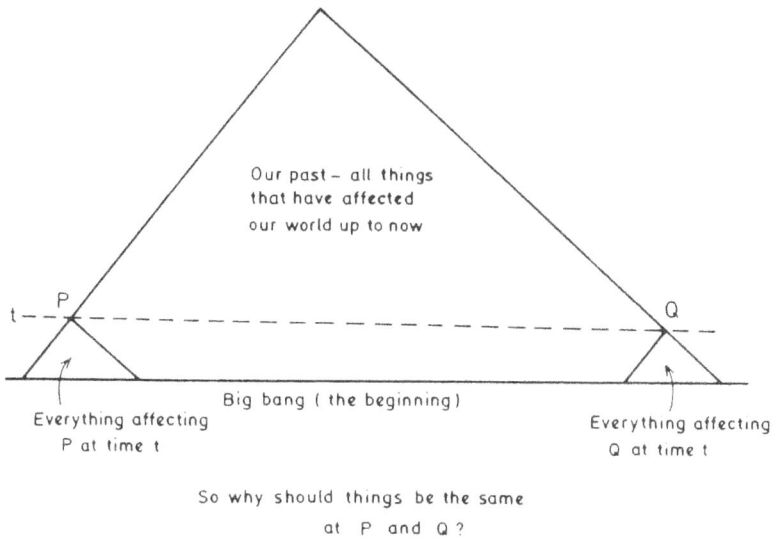

Our past – all things
that have affected
our world up to now

t —— P

Q

Everything affecting
P at time t

Big bang (the beginning)

Everything affecting
Q at time t

So why should things be the same
at P and Q ?

Fig. 18.2.

would then be attracted to regions of higher density from regions of lower density. But any initial inhomogeneities would tend to grow as a result of gravitational forces. Unless they were *very* small at the start they would grow too large (there is, however, some recent evidence that the universe is indeed "clumpier" than had been thought). Also, David Layzer of Harvard proposed in 1976 another mechanism for the origin of galaxies, which sprang from a slightly later time (some minutes after the beginning). According to Layzer's scenario, the universe went through a *solid* phase, and was shattered by expansion).

From the point of view of the inflationists it seems highly unlikely (whatever that means) that there would have been, at the beginning, such an *extremely* high degree of homogeneity, much greater than that of a gas in equilibrium, for example, to lead to the universe as it is.

18.3. The "Flatness Problem"

We do not yet know whether the universe will recontract (i.e. has positive curvature) or will expand forever (negative curvature). Up to the present time, study of the recession of the most distant galaxies appears to be quite consistent with the universe being flat (zero curvature). In such a case, the universe would be expanding so that its rate of expansion will keep decreasing toward zero, but that there will never be a recontraction. We know what the critical density, above which the recontraction would take place, is. All that we can say with confidence now is that the actual density is probably between 1/10 and 2 times that critical value. At first sight this seems to be a wide range of uncertainty, but a few years ago Peebles and Dicke showed that, as one goes back in time, that ratio must have been much closer to 1; in fact, to be in the range that we can now fix, it must, one second after the big bang, have differed from one by only one part in 10^{15}! Is it not extraordinary that the universe is so very close to being flat?

Once again, we could answer that it is as it is because it was made that way, and for that we have no explanation. It is a current prejudice that everything must have an explanation but that is ridiculous. There must always be some agreed basis for explanation, something à priori. But what then is the explanation of that? We are caught in an endless chain! However, we can push ourselves back from one unexplainable thing to another, and the inflationary model is attempting to do this in terms of the laws of physics as we know them today. It is a dangerous exercise, an immense extrapolation from the realm of our most sophisticated "experience". But it is surprising enough that we can, all the same, provide some sort of answers, that resolutions of the problems posed can be imagined.

How, then, does the inflationary model purport to provide an answer for the questions we have asked about the big bang? It is beyond the scope of this book to try to describe the new model, which is founded on as yet speculative "grand unified theories". What we can do however is affirm a consequence of quantum mechanics, that the "vacuum" is not nothing, but rather a very complex physical entity, containing a quantum residue of energy of all known fields. According to current theories, it is sufficiently complex that it can exist in more than one state. The "true vacuum" is its state of lowest energy (and is thus stable), but there is also a sort of excited state of higher energy, a "false vacuum" which is metastable. The motor for the expansion is the transition from the false vacuum to the true vacuum. This produces an expansion of the universe by a factor of perhaps 10^{20} in about 10^{-35} second, after which expansion it is still very small. In any case the enormous energy release can create all sorts of particles in abundance.

Obviously something very strange has happened. We started with some sort of "vacuum", and from this came the energy to create particles. Can we say that the particles came from "nothing"? No, vacuum is not "nothing". The question, "where did the matter in the universe come from" is replaced by another, "where did the false vacuum come from?"

The process still seems quite miraculous because particles represent mass energy, which is positive. It turns out that the gravitational energy, which is negative, almost exactly cancels out mass energy, so that the total energy is zero, or very close to it. What, then, is the process of creation? Is not its essential feature the creation of a pattern, of physical laws, rather than of substance?

Let us now see how the inflationary model explains the "problems" of the Big Bang model (remembering that these problems are optional and not compulsory).

So far as the "homogeneity problem" is concerned, before inflation takes place the universe is small enough for all its parts to be in equilibrium and so to be extremely homogeneous; this homogeneity is then maintained during the rapid expansion which follows.

As for the "smoothness problem", it is useful to think of the analogy, which we have used extensively, of the surface of an expanding balloon. Any unevenness existing in the pre-inflationary period may be thought of as wrinkles on the surface of the balloon which has shrunk; in the process of inflation the huge inflation flattens them out (it is really not quite so simple as that, but the image is useful to give a qualitative idea of the process).

As for the "flatness problem", it is claimed that the ratio of actual to critical density tends rapidly to unity in the process of inflation, whatever its value before, and that value of unity corresponds to a flat space. Nothing in the presently available evidence is inconsistent with a flat universe, though, as we have already noted, it does require that there must be hidden throughout the universe large amounts of energy in not yet detected form.

The inflationary model is not without its difficulties, and cannot at this point be taken as a definitive explanation of the qualitative features of the universe, even assuming that such an "explanation" is deemed necessary. But there is another question which is worth considering. Despite the claims of "creation from nothing", we live in a world which is describable in terms of a very sophisticated pattern of fields, related through sets of arbitrary rules. But why these particular fields? Why are there four interactions? Why do they have the strengths they have? Why have the basic parameters of nature their actual values? These, and a multitude of other questions, have no answer; we can only say: this is the way the world is. We use this cumbersome base, with all its arbitrariness, to explain notions of the utmost simplicity, the isotropy, the smoothness and the flatness mentioned above. It is as though the universe were made by the simplest possible recipe. Are these characteristics easier to accept *ab initio* than the complex and uncertain structure which we use to explain them?

We may not like to accept that there must always be, underlying our scientific theories, some things which must simply be accepted, and for which there is no base for rational explanation. Yet, as we have seen, this is inevitably so. Our present explanations are probably not final, but whatever may replace them as we gain more understanding will not enable us to escape this ultimate limitation. It is unlikely that we will ever be sure whether our speculations about the foundations of the universe are true or not. Science provides us with links between the elements of our knowledge, but does not reveal its origins to us. What human history teaches us is the wisdom of humility in facing the most fundamental problems of the universe. We have never been as clever as we like to think, so the future may yet reveal patterns which we cannot, at present, conceive.

Chapter 19

RADIATING BLACK HOLES: WHERE QUANTUM MECHANICS, THERMODYNAMICS AND RELATIVITY COME TOGETHER

Nature does not recognize the compartmentalization of physics into distinct subjects, such as relativity, thermodynamics, quantum mechanics, electromagnetic theory and the like. In the real world everything is mixed together. So it is appropriate to end with a discussion of a problem in which things get very thoroughly mixed up in a way which produces surprising results – that of the thermodynamics of black holes.

The black holes which we have discussed are truly black, which seems to leave little room for thermodynamics. From within their event horizon nothing, not even light, can escape. Thus, the black hole is an isolated, bubble-like subuniverse.

There is something disturbing about the idea, all the same, because it seems that the second law of thermodynamics is violated. All information associated with matter is lost when it is dumped into a black hole. Whether one dumps in a human being or the assortment of chemical substances (mostly water!) of which he/she is composed appears to make no difference to the final state entropy.

Can we assign an entropy to the black hole itself?

Stephen Hawking, in the early 1970's, tackled the problem of radiation emitted in the process of collapse of matter (a star) to the black hole state. There had to *be* radiation, because of the rapidly changing fields around the collapsing system. These would become so strong as to create particle-antiparticle pairs, a quantum process. What Hawking found, to his surprise, was that the intensity of the radiation predicted by the calculations did not decrease toward zero as the black hole state was approached. Rather, it approached a limit which was the black body radiation corresponding to a very specific temperature. The "black" body, having a temperature, was not really black. In fact, if it was small enough, the temperature could be very high.

Another feature of black holes, also demonstrated by Hawking, was that they showed a behaviour suggestive of that of entropy for interacting systems: If two black holes amalgamated, the surface area of the resulting hole was greater than the sum of the areas of the original ones. This is vaguely suggestive of the increase of total entropy associated with the thermal interaction of two initially separate bodies. It is certainly not evident, however, what the connection is between surface area and entropy, if indeed there *is* such a connection.

How can one go about assigning entropy to a black hole? The most obvious approach is to exploit the connection between entropy and information. Is there a change in entropy if a black hole of a certain size is introduced where previously there was empty space?

The answer is positive, and resides in the quantum mechanical concept of zero-point energy, which is found in the so-called vacuum, the state of lowest energy of fields. If we consider just the electromagnetic field, we can estimate the amount of information lost, due to the presence of the black hole, from the space which it occupies.

In infinite empty space, there is a zero-point energy of $\frac{1}{2}hf$ from a photon of frequency f. There are contributions of this sort for photons of every wavelength, however long. In the presence of the black hole, however, no photons of wavelength longer than the dimensions of the hole can be present. Thus, the information associated with such photons has been lost.

Our earlier discussions of the connection between entropy and information show that there are, associated with each particle, several bits of information. The total information is therefore proportional to the number of particles which could make up the black hole. Since the minimum particle energy allowed is, according to de Broglie's hypothesis, $hf_{min} = h\frac{c}{R}$, R being the black hole radius, whereas the total energy is Mc^2, the maximum number of particles is

$$\frac{Mc^2 R}{hc} = \frac{2GM^2}{hc} \ .$$

This gives, as a qualitative estimate of the entropy, or information lost,

$$S = k\xi \frac{GM^2}{hc} \ ,$$

where ξ is a constant whose correct value can be fixed by exact calculation.

We note that (assuming the black hole to be spherical) the surface area is

$$A = 4\pi \left(\frac{2GM}{c^2}\right)^2 = 16\pi \frac{G^2 M^2}{c^4} \ .$$

The connection between entropy and surface area appears to show that each is proportional to the square of the mass! In fact the relation between S and A is

$$S = k\xi \frac{c^3}{16\pi hG} A \ .$$

Detailed calculation gives $\xi = 8\pi^2$, so the entropy is

$$S = 8\pi^2 k \frac{GM^2}{hc} \ ,$$

k being the Boltzmann constant.

The quantum origin of the black hole entropy is clearly demonstrated by the presence of the Planck constant in the denominator. The result for a *classical* black hole can be obtained from the limit $h \to 0$, i.e. a classical black hole has *infinite* entropy. This is why it is so disturbing to the second law of thermodynamics. In a very real

and unexpected sense, quantum mechanics has come to the rescue
of the second law, which otherwise would have been menaced by a
consequence of the general theory of relativity.

We now have the basis for attributing a *temperature* to the black
hole, and an explanation of why it radiates a black-body spectrum.
The key relation is, defining U as the total energy of the hole

$$dU = TdS ,$$

where $U = Mc^2$. Thus we have

$$c^2 dM = T16\pi^2 kG \frac{M}{hc} dM ,$$

so that

$$\boxed{kT = \frac{hc^3}{16\pi^2 GM}} .$$

This is proportional to the surface gravity (the black hole equivalent
to the acceleration of gravity g at the earth's surface). This surface
gravity has been shown to be

$$K = \frac{c^4}{4GM} ,$$

so that

$$kT = \frac{hK}{4\pi^2 c} .$$

Values of the surface gravity were given earlier for black holes of
various mass.

Numerically, the temperature has the form

$$T = \frac{1.2 \times 10^{23}}{M(kG)} K .$$

Table 19.1.
Black hole temperatures.

Mass (kG)	Temp (K)	Peak blackbody wavelength	Kind of radiation
2×10^{33} (Sun)	6×10^{-11}	$\approx 10^5$ km	
1.2×10^{24} (Moon)	0.1	4.8 cm	microwave
2.4×10^{19}	5000	10^{-4} cm	near infrared
10^{15}	10^8	≈ 0.5 A	soft X-ray
10^{13}	10^{10}	5×10^{-3} A	
10^{11}	10^{12}	5×10^{-13} cm	γ-ray
10^9	10^{14}	5×10^{-15} cm	8BeV γ-ray

The rate of radiation U per unit area from the black hole at temperature T is given by Stefan's Law

$$U = \sigma T^4 ,$$

where the constant σ is

$$\sigma = \frac{\pi^2}{60}\frac{k^4}{\hbar^3 c^2} .$$

It follows that the rate of total energy output, obtained by introducing the previously calculated value of T and multiplying by the surface area, is

$$W = \frac{\beta c^2}{M^2}$$

where

$$\beta = \frac{1}{15360\pi}\frac{\hbar c^4}{G^2} .$$

With some simple calculus, starting from the equation for the rate of mass decrease,

$$\frac{dM}{dt} = -\frac{\beta}{M^2} \, ,$$

we find that, if M_0 is the initial mass and t_0 the total evaporation time,

$$\left(\frac{M}{M_0}\right)^3 = 1 - \frac{t}{t_0} \, ,$$

where

$$t_0 = \frac{M_0^3}{3\beta} \, .$$

If $M_0 \simeq 5 \times 10^{11}$ kg (the size of a fair-sized mountain), $t_0 \simeq 15$ billion years, so that black holes of this mass, formed early in the life of the universe, would be in the final stages of evaporation about now (on the time-scale of billions of years!).

If our universe is cyclic, and thus subject to ultimate recontraction, only small black holes would evaporate during its lifetime; black hole radiation would only have a slight effect on normal astronomical objects. But if it is open, and continues to expand forever, it would appear that *all* black holes would ultimately evaporate, emitting very long wavelength, low-energy radiation. The entropy would be greater than in the black-hole state, since the second law of thermodynamics would apply.

In Fig. 19.1 we show the relative rate of radiation R (in terms of the initial rate) as a function of the fraction of the lifetime elapsed, as well as the proportion of the mass remaining P. For the black hole which we are considering for illustrative purposes, the initial rate of radiation is approximately $1\frac{1}{2}$ thousand megawatts. As energy is radiated, the remaining energy (mass) of the black hole decreases, causing the temperature to rise and increasing the rate of radiation. At the end, the process becomes catastrophic. Over 20% of the total mass energy is radiated away in the last 1% of the lifetime; as the temperature increases, the frequency spectrum of the radiation shifts upward.

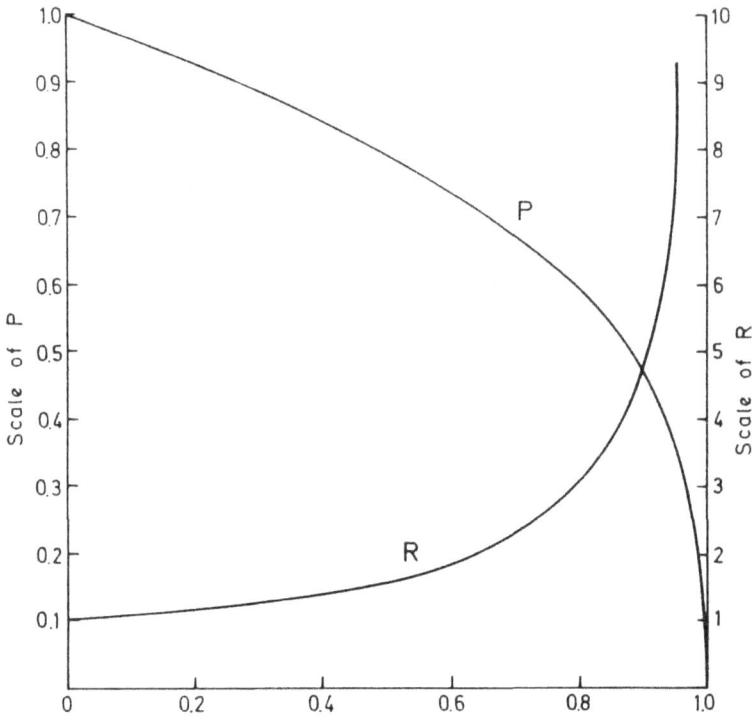

Fig. 19.1.

Where does the radiated energy come from? Since its emission causes the black hole to shrink, it comes ultimately from the rest-mass of the material (or energy) inside the event horizon of the hole. This is in flat contradiction to the classical statement that nothing escapes from a black hole.

The shrinking of the black hole may be understood by looking at a simpler situation. Imagine two parallel mirrors separated by a distance d. Between them there may be standing waves of light, but only for wavelengths $< 2d$ or of frequency $> c/2d$.

From quantum mechanics we know that every oscillator, and thus every photon, has a minimum energy of $\frac{1}{2}hf$, f being its frequency. In a region of space in which some frequencies are suppressed, this zero-point energy is not present; thus, the total density of zero-point energy between the mirrors is less than it would be in the region between them if the mirrors were not there. Thus, *even in a vacuum, a radiation pressure will be set up tending to push the mirrors together*. It appears that one can even squeeze some energy out of the vacuum! Furthermore, if the mirrors *do* come closer together, some frequencies of still shorter wavelength will be suppressed in the region between them, and the pressure to push them together will *increase*. This is very reminiscent of the situation with the black hole.

This phenomenon was first discovered many years ago by Casimir, when he studied the forces between molecules. It has been demonstrated more recently in the form in which we have presented it, in an experiment reported by Tabor and Winterton (*Proc. Roy. Soc.* **A312**, p. 435 (1969)) who were actually able to measure the force between the two mirrors. One is reminded of the statement by John Wheeler, that:

"No point is more central than this, that empty space is not empty. It is the seat of the most violent physics."

In the black hole, it is not mirrors, but the distortion of space by the gravitational field of the black hole which creates boundary conditions limiting the spectrum of photons. Energy is radiated at the expense of mass energy; the loss of mass produces an effect comparable to the approach of the two mirrors, and the number of particle degrees of freedom inside the black hole diminishes. When the black hole energy is exhausted, the geometrical constraints disappear and the vacuum is found once again in its "normal" state.

We have concentrated in this discussion on spherical black holes. If angular momentum is present, the black hole entropy, and hence its temperature, is changed. However, angular momentum is radiated away faster than energy; it follows that the later stages of collapse will take place in much the same way as in the spherical case.

Whether or not small black holes are ever demonstrated to exist, the mere possibility of their existence holds an important lesson for us. The requirements of relativity do not violate the second law of thermodynamics because of the quantum nature of energy. Our various fundamental physical laws appear to depend on each other in an unforeseen way. There is something very satisfying in this. Perhaps there are more profound links between the various facets of our knowledge of the physical world than we have yet suspected, and there lies yet to be discovered a deeper principle which assures the concordance of the various laws we know.

It seems appropriate to end with a statement of J. B. S. Haldane:

"The universe is not only stranger than we imagine, but stranger than we are *capable* of imagining."

Stranger, and more wonderful.

APPENDIX

Galileo, in a day when scientific communication was not so ritualized as it is nowadays, got the idea of revealing the mental processes by which new ideas were developed by imagining a dialogue between himself and a student who played the role of a foil and sometimes an *advocatus diaboli* (Dialogue Concerning Two New Sciences). It is hard for the newcomer to the subject to imagine the processes of thought by which one comes to have a certain understanding of quantum theory. The imaginary dialogue which follows may be similarly useful, at least in revealing some of the issues which might be raised and the questions that might come to mind.

It is hardly fair to Galileo to invoke his name in this context, since, Mr. G. (in the person of a grey-haired professor named Professor Gray) is clearly no Galileo. The student, Mr. S., is not unintelligent, but shall otherwise remain anonymous.

An Informal Dialogue Concerning One
Not-so-new Science

S: You promised to explain to me the strange features of the quantum theory, things like the uncertainty principle and the so-called wave-particle duality. Can we go back to where it all started, that is to say, to Planck's idea of the quantum of electromagnetic radiation?

G: Well, in the first place, it was not really Planck's idea, but Einstein's. Planck's guess was that radiation was only emitted in discrete energy bundles, but Einstein went further and said that it could only *exist* in such bundles.

S: Let me see if I understand that. Maxwell, building on ideas first enunciated by Michael Faraday, showed that light consisted of electromagnetic waves propagated through space at a definite speed.

G: Yes, and these waves were not visible things like water waves, but were waves of what we now call "fields". Faraday had an image of them; he talked about "lines of force". These were not *real* lines; they merely defined electric and magnetic *conditions* in space. Faraday put it this way. He said, "I desire to restrict the meaning of the term line of force so that it shall imply no more than the condition of force at any given place as to strength and direction..." What he meant was that if you picked an arbitrary position in space, anything put at that position would experience an electric or magnetic force pointing in the direction of his line of force at that point. So if you had a compass and kept moving it in the direction in which it was pointing, you would be following a "magnetic line of force".

S: That's a pretty abstract idea, but I guess I see the point. It seems to mean that you can make a sort of map of what will happen to a compass if you put it at a given place, but that the "condition of space" exists even if there is no compass there.

G: That is quite right; but what is more, Maxwell said that this "condition of space" can be propagated *in* space, the way a radio wave comes to our radio set from the antenna where it was emitted.

S: But what is it that is really being propagated?

G: I guess that, if you use the example of the radio wave, it is energy. Because if you stay out in the sun too long in the summer, you get burned by energy propagated out by the sun; and we know now that it is the same kind of thing as a radio wave, only more powerful. And of course the light by which we see things is also electromagnetic energy which is absorbed by the retina of our eyes.

S: Now I see that we are in trouble. Einstein has said that the energy comes in "bundles", but the energy of the electric and magnetic fields which constitute the light arrives continuously! It is a little difficult to imagine a bundle of wave energy!

G: Yes, but you shouldn't be misled by the word "bundle". This doesn't really mean that it is concentrated together in space. In fact, what you really mean is that you have a wave, but a wave whose energy comes only in specified amounts.

S: I don't understand. Einstein's "quanta of radiation" are supposed to behave sometimes like waves and sometimes like particles. Where do the *particles* come in?

G: Because that is the way they are absorbed. You can't absorb *part* of a quantum of electromagnetic field; you absorb all of it or none.

S: Oh, I think I misunderstood what physicists meant by "particle" in all this. It isn't at all like a speck of dust. But surely you've got to be able to say "There is a photon *there*, somewhere". Otherwise all you've got is just a weak wave, and I don't see why you talk about any *dual* character.

G: Well, strictly speaking you're right, but think about this. A quantum can come from a distant star, go through an astronomer's telescope, and be absorbed at a point on a photographic plate. That's pretty much "like a particle".

S: Now you have me *completely* confused. First the quantum is spread out like a wave and next thing you know it's delivered as a neat, compact package at the plate. Surely you can't have it both ways!

G: But that's the whole point of the paradox of duality.

S: Are you suggesting that there's something in all this that physicists can't explain? I've always thought that you fellows like to think yourselves smarter than you really are!

G: Oh, no, it's not like that at all. With quantum mechanics, everything we know can be *explained*. There are some things that can't be *predicted*, but that's a different story.

S: Well, then, *explain* how that photon which goes through space like a wave suddenly does itself up in a neat package to get absorbed at one point!

G: Well, of course, it isn't *really* a point, not exactly. But since it can't be split or shared, and that's the essential thing, it must be absorbed by one *electron* on the photographic plate. Of course, electrons also are sometimes wave, sometimes particle. So it's not quite right to say that, because it is absorbed by one electron, it is absorbed *at a point*. It just looks like a point to us, because we can't see what is going on down at the level of single atoms. They are just too small.

S: This is like Alice in Wonderland; it keeps getting curiouser and curiouser! But there are two things I'd like you to explain to me. First, how does the electron decide when to be a "particle". (which we agree is not *really* a particle) and when to be a "wave". I think I'd better be more scientific and ask the question in less flippant terms: under what conditions is it spread out like a wave and under what conditions does it bundle itself up and behave like a sort of particle? And secondly, if it is really a "quantum of a wave" being absorbed, doesn't it take a fair bit of time to make the transition? Surely it can't contract into a "bundle" instantaneously!

G: First question: it's wavelike till it's observed or detected; then it becomes particle-like. Second question: if it is only detectable by being totally absorbed, we can't really see whether the process takes time or not. We can't even predict when the process is going to be observed, much less what led up to the observation!

S: It seems that my two questions are really part of *one* question! But it does seem to me that, using the same line of argument, there's no way you can *know* that it behaves like a wave when its's not being detected, because then you can't detect it! So by what logic can you claim to say *anything* about what it's like then?

G: You are trying my patience!

S: Well, you are certainly trying mine. First, you say that the light quantum is a weak wave with a definite amount of energy, and now you cannot tell me how you know that it *ever* appears like a wave; whenever you see it, it's particle-like.

G: I'm sorry; the answer to that one is really simple. We know that is is wavelike because it undergoes diffraction. When light passes through two slits, the parts going through the two slits interfere with each other (the waves are just added one to the other). You can get the same effect with water waves, if you let them go through two adjacent holes in a breakwater.

S: Is that true even when it's so weak that only one quantum goes through at a time?

G: Obviously not! When a quantum is absorbed, it is absorbed at a point (or at least, what looks like a point to us monsters). But if you let enough single quanta through, we can't tell where any *one* will be absorbed; we only know that the more points we get, the more obvious it is that the overall pattern is that of ordinary diffraction.

S: If one of Maxwell's electromagnetic waves goes through the two slits, you get an interference pattern. If the photon is simply a quantum of such a wave, doesn't each photon have to go through both slits, so shouldn't each photon make the diffraction pattern?

G: But it can't, because it all has to be absorbed by one electron on the detecting screen.

S: No wonder each photon doesn't make a diffraction pattern! Here we have countless trillions of electrons in the screen all waiting hopefully to absorb the photon, but only one can win. It's even worse than our state lotteries! But at least the most likely winners are the electrons near those points where the diffraction field is strongest, the well-known peaks which characterize the pattern. I think I'm beginning to believe that each photon *does* diffract, only the fact that they must be absorbed whole hides the fact until we've seen a lot of them.

G: Well, I suppose you're right if you consider the wave-like aspect of photons, but not if you look at the "particle" aspect.

S: But you're only detecting them on the screen, so it's only there that you have to worry about that aspect. And after all, you were trying to convince me that you sometimes have to treat them as waves, and you used diffraction as a proof. You have been very convincing with that part of the argument, so can we agree on that and go on?

G: I'm getting a headache; I think we'd better continue tomorrow.

S: As for me, I think I'm finally beginning to see a bit of light. But I've got a lot more questions to ask. *A demain!*

★   ★   ★

G: Well, have you had any afterthoughts about yesterday's discussion?

S: Only that I believe I had it right yesterday. Especially about individual photons each going through the two slits. Each one has got to carry the whole message, because each one is a sort of sample of the whole message.

G: What message are you talking about?

S: The message is the information required to make a diffraction pattern. After all, it's just a consequence of what you told me a quantum of radiation, what you call a photon, was.

G: I guess I can't quarrel with that, though your way of arguing from it is rather unusual. What we usually say is that the wave is a wave of probability, and so determines the probability that the photon gets absorbed at one or another point.

S: Really, now, this is too much. How on earth can you have a "wave of probability"? Probability is a concept, and I don't think it makes sense to talk of a wave of a concept!

G: Well, quantum mechanics *is* a bit unusual, to put it mildly. So a quantum probability is not quite like an ordinary probability.

S: I don't see why it's necessary to introduce something as weird as that. There wasn't anything like that in *my* explanation of what was going on, that there was a sort of lottery between electrons to see which one would absorb the photon. That seems to be a case of perfectly *ordinary* probability. What is wrong with that way of explaining things?

G: I'm not quite sure. I'll have to think about that a bit. But after all, probability is determined by something random happening repeatedly, like flipping a coin, and determining the relative frequencies of the different outcomes, and it seems that is what is happening here. The "trials" are the successive photons coming through the slits. So it's not so weird to talk about probabilities.

S: OK, but if you talk about it that way, it seems to me that you're not talking about some special kind of probability; it seems like perfectly ordinary probability to me.

G: What is not ordinary is having a probability propagated like a wave.

S: I get the feeling that your "wave-particle duality" enables you to keep shifting your ground to respond to anything I say. You've just given an example of applying probability to particles, which I suppose is all right because observation is in question. But before the observation, the quanta should have behaved like *waves*, so you then put the probability back into the waves. It seems to me as if you are implying that, until they are observed, quanta are only conceptual entities, but as soon as you observe them they materialize, so to speak. I get the feeling that there's a philosophical ambiguity here, that you shift gears constantly between epistemic and ontic elements in your arguments.

G: Please don't throw philosophical complications at me; I'm only a poor physicist.

S: Well, I don't think *I'm* responsible for your arguments. It was *you* who did it.

G: I think we're getting a bit testy again. Let's pause for a cup of coffee so that we can cool off a bit.

*   *   *

S: May I change the subject, or rather, talk about a related problem which intrigues me?

G: That seems a good idea. What is it?

S: Well, it has to do with the question of the possibility of subdividing a quantum. It is that two-slit experiment which brings it to mind. Suppose that, on the far side of the screen with the slits, you put in a barrier to completely isolate the waves going through them from each other. Haven't you got *one* quantum on the source side of the barrier, and two on the other side? And just to make sure that we're eliminating any prejudices carried over from classical physics, let's use electrons rather than photons as an example.

G: That's easy. You just have 50% probability of having an electron on one side and 50% on the other. That, you see, is the advantage of the "probability interpretation".

S: Well, I never said that the probability interpretation wasn't handy! But is it really as simple as that?

G: Why not? What's wrong with it?

S: I guess it's that question of epistemology versus ontology again. That is, to translate it into a physicist's language, is the wave a real physical thing or only a device for organizing our knowledge? To put it differently, is there really something on both sides of the barrier, or just the *probability* of something on one *or* the other side?

G: Until you do a measurement, it is just a probability. When you set up an experiment to detect it, you find it on one side.

S: So that the measurement has suddenly transferred something from one side to the other? And on the first side, nothing is left, neither something material, nor a probability of something being there?

G: That's the normal interpretation.

S: Fine, but I think this opens a number of questions. Am I being naive, or does this have one of the following consequences:

1. that the particle was only on one side all the time, which means that it only went through one hole, or

2. that the measurement made something go from one side to the other, that something being either part of the particle or the probability of the particle being on the other side, or, I suppose

3. that putting the barrier in beyond the slits somehow constrained the particle to go through only one slit? Can you tell me which is right?

G: Aren't options 1 and 3 pretty much the same thing? I think 3 is the better option of the two. Option 1 gets into difficulty because it seems to be inconsistent with the existence of a diffraction pattern, while 3 says that the pattern was somehow destroyed by the introduction of the barrier on the far side of the slots. The conventional interpretation is neither; it is option 2, but it being the probability, not part of a material particle, which is affected by the measurement. And the *probability*, as I have said, is what is involved in the diffraction pattern, so everything hangs together nicely.

S: Let me see if I've got this straight. The particle is definitely on one side or the other, but we don't know which; that is determined by the wave function, which determines the probability of one or the other possibility. And the probability, and nothing material, is what is affected by the measurement, because once we know where the particle is, the probabilities have been replaced by actualities.

G: Precisely. And that is what is usually referred to as "the collapse of the wave function".

S: Oh. But I'm sure I heard that Richard Feynman once said that there was no such thing as the collapse of the wave function.

G: Feynman never could see things the way other people do. I have no idea what was in his mind if he said that.

S: Well, anyway, I'd like to come back to my question of "splitting" a particle, because I'm not sure that it's been resolved. Suppose that you had a particle in a box, and then put a barrier down the middle of it, and took the two halves off as far as you want from each other, so that there could be no doubt as to whether we had accomplished the splitting or not.

G: I thought I'd made it clear; you can't split particles that way. Furthermore, the example is a bit ridiculous, boxes are made of electrons, among other things, and electrons are indistinguishable, so the "one electron in a box" doesn't make much sense. But I think there's another example which involves the same issues. There's something called the "hydrogen molecule-ion" which consists of one electron and two hydrogen nuclei which are held together. Suppose that somehow (I'm not sure how!) the two nuclei could be separated, by as large a distance as you like. Then you would see that one nucleus would carry the electron with it, and the other would be bare; in other words, you'd get a hydrogen atom and a hydrogen ion.

S: Let me anticipate your answer on this one: one hydrogen nucleus, a proton, has the electron around it, but we don't know which until we make a measurement, or observation, or whatever on one of them, in which case we will either find the electron on that one or not; in the latter case we can deduce that it will be found on the other. But what about that epistemological appendage, the wave function, which is supposed to be the key to the physical realities of the situation. I expect that it will be split, presumably 50:50, between the two protons. So what happens when the measurement is made? Presumably all the wave function is suddenly found around one of them. Now if they are a light year apart, that seems to me to create some sort of difficulty.

G: You have learned very well; your analysis is impeccable. As for the instant influence of one proton on the other at an arbitrarily large distance, that has worried philosophers of quantum mechanics. You may ask, doesn't this contradict the theory of relativity, which says that no influence can be propagated faster than the speed of light? And the answer is negative, because only probabilities are involved, and there is no transmission of information.

S: But probabilities constitute information!

G: No, I think that only the actualities which accompany measurement might. If you find the electron on one side, the deduction that it is not on the other follows from the law that the sum of all probabilities must be 1.

S: So please explain to me exactly what *does* happen to the wave function when the measurement or observation is made on one of the components of the original system and the electron is found there.

G: The wave function on the other vanishes!

S: Yes, but when?

G (after a long silence): I was going to say, at the instant of the measurement. But to tell the truth, there's something here that hadn't occurred to me. Relativity says at least that "simultaneous" has no unambiguous meaning, so I'm stuck for a clear answer at the moment. I think, though, that it mustn't really make any difference. In fact, perhaps this is the answer to the old worry about propagation faster than light; perhaps the influence *is* carried at the speed of light, though I'm not quite sure how.

S: Well, I see a problem here. Doesn't that mean that there is a period to time, which may be long, during which the probabilities don't add up to 1? And that means, in fact, that if one made a measurement at the proton which hasn't got the electron, while the signal between them was on the way, there's a finite probability that an electron would be found there! So suddenly we'd have two electrons instead of one. I don't see how a signal arriving later could get rid of it.

G: I guess that my thought about the signal being propagated with the speed of light was wrong.

S: Yes, but giving way on that doesn't get us out of the hole; not if I've understood correctly. Because in the period in which light signals can't get between the two...

G: ... in that case we say there's a space-like separation between the protons...

S: ... in that period, if the vanishing of one wave function is simultaneous with the measurement of the other, it will, in other frames of reference, be seen to happen before, or after, the measurement. In either case, the probabilities won't add up to 1, which seems to throw the whole "probability interpretation" out of whack. If the probabilities *don't* add up to 1, there's always a finite probability that for a time there will be two electrons, or none. That will be true in all frames of reference but one, and that one is arbitrary!

G: I don't, unfortunately, see any flaw in your argument, but I can't believe that no one has seen this before. I can only think of two ways out. One is that this situation can only *arise* in systems that are unreal physically. I believe, in fact, that this hypothesis had been proposed by a prominent physicist, though I can't remember who at the moment. The other possibility is that the very workings of quantum mechanics don't *allow* a wave function which is "split" in the way that we have assumed. And in fact, this doesn't seem to me to be such a crazy idea. Most of the examples I can think of involve a symmetry in the system, as that between the two protons. But of course physicists are very conscious these days of the phenomenon of "spontaneous symmetry breaking", where things don't balance in the middle between falling one way or the other, but go one way or the other depending on minute random asymmetries. It seems that this is a problem worth looking at further.

S: I didn't realize that my naive questions would stir up such a tempest... I think I'd like to continue the discussion in at least one more session; my question bag is not exhausted yet. I've got a great curiosity about this purported phenomenon of "observer-created reality". I think we've sort of stumbled on it in our discussion today.

G: All this is getting tougher than I anticipated. Why can't the world be simple? Or perhaps the question is, why can't we *see* that it's simple!

*   *   *

G: You said you wanted to talk about "observer-created reality". As you also said, we skirted around it in our discussion which came to the conclusion that, in one of its strong forms at least, it didn't hold up. Because if we assume that we can create a *distant* reality by doing an experiment at home, it was shown not to be true. The idea, though, is usually put something like this: a system is said to be in a "mixture of states", which collapses into one state when we make a measurement on it. It wasn't in that state before, so we made that state.

S: I don't see what the big deal is here; surely with *anything* we do we make a condition, a state if you like, that wasn't there before. Nobody's going to argue about that, and it isn't peculiar to quantum mechanics.

G: Well, let me tell you about "Schrödinger's cat". Schrödinger wanted to see if the "queer" features of quantum mechanics, which was meant to deal only with atomic or atomic-scale phenomena, had some repercussions in macroscopic physics. So he thought of this idea: suppose we put a cat in a box with a radioactive source and a detector such that a decay of a nucleus would register on it, and that that would be triggered to release a deadly poison from its container which would kill the cat. One could find such a kind and amount of radioactive matter that the probability of a decay in one hour would be exactly one half. Now conventional quantum mechanics says that the radioactive nucleus would be neither in the unradiated state nor the decayed state, since we would only know the *probability* of the decay taking place at any arbitrary time. We say then that it is in a "mixed" state, i.e.

$$(\text{decayed state}) + (\text{non-decayed state}).$$

But this would have consequences for the cat, which we could then write in a similar way: (dead state of cat) + (live state of cat). Until we make the observation an hour later, we don't know which; but the moment we look, the state collapses, and we know that the cat is either dead or alive.

Now there are two ways of reacting to this picture. One is to say that although we didn't *know* until we looked, the outcome was already *decided* by that time; according to the other, the cat was neither definitely dead nor definitely alive before we looked. Philosophers may quibble about this, but scientists *must*, it seems to me, assume that there is a real physical world out there; otherwise one could not justify their function, which is to understand it. It is the second sort of outlook, however, which opens the door to ideas like "observer-created reality".

S: Excuse me for interrupting at this point, but I must get one thing clear before we go on: how, in scientific terms, can one make a one-to-one correspondence between states of a simple thing like a chunk of radioactive matter and something as complicated as a cat, to say nothing of the system which amplifies the radioactive decay into the release of the poison, or the process of poisoning itself? The cat is, after all, a *macroscopic* object, and the whole "observation" is a macroscopic observation, and I can't see that we get any deeper into this question by trying to explain all sorts of macroscopic stuff by quantum mechanics. I doubt that it's even possible, but if it is, it surely is no more useful than trying to describe the sound of my voice by trying to follow the motion of each molecule of the air through which it passes. And even if this pointless task could be accomplished, it seems to stretch credulity to claim that one could transform the two-state system of a radioactive nucleus into a two-state system of a cat.

G: I think there's some merit in what you say, but still, we are only concerned with minimal knowledge about the cat, its aliveness or deadness. Aside from that, it could equally be anything from an owl to an elephant. And it's not even that aliveness or deadness is essential to the point; some other irreversible selection between two possibilities would do equally well.

S: Well, isn't there another way of looking at it? The cat is, after all, only a macroscopic indicator of something that has happened, a crude measuring instrument, if you like. Some thing inanimate, but still macroscopic, would do equally well. It isn't even important, is it, whether we or some instrument detects what has gone on in the box, which seems to dispose of the issue of human consciousness. And the instrument could have one useful feature which the cat hasn't got, it could make an automatic record of the *time* that the radioactive decay took place, if in fact it did. That of course would have nothing to do with when we looked. It seems to me that the whole issue here is the relation between measuring instruments and what they measure.

G: Of course, that's what is known as the "problem of measurement" in quantum mechanics, and volumes have been written about it.

S: So what was the point of the whole "Schrödinger's cat" fable? Wasn't it supposed to embody a paradox?

G: Well, Schrödinger's idea was to ask whether what we consider as characteristically quantum behaviour might be transposed into the macroscopic realm. Since it seemed to him that this had unacceptable consequences, if it could be done, he felt that the whole foundations of quantum mechanics could be thrown into doubt. Which would not have displeased him at all.

S: Well, it doesn't seem to me that all this has proven much of anything, so I'd like to get back to the matter of quantum states, *real* quantum states, I mean, and not cat ones!

G: I'm sorry. I dragged in the cat, so to speak, to illustrate the idea of quantum states, and mixtures of states. It wasn't a good example, so let's talk about the hydrogen atom, because this illustrates something that *is* important. Before we get down to that discussion, there are a lot of other things that have to be put in place. Although this will involve saying some things that you already know, such as that electrons can behave as waves or fields, I think I should go back over the history of how all this came about in order to put quantum phenomena in their proper perspective. After all, between classical and quantum mechanics there was a major conceptual revolution, and we have to be clear what it involved. So, to start with, let me explain that the mechanics of Newton regarded the goal of physics as being to follow the motions of objects, particles if you like, in space and time. Or rather, in space as a function of time. It was a matter of predicting the trajectories or paths of these particles by the use of physical laws, his laws, in fact.

S: But the things scientists wanted to explain did not concern "particles", but *real* things, like the orbits of the planets, for example.

G: Right. But that was the beauty of Newton's vision. If you had to make separate laws for each sort of thing you studied, the life of a scientist would be very frustrating. But Newton saw that if you considered everything to be made up of particles, his laws could be used to deduce the behaviour of things like planets. In fact, it turned out rather nicely, because if planets were considered as rigid they had only two kinds of behaviour. On the one hand, they traced out orbits just as though they were particles, and in addition had independent motions of rotation, and one did not affect the other. So the astronomers could concern themselves with the orbits while they ignored the rotations.

Now in the 19th centuries a lot of work was done on the radiation of light, or more generally, electromagnetic waves, when various substances were heated to high temperatures. And then there was related work on the *absorption* of radiation, especially sunlight. Something unexpected was found, that radiation was neither emitted nor absorbed at all frequencies, but rather at sequences of very particular frequencies. This generated the science of spectroscopy. But for a long time, one had no explanation for these particular frequencies, even though there were simple mathematical formulae for many of them.

After Planck and Einstein came up with their idea of electromagnetic quanta, i.e. photons, Ernest Rutherford demonstrated that atoms seemed to be something like solar systems, with a very heavy object, the nucleus, at the centre and light electrons circling around it. But the trouble with the solar system model was that the particles had electric charge and were held together by electric forces. Now it was known since Maxwell's time that when electrically charged bodies were accelerated, they radiated away electromagnetic energy – at a continuous range of frequencies. But the important thing was, that if one calculated this radiation, it was so intense that the electrons would radiate away *all* their energy in a fraction of a second, and so spiral into the nucleus. This was disastrous, because theory appeared to predict that atoms should not be stable!

S: I was already aware of all this, but thanks all the same for recalling it to me. Isn't this where Niels Bohr came in?

G: Exactly. Bohr "fixed up" the orbit theory by decreeing that not all orbits were allowed by nature, but only those with integral numbers of "quanta" of angular momentum (the quantum was Planck's constant, by the way). These "allowed" orbits then had discrete energies. Bohr reasoned, then, that if an atom made an abrupt transition from one of these states to one of lower energy, the remaining energy would be radiated away as a light quantum. Because these photons could then only have specific energies, Planck's law said that this radiation could only have specific, discrete frequencies, which was very satisfactory to the spectroscopists. Not only that but, for hydrogen at least, the predicted frequencies were exactly those that had been observed.

S: But surely it isn't enough, for a scientist, to be able to reproduce numbers from the real world. Don't your theories have to be compatible with each other, and consistent with all that is already known? How did Bohr explain why atoms didn't radiate continuously and collapse as Maxwell would predict?

G: He didn't. He simply felt that there must be *some* significance in the agreement between his theory and experiment, and that the existence of stable orbits was due to some as yet unexplained factor. · What he didn't realize was that the trouble lay in Newton's whole program of following the trajectories of particles.

S: From that statement it is obvious that the next step had to be a quite revolutionary one.

G: Yes. It started with Louis de Broglie in Paris. He made the absurd suggestion that if light could behave in a particle-like manner (that is, like photons), perhaps electrons could behave like a *field*, or, as he put it, like waves. They were rather peculiar waves, though, because they had to be charged and so would be attracted to the nucleus.

S: I presume that the idea was that waves, unlike orbiting particles, wouldn't radiate continuously and spiral into the nucleus. I don't see why, though; aren't they still accelerated charges?

G: In matters like this, you usually try to draw on your previous experience; you ask, haven't I seen something like this before? And Erwin Schrödinger got the key idea in this way. There was a problem which attracted a great deal of attention in the 19th century. It was this: if you had a continuous system and it was constrained in some way, it could only vibrate at very particular frequencies. Because Planck had linked frequency to energy, this appeared to give a way to explain discrete *energies*. Musical instruments provide a good example, whether it's a violin string, an organ pipe, a flute or a drum, all music depends on this phenomenon. Of course, nowadays we know other examples, like electromagnetic wave cavities.

S: But this always involves constructing a material barrier to reflect the waves. What plays that role in Schrödinger's atom?

G: Actually, it's the nucleus which, by electrically attracting the electron (even if it's an electron field), prevents it from escaping from the nucleus. Schrödinger set up a "wave equation" to describe it, and he found that the energies (or frequencies) were exactly what were needed to fit what we knew about spectroscopy.

S: You mean he got the same formulae for the energies that Bohr did?

G: Yes, and better, because he found that there could be states which didn't have *any* angular momentum about the nucleus, so they weren't in *any* respect like orbits. Quite simply, there were no orbits any more; because you were dealing with a field, you couldn't follow it from point to point. So the whole program of Newtonian mechanics had to be abandoned.

S: That is to say, the job was not to follow objects from point to point, but rather to determine the possible *states* of systems! But this raises some other questions, doesn't it? First, what is the 'wave' a wave *of*? And secondly, how, in this system, do you describe *change* in a system? The states seem to be rather frozen things. Surely in the physical world one still has a history of things?

G: We've covered quite a bit of ground, so let's save these questions for one more session tomorrow.

★   ★   ★

G: The first question that we left unanswered last time was, what were the waves in Schrödinger's equation waves of? I'm rather disappointed that you asked that question, because it is the same one that was asked about Maxwell's electromagnetic waves at the beginning, and gave rise to several decades of nonsense about the so-called "aether", which of course didn't exist. So the answer is, I guess, that it's a wave of electrons. But the second question, namely, how, in quantum physics, does one describe change, is a good one, and shows how utterly different quantum mechanics is from classical mechanics. Change, in quantum mechanics, is described simply by systems making transitions from one state to another.

S: But what causes the change, or, how is it brought about?

G: The answer is, by interaction with other fields. For instance, if you consider the problem of the electron spiralling in the hydrogen atom, it doesn't follow a path, but because of its interaction with the electromagnetic field it can start in a state which is weakly bound to the nucleus, and then can "decay" through a series of states of decreasing energy. However, and this is of capital importance, there is a lowest state, and when the electron gets there it can't go further, and remains stable in that state. The state, reasonably enough, has no angular momentum, so there's no question of an orbit or even of rotation. Once in that state, there's no acceleration and so no radiation.

S: But what holds it that way in the face of the inward electrical force on it?

G: The very first principle enunciated by de Broglie, whose "electron waves" had a wavelength h/p, where p is the electron momentum. If it went faster, the wavelength would get smaller and the frequency bigger, which would cause an increase in the energy. This, it turns out, more than compensates the energy it would lose by getting closer to the nucleus. So it settles down with everything in nice balance, adjusting itself to get as close to the nucleus as it can without invoking too strongly this opposing quantum effect.

S: How can you talk about the wavelength of something which isn't wavy?

G: It is just as much a wave as the "wave" on the vibrating drumhead. All the same, I'm afraid that what I've just said is rather oversimplified. The detailed solution of the problem is quite difficult. But let me try another way of approaching it. If I go back to my analogy with vibrating systems, for each of those allowed frequencies of which we spoke, there is a pattern of the field. The simplest case is that of a violin string. The "field" is the sideways displacement of the string which has a definite value at each point. Here we have real periodic waves, but the constraints of fixing the ends ensure that the wave "fits" at the two ends. The longest wave you can get in is half a wavelength, or a wavelength twice the length of the string. The longest wavelength corresponds to the lowest frequency; this is the "fundamental" vibration, and would correspond to the lowest energy state of the hydrogen atom. Or, to take something a bit more complicated, consider a drumhead. It too has a lowest frequency, in which it fits between the boundary points all around the circumference. If you took a cross-section of the drumhead going through its centre, it would again rise from zero to a maximum and then go back to zero at the opposite side. But the important thing is that there always *is* a state of lowest frequency, and so, in the atom, a state of lowest energy.

S: So your description of the atom, for instance, involves giving these vibration patterns for each state, and somehow seeing how interaction with something else gets you from one to another.

G: Actually, it's simpler than that; you don't need the vibration patterns. Dirac showed that the state can be *completely* specified by giving values of a small number of quantities which in fact are related to the symmetries of the system. What I am saying is that giving these constitutes a complete physical description of the system. And these must be *compatible* quantities. For instance, you can't give an answer to the two questions: "*where* is the electron", and "what is its energy (or momentum)" because if it has an energy or momentum these are determined by the properties of a *field*, which is defined from point to point, and so has no "position". Compatible quantities are quantities whose values can all be specified *simultaneously*. These quantities can be measured without affecting the state of the system and so without affecting each other's measurement.

S: This is strange. Even in classical physics, you can't measure *anything* without changing it at least a little bit, and it is usually said, isn't it, that in quantum mechanics this feature is built in in a very fundamental way?

G: This is all very true for *incompatible* variables, but not for the basic "quantum numbers" which describe the system.

S: But *isn't* it strange, that you can measure something without ever touching it, so to speak?

G: Well, you might think so at first sight, but let me give you an example which is often used in teaching. It involves photons. One thing that it illustrates is that there isn't only *one* choice of compatible variables. One way of describing photons is to specify their *polarization*, which is simply the direction of the electric field vector in the quantized wave. There are only two possibilities; it can either be along some arbitrary direction or in a direction perpendicular to it. We can call these directions x and y respectively. But another property which can be used to describe them is related to the fact that photons have angular momentum about their direction of propagation, which can be either clockwise or anticlockwise and whose magnitude is proportional to the Planck constant. These two properties, polarization and angular momentum, are mutually incompatible; you can't specify both of them at once. In fact, if you specify one (polarization, say), the photon is described by a linear combination of the states of the other, the angular momentum; or vice versa. If you try to measure the angular momentum then, i.e. when the polarization has been determined, you may get one or the other value, with equal probability.

S: I'm not worried by the last statement, which illustrates what you have claimed to be the essence of the quantum theory. But I'm still bothered by the possibility of measuring anything without disturbing it in the least.

G: Yes, well, I was just coming to that. You see, there are filters, partially transparent pieces of plastic, which can let through just photons of a certain polarization or, alternatively, just photons of a certain angular momentum. For the waves that get through, you then *know* whichever of the properties you have chosen, but you haven't changed that property for these photons.

S: Ah, I see. You *have* had an influence on the system; you didn't get the information free, so to speak.

G: Well, if you put the outgoing wave through another filter of the same polarization, it would get through with no disturbance.

S: Yes, but you'd have already known the value of the polarization, so at this point you weren't determining it.

G: Quite true.

S: Look, it seems to me that your "measurement" just corresponds to inducing a transition in the system. But transitions are going on all the time, they are essential for all physics! So what is special about measurement by humans? You see, of course, that I'm getting at the question of "observer-created reality". The physical world, it seems to me, is constantly changing its own realities; we just stick our oar in now and then. After all, we are *part* of the physical world!

G: But the transitions you say we induce by our measurements are of a special sort; we call them "non-unitary", which signifies that the system at the end of a measurement has lost something relative to its initial state. We have, in a sense, only part of the initial system.

S: But actuallly nothing has been lost, except that we've lost interest in the part of the wave that didn't get through the filter. And the measurement certainly has involved a transition in the system.

G: But you forget that this can be brought down to the level of individual photons. Then, they either get through or they don't.

S: So if they get through, nothing is lost and if they don't, we're not interested.

G: But then, hasn't the measurement created the reality for that photon?

S: That's a strange statement. If there weren't some photons in the original beam with one polarization and some with the other, what would be the point of a filter to isolate one from the other?

G: You could say that the "filter" isn't a filter at all, that it can *create* the photons with one or the other polarization, with equal probability. Well, what I mean is that it can *process* the incoming photons this way, and that's different from filtering them.

S: All right, now tell me this: Do all beams of light necessarily have the same numbers of photons of the two polarizations, and does "filtering" a beam always produce them in equal numbers? Because if the result of the measurement doesn't always produce the same result, but depends on what went in, you can hardly say that it's the filter, to be neutral let's call it "the apparatus", if you want. Anyway, you can hardly say it's the apparatus which created the outcome and that it wasn't just a function of what was fed in. In other words, it would be logical to conclude that there was a pre-existing "reality" which we were determining by the measurement; the apparatus wouldn't be *creating* the reality. In fact I should think that there would be a simple test of whether "it's what came in that came out". You could create a beam with, say, a 2:1 mixture of the two polarizations, and then verify by an experiment that that was indeed what the beam contained.

G: How would you produce the 2:1 mixture?

S: With two different instruments, you could create two beams of opposite character, one twice as intense as the other, and then combine them with totally reflecting mirrors.

G: But you see, you're then using the filters to verify the properties of the filters! Isn't this a circular argument?

S: I think you've slipped through my net again, but I still think that most reasonable people would not quibble that way. And in fact there's some of my argument left, despite it. Because if you take a lot of random beams, that is, beams that have been produced under diverse conditions, and on putting them all through the same apparatus get different results, it's pretty hard to see how you could attribute those results to the apparatus.

G (tongue in cheek): Couldn't you still attribute it to "human consciousness"?

S: No doubt, but I'd have thought that I'd have a hard time convincing anyone.

G: Well, I think there's only one item left on our agenda. You wanted to know what it meant to say that before a measurement a system was in a "linear combination of states", but that the measurement made it "collapse to a definite state".

S: Quite correct. The question is, when is a state a state and when is it a *collapsed* state?

G: Let me explain. An experiment is set up to determine a specific property of a system. In general, the system will not be in a state in which that property has a definite value. It has different probabilities of being found to have one or another value. That's essentially what we mean by saying that the system being measured is in a "linear combination of states". That is with respect to a given property. When the measurement is made, one gets a definite value. And that is what we mean by the "collapse of the wave function".

S: Aha! But surely the system was, in a literal sense, in a "definite state" both before and after the measurement, or transition, or whatever you want to call it. As you yourself said, the "linear combination" is defined with respect to a specific and arbitrarily defined property. But the same state could be a "definite state" in a different situation. In fact, if it exists in nature, doesn't it have to be? It seems to me that a "linear combination of states" is a definite state, just as a "definite state" can also be a linear combination of states!

G: In a sense, certainly this is true. The distinction depends on the circumstances.

S: Good. So what has been said up to now has convinced me that a "measurement" was just an experimentally induced transition from one state to another. Am I wrong about that?

G: I suppose that I can't quarrel with that, though the statement seems to me to need a fair bit of elaboration. For example, the measuring apparatus is always *macroscopic*, so one gets into the problems of how we get microscopic information from a macroscopic apparatus. Your statement presumably means that a measurement is "just" a microscopic quantum transition, presumably amplified or somehow detected by a macroscopic measuring apparatus. I'm not happy about the word "just", because that statement has a lot of implications. For instance, when you use the word "measurement", you normally think of some fixed property which the microscopic system has, yet you are describing it in terms of transition, i.e. change.

S: Let me see if I can unscramble this. I find it easier to reason with a particular example rather than in terms of generalities.

G: So you're going to reason inductively, like a good physicist.

S: So be it. A nice simple example to take is the polarizing filter. Photons impinge on the filter, and they either get through or are absorbed or perhaps scattered back. If they get through, we have determined that they have the polarization in question, both when they arrive at the filter and when they emerge on the other side. On the other hand, if they don't get through, they undergo a transition and do not emerge. So a beam of photons is partially scattered and partially transmitted. And that, I contend, is a transition in the beam.

G: Ah, but there you go again; the problem starts when you consider *single* photons. Then some are measured to have a definite polarization, because they get through; they are measured but do not make a transition. They are filtered through. And the others make a transition.

S: But surely in both cases they are in some way being measured!

G: I guess it's a bit like measuring the lifetime of a radioactive nucleus. In a given period of time it may decay or it may not, and it is this discrimination which constitutes the measurement. To take another instance, in Rutherford's famous experiment on the scattering of alpha-particles, there were not two but many different outcomes for a given particle. In both this case and that of polarizing filters, though, the measurement is only complete when another transition takes place, involving the *detection* of the particle which is the object of measurement. And this detection involves a framework of theoretical interpretation. So you see, things get quite complex.

S: I think it comes back to whether there's anything "paradoxical" about the so-called "collapse of the wave function". If a measurement just involves transitions from one state to another, isn't it a perfectly normal process of the sort that is necessary to describe *any* change within the framework of quantum theory? And if so, what is all the fuss about "collapse of the wave function", as though it posed some special problems?

G: It is a form of words to describe the process which I spoke about a moment ago. But what distinguishes a "measurement" is that it is a non-unitary transition.

S: Well, that's what "collapse" appears to mean. I thought, though, that we'd agreed that it wasn't non-unitary. It still seems to me that, if I've understood this idea of non-unitarity properly, it only appears non-unitary because we don't take account of everything, we ignore a part of the system if the system contains more than one photon, or some information if it is only a single photon.

G: All the same, it still seems to me that when you talk of systems of many photons, what you say is quite reasonable, but the problems come when you deal with *single* photons.

S: It's certainly true that things seem stranger, but is that anything more than a subjective reaction to new discovery? In any case, I can't see any special role for the "human factor". Anything that a human can do, anything that springs from his thoughts and which leads him to set up an apparatus to do a particular experiment, can't affect the outcome of the experiment itself; it's that which is governed by physical laws. Of course humans leave their mark upon the physical world on everything they do in life; in that sense, "human consciousness" has an effect on physical reality. But so does a windmill or a sunspot. The only thing that is special about *our* interaction with the physical world outside us is that we make deliberate choices about exactly *how* we will play our part as an element in it. And it's not very clear how the laws and actions of the physical world bear on the choices that we make.

G: I think that this, at least, is one point on which we can agree. But there is something else that still bothers me. A lot of our argument has had to do with how we *interpret* reality. And time and again it has seemed to me that these differences of interpretation have no real bearing on reality itself. They just don't seem to matter. Still, that's an idea that I don't feel entirely comfortable with. But perhaps whether this is really true or not is the most important question which has come out of the whole discussion.

INDEX

Abelard, Pierre   67
absolute space (*see* space)
acceleration   23, 25, 75
accelerators   104, 130, 460, 493
accretion disk   233–234, 250
aether   6, 66–69, 71–73
agriculture   7
air table   164–165
alkali metals   419
alpha particles   320
alpha rays   317
Anaxagoras   307
Anderson, C. D.   424–426
angular momentum   28–30, 339, 377–378, 394, 401–404, 421, 442
    – conservation of   30, 32–33
    – orbital   395, 444
Annalen der Physik   264, 266
annihilation of
    – operators   382
    – particles   96
antiparticles   96, 193, 421–433
Arecibo   191
arrow of time, cosmological   305–306
astronomical distances, measurement of   177–181
astronomy   15

www.ingramcontent.com/pod-product-compliance
Lightning Source LLC
Chambersburg PA
CBHW052115230326
41598CB00079B/3699